Blue Sky, Blue Water

Blue Sky, Blue Water is a timely and comprehensive resource that explores the dynamic intersection of air pollution, climate resilience, and water governance in the 21st century. This scholarly book brings together cutting-edge research, case studies, and innovative strategies aimed at promoting environmental sustainability and public health. Spanning 18 chapters, it offers both conceptual and real-world perspectives across diverse regions and scientific disciplines.

Key themes covered in this volume include an in-depth analysis of air pollution's health effects, microplastic risks, climate extremes, and advanced wastewater treatment techniques such as nanotechnology and blockchain-based governance. This book also highlights the value of traditional knowledge and community-based water conservation, bridging science with practice. Readers will benefit from multidisciplinary insights and practical recommendations for sustainable policy-making and environmental stewardship.

This book is an essential resource for researchers, environmental scientists, policymakers, urban planners, and students of environmental studies and public health. Its broad thematic coverage and global relevance make it suitable for academic institutions, government agencies, and non-governmental organizations focused on creating a sustainable and equitable future.

Blue Sky, Blue Water

Advances in Air Pollution, Climate Resilience, and Water Governance for Sustainable Futures

Edited by
Sneha Gautam, Suneel Kumar Joshi,
and Balram Ambade

CRC Press
Taylor & Francis Group
Boca Raton London New York

CRC Press is an imprint of the
Taylor & Francis Group, an **informa** business

Designed cover image: A breathtaking view of the tranquil waters and misty hills of Shillong, Meghalaya, India, captured at dawn with soft cloud reflections and forest silhouettes. © Sneha Gautam

First published 2026
by CRC Press
4 Park Square, Milton Park, Abingdon, Oxon, OX14 4RN

and by CRC Press
2385 NW Executive Center Drive, Suite 320, Boca Raton FL 33431

CRC Press is an imprint of the Taylor & Francis Group, an informa business

British Library Cataloguing-in-Publication Data
A catalogue record for this book is available from the British Library

ISBN: 9781041011101 (hbk)
ISBN: 9781041011118 (pbk)
ISBN: 9781003613183 (ebk)

DOI: 10.1201/9781003613183

Typeset in Times
by codeMantra

Contents

About the editors

Sneha Gautam is an Associate Professor in the Division of Civil Engineering at the Karunya Institute of Technology and Sciences, Coimbatore, India. He earned his PhD (Engineering) from the Indian Institute of Technology Kharagpur and his MTech (Environmental Science and Engineering) from IIT-ISM Dhanbad. After his PhD, he joined INCLEN and the University of California, Berkeley, as a researcher, followed by a Postdoctoral Fellowship at the Ministry of Science and Technology, Taiwan, and pursued postdoctoral research at NCTU, Taiwan. His name has been featured consecutively (2021–2024) in Stanford University's Top 2% Scientists of the World list.

He has received numerous awards, including the Young Scientist Award 2023 by the Academy of Sciences, Chennai, the Young Scientist Award 2022, the IEI Special Appreciation Award 2021, the IEI Young Achiever Award 2020, the IEI Young Scientist Award 2019, the Outstanding Reviewer Award from Elsevier, and an International Travel Grant from the Government of India. He also serves as an Editor, Associate Editor, Editorial Board Member, and Guest Editor for several high-impact journals. With over 175 articles in leading journals (h-index 41, i10-index 104, and over 5923 citations), he has developed a global network of collaborators, and his research has been featured in prominent media outlets such as Al Jazeera, *The Times of India*, and Trinity.

Suneel Kumar Joshi is a distinguished hydrogeologist and landscape restoration expert with over 15 years of multidisciplinary experience across hydrology, groundwater modeling, river science, remote sensing, and sustainable water management. Currently serving as the Director of Landscape Restoration Projects at Geo Climate Risk Solutions Pvt. Ltd., he leads a broad spectrum of initiatives in flood risk management, water stewardship, lake rejuvenation, and geohazard assessments. Dr. Joshi holds a PhD from the Indian Institute of Technology (IIT) Kanpur, specializing in hydrogeology, groundwater modeling, stable isotopes, and GIS applications. He further pursued an MTech (Remote Sensing and GIS) from SRM University, as well as an MSc (Earth Sciences). His academic and research affiliations include prestigious institutions such as Imperial College London and the National Institute of Hydrology, Roorkee.

He has led or contributed to more than 225 groundwater and hydrological modeling projects across India and abroad, supporting mining, industry, and infrastructure sectors. His technical expertise spans tools like ArcGIS, Visual MODFLOW, HEC-HMS, HEC-RAS, SWMM, and ERDAS. His work emphasizes sustainable development, having developed artificial lakes, early warning systems, and integrated water resource plans for several landscapes in India and globally. As an active scholar, he has authored multiple peer-reviewed publications and serves as a Guest Editor and Editorial Board

Member for top journals. He is affiliated with international bodies, including the AGU, GSA, and International Water Association. His impact extends through mentorship, capacity-building workshops, and public policy advising. His leadership in hydrological sciences and commitment to climate resilience and sustainable landscape restoration make him a leading expert in India's environmental and water resource management landscape.

Balram Ambade is an Associate Professor and former Head of Department in the Department of Chemistry, National Institute of Technology Jamshedpur, India. He has been in academia for over 20 years and has contributed prolifically to research in atmospheric and water chemistry, environmental remediation of toxic metals, microplastics, and PFAS. He is the recipient of several prestigious awards, including the Early Career Research (ECR) award by the Science and Engineering Research Board (SERB); the Marie Curie Award from Hyytiälä Forest Field Station, Helsinki, Finland; the Groningen Biomolecular Sciences and Biotechnology Institute (GBB) Award from the University of Groningen, Netherland; the Young Scientist Award from China in 2014; and the NASA Earth Science Division Modeling, MAP Award by Météo-France in Toulouse, France. He has authored seven books, served as a guest editor for many reputed journals, and published over 100 Science Citation Indexed journal articles and ten book chapters. He has supervised nine PhD students, 11 MTech students, and 48 MSc students. Currently, he is mentoring and supervising six PhD students. Further, he was listed among the Top 2% of the most cited scientists across various disciplines globally in 2023 and 2024.

Contributors

Reyana A
Division of Computer Science and Engineering
Karunya Institute of Technology and Sciences
Coimbatore, Tamil Nadu, India

Insha Ali
Division of Agronomy
Faculty of Agriculture
SKUAST
Sopore, Jammu and Kashmir, India

Maha K. Al-Mishrey
Department of Biology
College of Science
University of Basrah
Basra, Iraq

Hasan Umur Alsancak
Ministry of Transport and Infrastructure
Ankara, Türkiye

Hamid T. Al-Saad
College of Marine Sciences
University of Basrah
Basra, Iraq

Aagam Bamb
Biochemical Science Division
National Chemical Laboratory
Pune, India

Sudarshana Banerjee
Department of Physics
Srinath University
Jamshedpur, India

M Auyoub Bhat
Division of Soil Science and Agricultural Chemistry
Faculty of Agriculture
Wadura SKUAST-K
Sopore, Jammu and Kashmir, India

Tauseef Ahmad Bhat
Division of Agronomy
Faculty of Agriculture
SKUAST
Sopore, Jammu and Kashmir, India

Lalita Bisht
Department of Environmental Science
University School of Vocational Studies and Applied Sciences
Gautam Buddha University
Greater Noida, India

Laxmi Lahalya Boddeti
Buro Happold Engineers India Private Limited
Bengaluru, India

M. H. Chesti
Division of Soil Science and Agricultural Chemistry
Faculty of Agriculture
SKUAST
Sopore, Jammu and Kashmir, India

Murali Dadi
Department of Chemistry
School of Mathematics and Natural Sciences
Copperbelt University
Zambia

Esther Daniel
Division of Computer Science and Engineering
Karunya Institute of Technology and Sciences
Coimbatore, Tamil Nadu, India

Khursheed Ahmad Dar
Division of Agronomy
Faculty of Agriculture
SKUAST
Sopore, Jammu and Kashmir, India

Amutha Dhanaraj
Division of English
Karunya Institute of Technology and Sciences
Coimbatore, Tamil Nadu, India

Rifat Fayaz
Division of Soil Science and Agricultural Chemistry
Faculty of Agriculture
Wadura SKUAST-K
Sopore, Jammu and Kashmir, India

Tajalee Gulshan
Division of Agronomy
Faculty of Agriculture
SKUAST
Sopore, Jammu and Kashmir, India

Vidhu Gupta
Department of Environmental Science
H.N.B. Garhwal University
Srinagar Garhwal, India

Umma Salma Hashe
Department of Environmental Science
Bangladesh University of Professionals (BUP)
Dhaka, Bangladesh

Kaan Isinkaralar
Department of Environmental Engineering
Faculty of Engineering and Architecture
Kastamonu University
Türkiye

Oznur Isinkaralar
Department of Landscape Architecture
Faculty of Engineering and Architecture
Kastamonu University
Türkiye

Bisma Jan
Division of Agronomy
Faculty of Agriculture
SKUAST
Sopore, Jammu and Kashmir, India

Syed Andleeba Jan
Division of Soil Science and Agricultural Chemistry
Faculty of Agriculture
Wadura SKUAST-K
Sopore, Jammu and Kashmir, India

Ruqqaya Javid
Division of Agriculture Entomology
Faculty of Horticulture
SKUAST
Sopore, Jammu and Kashmir, India

Dipankar Jena
Department of Chemistry
Fakir Mohan University
Vyasa Vihar, Balasore, Odisha

Amol Kahandal
School of Bioengineering Science and Research
MIT ADT University
Pune, India

Inayat Mustafa Khan
Division of Soil Science and Agricultural Chemistry
Faculty of Agriculture
Wadura SKUAST-K
Sopore, Jammu and Kashmir, India

Diwesh Kumar
Department of Physics
Srinath University
Jamshedpur, India

Parth Sarathi Mahapatra
International Centre for Integrated Mountain Development (ICIMOD)
Kathmandu, Nepal

Kazi Khaled Mahmud
Department of Environmental Science
Stamford University Bangladesh
Dhaka, Bangladesh
and
Department of Port and Shipping Management
Bangladesh Maritime University
Bangladesh

Rakesh Kumar Maikhuri
Department of Environmental Science
H.N.B. Garhwal University
Srinagar Garhwal, India

Ahmad Kamruzzaman Majumder
Center for Atmospheric Pollution Studies (CAPS)
Dhaka, Bangladesh
and
Department of Environmental Science
Stamford University Bangladesh
Dhaka, Bangladesh

Aamina Manzoor
Division of Soil Science and Agricultural Chemistry
Faculty of Agriculture
Wadura SKUAST-K
Sopore, Jammu and Kashmir, India

Aneesa Manzoor
Division of Soil Science and Agricultural Chemistry
Faculty of Agriculture
SKUAST
Sopore, Jammu and Kashmir, India

Kenneth Maseka
Department of Chemistry
School of Mathematics and Natural Sciences
Copperbelt University
Zambia

Shakeel A. Mir
DEAN
Faculty of Agriculture
SKUAST
Sopore, Jammu and Kashmir, India

Prakash Chandra Mishra
Department of Environmental Science & Centre of Excellence for Bio-resource Management and Energy Conservation Material Development
Fakir Mohan University
Vyasa Vihar, Balasore, Odisha

Mohammad Moniruzzaman
BCSIR Laboratories
Dhaka, Bangladesh

Idrees Mukhtar
Division of Soil Science and Agricultural Chemistry
Faculty of Agriculture
Wadura SKUAST-K
Sopore, Jammu and Kashmir, India

Ishrat Mumtaz
Division of Agronomy
Faculty of Agriculture
SKUAST
Sopore, Jammu and Kashmir, India

Tahira Mumtaz
Division of Plant Pathology
Faculty of Agriculture
SKUAST
Sopore, Jammu and Kashmir, India

S.M. Shiva Nagendra
Department of Civil Engineering
Indian Institute of Technology Madras
Chennai, Tamil Nadu, India

Nishtha Naudiyal
Department of Environmental Sciences
College of Basic Sciences and Humanities
G. B. Pant University of Agriculture and Technology
Uttarakhand, India

Abdullah Al Nayeem
Department of Environmental Science
Stamford University Bangladesh
Dhaka, Bangladesh
and
Center for Atmospheric Pollution Studies (CAPS)
Dhaka, Bangladesh

Jibin Jose P
Division of English
Karunya Institute of Technology and Sciences
Coimbatore, Tamil Nadu, India

Smaranika Panda
Department of Civil Engineering
Sardar Vallabhbhai National Institute of Technology
Surat, India

Amit Passi
Department of Civil Engineering
Indian Institute of Technology Madras
Chennai, Tamil Nadu, India

Md Nasir Ahmmed Patoary
Department of Environmental Science
Stamford University Bangladesh
Dhaka, Bangladesh
and
Center for Atmospheric Pollution Studies (CAPS)
Dhaka, Bangladesh

Vikas Poonia
Centre of Excellence in Water Management
MANIT
Bhopal, Madhya Pradesh, India

Bidya Banmali Pradhan
International Centre for Integrated Mountain Development (ICIMOD)
Kathmandu, Nepal

Siva Praveen Puppala
International Centre for Integrated Mountain Development (ICIMOD)
Kathmandu, Nepal

Marziat Rahman
Center for Atmospheric Pollution Studies (CAPS)
Dhaka, Bangladesh

Chaitanya Raj
Department of Earth and Environmental Sciences
IISER
Bhopal, Madhya Pradesh, India

Ahmed Rushdi
College of Earth, Ocean, and Atmospheric Sciences
Oregon State University
Corvallis, Oregon

Vaishnavi Sahu
Department of Earth and Environmental Sciences
IISER
Bhopal, Madhya Pradesh, India

Akshay Saini
Department of Environmental Science
H.N.B. Garhwal University
Srinagar Garhwal, India

Nader A. Salman
Al-Manara College for Medical Sciences
Maysan, Iraq

Asma Shakeel
Division of Soil Science and Agricultural Chemistry
Faculty of Agriculture
Wadura SKUAST-K
Sopore, Jammu and Kashmir, India

Mehnaz Shakeel
Division of Plant Pathology
Faculty of Agriculture
SKUAST
Sopore, Jammu and Kashmir, India

Rehana Shakeel
Division of Soil Science and Agricultural Chemistry
Faculty of Agriculture
SKUAST
Sopore, Jammu and Kashmir, India

Stanford Siachoono
Department of Zoology and Aquatic Sciences
School of Natural Resources
Copperbelt University
Zambia

Surya Pratap Singh
Department of Civil Engineering
Srinath University
Jamshedpur, India

Aruna Sivaram
School of Bioengineering Science and Research
MIT ADT University
Pune, India

Rajeev Kumar Srivastava
Department of Environmental Sciences
College of Basic Sciences and Humanities
G. B. Pant University of Agriculture and Technology
Uttarakhand, India

Somil Swarnkar
Department of Earth and Environmental Sciences
IISER
Bhopal, Madhya Pradesh, India

Mathu T.
Department of Computer Science and Engineering
Nehru Institute of Technology
Coimbatore, Tamil Nadu, India

Chandrakant K. Tagad
Department of Biochemistry
S.B.E.S. College of Science
Chhatrapati Sambhajinagar (Aurangabad), India

Alpha Thapa
International Centre for Integrated Mountain Development (ICIMOD)
Kathmandu, Nepal

Chithra V. S.
Ecology and Environment Research Group
Centre for Water Resources Development and Management
Calicut, India

Deepthi Yaparla
Department of Civil Engineering
Indian Institute of Technology Madras
Chennai, Tamil Nadu, India

Preface

The idea for *Blue Sky, Blue Water: Advances in Air Pollution, Climate Resilience, and Water Governance for Sustainable Futures* was born out of an urgent recognition: the interconnected crises of air pollution, climate change, and water scarcity are no longer distant environmental threats. They are present, intensifying, and deeply interwoven with public health, urban life, and global development. As editors and environmental scholars, our motivation in curating this volume stems from a shared commitment to fostering interdisciplinary dialogue and promoting science-backed, community-driven solutions.

This book originated from our collective engagement with researchers, practitioners, and institutions across disciplines who are working on the frontlines of environmental change. From conferences and collaborative projects to classroom discussions and policy roundtables, we encountered a recurring need for a consolidated platform that integrates advanced research in air and water systems with perspectives on climate resilience and sustainable governance. This volume is our response to that need.

Air Pollution and Human Health

Risks, Mechanisms, and Mitigation Strategies

1

Asma Shakeel, Inayat M. Khan,
Mehnaz Shakeel, Syed Andleeba Jan,
M. Auyoub Bhat, Shakeel A. Mir, Rifat Fayaz,
Aamina Manzoor, and Idrees Mukhtar

INTRODUCTION

Pollution is the introduction of harmful substances into the environment, posing risks to humans, animals, plants, and ecosystems (Ukaogo et al. 2020). Pollutants, which may be solids, liquids, or gases, often exceed natural concentrations and degrade the environment (Manisalidis et al. 2020). These originate from industrial processes, vehicular emissions, agriculture, and household waste, affecting air, water, and soil quality. Air pollution, a major environmental issue, arises from anthropogenic sources like industrial emissions, vehicular exhaust, agricultural practices, and natural events like wildfires and volcanic eruptions. Harmful pollutants, including particulate matter (PM), nitrogen oxides (NO_x), sulfur dioxide (SO_2), carbon monoxide (CO), ozone (O_3), and heavy metals, are linked to severe health risks (Khajeamiri et al. 2021). Indoor air pollution results from combustion sources, tobacco smoke, building materials emitting volatile organic compounds (VOCs), and biological contaminants like mold and

DOI: 10.1201/9781003613183-1

dust mites. Poor ventilation worsens pollutant accumulation, leading to respiratory and cardiovascular diseases, neurological effects, and cancer. Outdoor pollution, primarily from industry, vehicles, and agriculture, releases PM2.5, SO_2, and NO_x, causing respiratory and cardiovascular issues, cancer, and pregnancy complications while contributing to climate change. Addressing these impacts requires targeted mitigation strategies. Air pollution significantly affects health, even at low levels. Short-term exposure is linked to chronic obstructive pulmonary disease (COPD), asthma, respiratory infections, and hospitalizations (Izah et al. 2024). Long-term exposure exacerbates chronic asthma, pulmonary insufficiency, cardiovascular diseases, and mortality (Warburton et al. 2019). Studies suggest prolonged exposure may also contribute to diabetes (Clark et al. 2017) and various early-life disorders, including respiratory, cardiovascular, and perinatal issues (Kim et al. 2020). Global research correlates daily fluctuations in PM concentrations with mortality rates (Schraufnagel et al. 2019). Climate change is expected to intensify these health impacts, increasing hospitalizations among vulnerable populations. Fine and ultrafine particles penetrate the lungs and bloodstream deep, leading to severe illnesses, especially in urban areas where road emissions are a major contributor (Piracha and Chaudhary 2022). Industrial accidents further exacerbate pollution, with dispersion influenced by atmospheric stability and wind patterns (Popov et al. 2020).

Developing nations face heightened air pollution challenges due to overpopulation, rapid urbanization, and unchecked industrialization, compounded by socioeconomic disparities and limited environmental awareness. Many households rely on solid fuels like wood for cooking, exposing three billion people globally—especially women—to harmful indoor air pollution (WHO 2016; Austin & Mejia 2017). Fine PM is strongly linked to lung cancer mortality in China (Li et al. 2020). India ranks among the most polluted countries, with cities like New Delhi frequently experiencing hazardous air quality (Guttikunda et al. 2014). Industrialization, urbanization, and motorcycle use further deteriorate air quality, while biomass combustion remains a significant household pollutant (Paul 2023). Northern Indian states, with colder climates, report significantly higher PM2.5 levels (557–601 $\mu g/m^3$) compared to southern states (183–214 $\mu g/m^3$) due to increased indoor heating (Bartaria et al. 2023). Addressing air pollution requires region-specific strategies to mitigate its health effects and enhance global quality of life (Figure 1.1).

Major Air Pollutants

The World Health Organization (WHO) identifies six major air pollutants: PM, ground-level ozone (O_3), carbon monoxide (CO), sulfur oxides (SO_x), nitrogen oxides (NO_x), and lead (Pb). These pollutants contribute to environmental degradation, acid rain, global warming, and climate change.

Particulate Matter (PM)

PM consists of fine solid and liquid particles suspended in air. PM10 ($\leq$10 μm) and PM2.5 ($\leq$2.5 μm) are most concerning (Ahmad et al. 2023). PM10 can reach the lungs, while PM2.5 penetrates deeper, entering the bloodstream (Zhang et al. 2021).

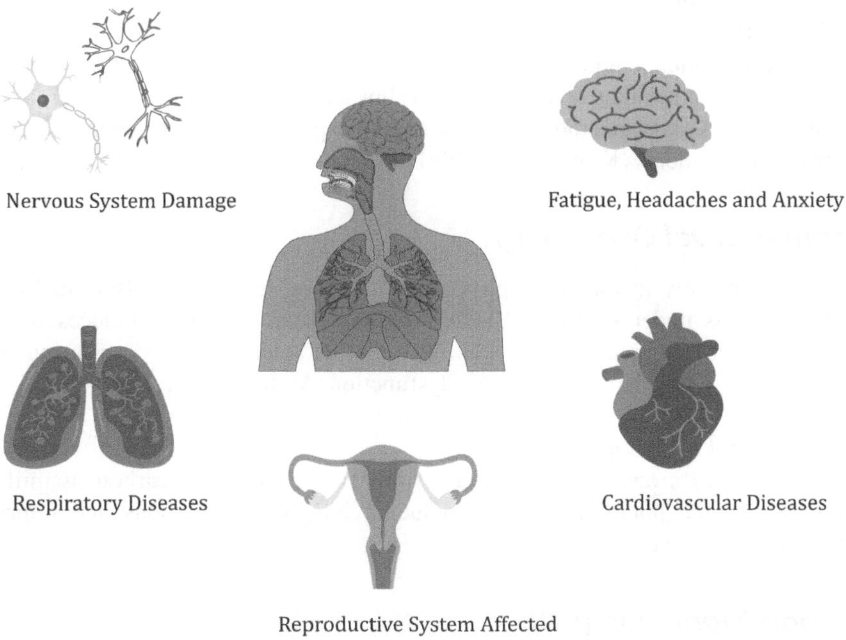

FIGURE 1.1 The effect of air pollution on human health.

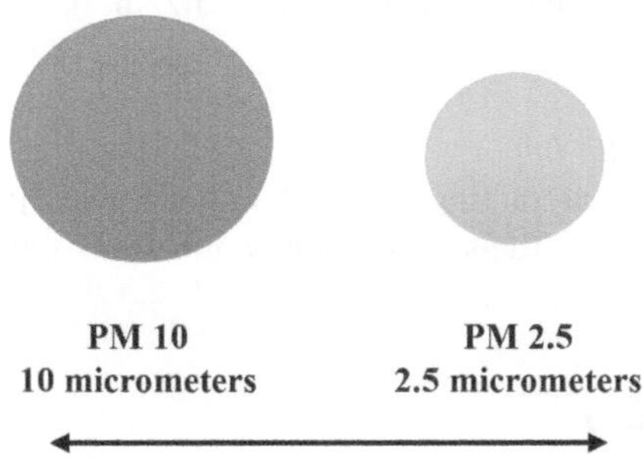

FIGURE 1.2 Particle size diameter of particulate matter.

Its composition includes compounds in organic (e.g., benzene, dioxins) and inorganic (e.g., sulfates, nitrates, metals) (Figure 1.2).

Categories of PM:

1. **Gas Contaminants**: Present in aerial masses.
2. **Particulate Contaminants**: Smog, soot, tobacco smoke, fly ash, and cement dust.
3. **Biological Contaminants**: Bacteria, fungi, viruses, and allergens.
4. **Dust Types**: Suspended atmospheric dust, settling dust, and heavy dust.

Environmental and Health Impacts
PM2.5 contributes to haze, affecting visibility and degrading air quality (Chauhan et al. 2024). It alters aquatic nutrient balances, damages vegetation, and acidifies water bodies. Health risks include respiratory and cardiovascular diseases, lung cancer, and premature death. A Harvard study improved PM2.5 exposure models using remote sensing for better assessments (Knibbs et al. 2018).

Ground-Level Ozone (O_3)

O_3 forms naturally in the stratosphere, but at ground level, it results from NO_x and VOC reactions under sunlight (ESCAP 2024). Ozone inhalation induces oxidative stress, depleting antioxidants like vitamins C and E (Thiele et al. 1997), causing airway inflammation, DNA damage, and lung dysfunction (Mohania et al. 2017).

Environmental Impact and Mitigation
Ozone damages crops, forests, and ecosystems by disrupting carbon assimilation. Mitigating ozone pollution includes reducing NO_x/VOC emissions and improving urban air monitoring.

Carbon Monoxide (CO)

CO results from the incomplete combustion of fuels like gasoline, diesel, and wood (Kumar et al. 2023). It reduces OH radicals, prolonging atmospheric methane and VOC lifespans, contributing to global warming (Singh & Yadav 2021). CO binds to hemoglobin, reducing oxygen transport causing dizziness, cognitive impairment, and asphyxiation.

Environmental and Health Impacts
CO exacerbates smog by interacting with PM and NO_x to amplify pollution. Mitigation strategies include cleaner fuels, emission regulations, and efficient combustion technologies.

Sulfur Oxides (SO_x)

SO_2 originates from fossil fuel combustion, industrial processes, and volcanic eruptions (Thanvisitthpon et al. 2024). It irritates the respiratory system, worsens asthma, and triggers bronchospasms. Prolonged exposure leads to chronic bronchitis and increased mucus production.

Environmental Impact and Mitigation
SO_2 forms acid rain, leading to soil acidification, vegetation damage, and aquatic ecosystem disruption. Solutions include shifting to renewable energy, flue-gas desulfurization, and stricter industrial regulations.

Nitrogen Oxides (NO_x)

NO_2, mainly from vehicle emissions, causes respiratory diseases, cardiovascular issues, and chronic lung damage (Theodorakidou & Lambrou 2017). It impairs immune function by affecting T-lymphocytes and natural killer (NK) cells (García-Ortiz & Serrador 2018).

Environmental Impact and Mitigation
NO_x reduces crop yields, contributes to haze, and damages materials. Reducing emissions through catalytic converters, cleaner fuels, and air quality regulations is crucial.

Lead (Pb)

Pb pollution arises from industrial emissions, batteries, and contaminated soil/water (Levin et al. 2021). It accumulates in the brain, liver, and bones, leading to neurological disorders, developmental delays, and cognitive impairment (Iqubal et al. 2020; Collin et al. 2022).

Environmental Impact and Mitigation
Pb disrupts photosynthesis and causes neurotoxicity in animals. Measures include phasing out leaded fuels, industrial emission controls, and public awareness campaigns.

Microplastics as Air Pollutants

Microplastics (<5 mm) originate from plastic degradation, textiles, and tire wear. They cause respiratory inflammation, oxidative stress, and lung impairment. Inhaled microplastics also carry toxic chemicals, exacerbating health risks.

Environmental Impact and Mitigation
Microplastics contaminate soil, water, and remote ecosystems. Reducing plastic use, improving waste management, and implementing filtration technologies can mitigate their impact.

Nanomaterials as Air Pollutants

Nanoparticles (<100 nm) are released from industries, combustion, and consumer products. They penetrate deep into the lungs, cause neurotoxicity, and accumulate in vital organs.

Environmental Impact and Mitigation
Nanoparticles disrupt ecosystems and bioaccumulate in plants and animals. Regulatory oversight, advanced filtration, and green nanotechnology can reduce risks.

Other Air Pollutants

- **Polycyclic Aromatic Hydrocarbons (PAHs)**: From coal, tar, and forest fires, PAHs like benzopyrene are carcinogenic (Mallah et al. 2022).
- **Volatile Organic Compounds (VOCs)**: Indoor pollutants linked to cancer, affecting mucous membranes (Maung et al. 2022).
- **Dioxins**: Industrial byproducts accumulate in animal tissues, leading to immune suppression and cancer (Wei et al. 2021).
- **Fossil Fuels and Air Contamination**: Major pollution sources include agriculture, industry, and transportation (Li 2020). European Air Quality Directives remain weaker than WHO standards, emphasizing the need for stricter measures (Table 1.1).

TABLE 1.1 The primary sources, health effects, and mechanisms of air pollutants affecting human health

AIR POLLUTANT	*SOURCES*	*HEALTH EFFECTS*	*MECHANISM OF ACTION*	*MITIGATION*
Particulate matter (PM)	Vehicles, industries, and wildfires	Respiratory and cardiovascular diseases, lung cancer	PM10/PM2.5 penetrates lungs, causing inflammation and oxidative stress	Clean energy, emission controls, and air filtration
Ground-level ozone (O_3)	NO_x and VOCs reacting under sunlight	Asthma, reduced lung function	Induces airway inflammation and oxidative damage	Reduces NO_x/VOC emissions and improves transport and monitoring
Carbon monoxide (CO)	Fossil fuel combustion, vehicles, and heating	Dizziness, cognitive impairment, and asphyxiation	Binds to hemoglobin, reducing oxygen transport	Improves combustion and enforces emission controls
Sulfur oxides (SO_x)	Coal and oil burning industries, and volcanoes	Bronchitis, asthma, and lung diseases	Forms sulfuric acid, irritating the respiratory tract	Flue-gas desulfurization and renewable energy
Nitrogen oxides (NO_x)	Vehicles, power plants, and industries	Asthma, lung infections, and cardiovascular issues	Forms nitric acid, causing oxidative stress and inflammation	Catalytic converters, electric vehicles, and emission regulations
Lead (Pb)	Batteries, industries, and contaminated soil/water	Neurological damage and developmental issues	Accumulates in the brain and bones, disrupting neurotransmission	Ban leaded fuels and remediate contaminated sites

Health Effects of Air Pollution

Respiratory Health Impacts

Air pollution has profound effects on the respiratory system, primarily due to the inhalation of harmful pollutants such as PM (PM10 and PM2.5), nitrogen oxides (NO_x), sulfur dioxide (SO_2), and ozone (O_3) (Gupta 2024).

- **Particulate Matter (PM)**: Fine particles (PM2.5) penetrate deep into the alveoli of the lungs, triggering inflammation, reduced lung function, and respiratory diseases such as asthma, bronchitis, and COPD. Long-term exposure to PM increases the risk of lung cancer (Perumpully et al. 2024).
- **Nitrogen Oxides (NO_x)**: NO_x gases are strong irritants, penetrating the lower respiratory tract, leading to symptoms like coughing, wheezing, and dyspnea. Chronic exposure can cause pulmonary edema and impair immune response.
- **Ozone (O_3)**: Ground-level ozone forms through photochemical reactions and is a potent irritant. It exacerbates asthma, decreases lung function, and increases the risk of respiratory infections by damaging epithelial cells in the respiratory tract.
- **Sulfur Dioxide (SO_2)**: SO_2 irritates the mucous membranes of the nose and throat, causing bronchoconstriction and acute respiratory symptoms.

Children and individuals with pre-existing respiratory conditions are particularly vulnerable, experiencing more frequent hospitalizations and reduced quality of life due to air pollution.

Cardiovascular Health Impacts

Air pollution significantly contributes to cardiovascular morbidity and mortality (Rajagopalan et al. 2018), especially due to exposure to PM2.5 and gaseous pollutants like CO, NO_x, and SO_2.

- **Particulate Matter (PM)**: Fine particles enter the bloodstream via the lungs, triggering systemic inflammation, oxidative stress, and endothelial dysfunction. These effects accelerate atherosclerosis, increase blood pressure, and raise the risk of heart attacks and strokes (Sellami et al. 2025).
- **Carbon Monoxide (CO)**: CO binds to hemoglobin with greater affinity than oxygen, forming carboxyhemoglobin. This reduces oxygen delivery to tissues, leading to hypoxia and increased strain on the heart, especially in individuals with pre-existing cardiovascular conditions (Gautam et al. 2019).
- **Ozone and NO_x**: Both pollutants contribute to arrhythmias, vascular dysfunction, and plaque instability in coronary arteries, increasing the risk of cardiac events (Chelani et al. 2023).

Long-term exposure to polluted air significantly raises the incidence of ischemic heart disease, heart failure, and hypertension, with higher mortality rates in urban areas.

Nervous System Effects

Recent studies have highlighted the neurotoxic effects of air pollution (Shabani 2021), particularly cognitive decline and neurological diseases.

- **Neurotoxicity of Pollutants**: Fine PM and toxic gases like NO_x can cross the blood–brain barrier or reach the brain through the olfactory bulb, causing neuroinflammation and oxidative stress.
- **Cognitive Decline**: Prolonged exposure to PM2.5 is associated with reduced cognitive function, memory impairment, and accelerated brain aging.
- **Neurological Disorders**: Air pollution has been linked to a higher risk of Alzheimer's disease, Parkinson's disease, and other neurodegenerative disorders. Children exposed to polluted air are more likely to exhibit developmental delays and behavioral problems such as attention deficit hyperactivity disorder (ADHD).
- **Immune Impact on the Brain**: Pollutants like NO_2 affect the immune response, altering CD8+ T-lymphocytes and NK cells, which can lead to nervous system dysfunction.

Reproductive and Developmental Effects

Air pollution adversely affects reproductive health and fetal development (Luderer 2024), with long-term consequences for future generations.

- **Effects on Fertility**: Exposure to pollutants like VOCs, heavy metals, and PM reduces fertility in both men and women by impairing gamete quality, hormonal balance, and reproductive organ function.
- **Fetal Development**: Pregnant women exposed to high levels of air pollution face an increased risk of miscarriage, preterm birth, low birth weight, and developmental delays in infants.
- **Neurodevelopmental Disorders**: Transplacental transfer of pollutants, such as lead, impacts the developing fetal nervous system, leading to cognitive impairment, hyperactivity, and learning disabilities.
- **Endocrine Disruption**: Certain air pollutants act as endocrine disruptors, interfering with hormonal signaling pathways critical for reproductive and developmental processes.

Skin and Ocular Effects

Air pollution also affects external organs such as the skin and eyes, often acting as irritants or causing chronic conditions (Lin et al. 2022).

- **Skin Damage**: Pollutants like ozone, PM, and UV radiation degrade the skin barrier, leading to dryness, eczema, premature aging, and inflammatory conditions like acne and psoriasis (Ali et al. 2020). Long-term exposure to dioxins and PAHs is linked to dark spots and skin lesions.

- **Eye Irritation**: Pollutants such as SO_2, NO_x, and ozone irritate the eyes, causing redness, watering, itching, and, in severe cases, corneal damage. Prolonged exposure can exacerbate conditions like conjunctivitis and dry eye syndrome.

Cancer Risk

Several air pollutants are known carcinogens, increasing the risk of various types of cancer (Ma 2022).

- **Lung Cancer**: Long-term exposure to PM2.5, benzene, and PAHs (e.g., benzo[a]pyrene) significantly raises the incidence of lung cancer, even in non-smokers.
- **Skin Cancer**: Dioxins and UV radiation exacerbate skin cancer risk by causing DNA damage and impairing repair mechanisms.
- **Other Cancers**: Exposure to VOCs and heavy metals like lead is associated with cancers of the bladder, liver, and other organs.

Mental Health Effects

Air pollution has emerging links with mental health disorders, particularly in urban populations (Bhui et al. 2023).

- **Anxiety and Depression**: Chronic exposure to polluted air increases systemic inflammation, which is linked to mood disorders such as anxiety and depression.
- **Psychological Stress**: Living in highly polluted areas contributes to a sense of vulnerability and stress, affecting overall well-being.
- **Cognitive Decline**: Pollutants, especially PM and NO_x, impair cognitive function, contributing to memory loss and reduced executive function.

Air pollution's effects on mental health remain an underexplored area, requiring further investigation to mitigate its psychological burden.

Mitigation Strategies for Air Pollution on Human Health

Air pollution is a critical global issue that poses severe risks to human health. Effective mitigation requires a multifaceted approach that combines innovative technologies, policy frameworks, and public health interventions. These strategies aim to reduce pollutant emissions, minimize human exposure, and safeguard public health.

Innovative Technologies

Advancements in technology play a pivotal role in mitigating air pollution. Key innovations include:

- **Emission Control Technologies**
  Devices such as catalytic converters in vehicles reduce harmful emissions like carbon monoxide, hydrocarbons, and nitrogen oxides. Industrial facilities use scrubbers and electrostatic precipitators to capture pollutants before they are released into the atmosphere.
- **Air Purification Systems**
  High-efficiency particulate air (HEPA) filters and activated carbon technologies in homes, offices, and public spaces help to remove fine PM (PM2.5) and gaseous pollutants, providing cleaner indoor air.
- **Clean Energy Solutions**
  The transition to renewable energy sources, such as wind, solar, and hydroelectric power, reduces dependence on fossil fuels, thereby cutting sulfur dioxide (SO_2) emissions, nitrogen oxides (NO_x), and PM.
- **Smart Monitoring Systems**
  Internet of Things (IoT)-enabled air quality monitors provide real-time data on pollution levels, helping policymakers and individuals take timely actions to minimize exposure.

Policy Frameworks

Robust policy frameworks are essential for long-term mitigation. These frameworks can vary significantly across regions, with successful models offering valuable lessons:

- **National Standards and Regulations**
  Countries like the United States enforce stringent air quality standards through the Clean Air Act, which mandates reductions in emissions of key pollutants. Similarly, the European Union's Air Quality Directive sets binding limits for pollutant concentrations.
- **Global Agreements**
  International frameworks, such as the Paris Agreement, emphasize reducing emissions of greenhouse gases, indirectly addressing air pollutants like methane and black carbon.
- **Comparative Analysis of Policies**
  - **China**: The "war on pollution" includes restrictions on coal use, vehicle emission standards, and public transport expansion, significantly improving air quality in urban areas.
  - **India**: Programs like the National Clean Air Programme (NCAP) aim to reduce PM2.5 and PM10 concentrations through city-specific plans.

- **Scandinavian Countries**: Policies promoting electric vehicles (EVs), extensive urban greening, and renewable energy adoption offer replicable models for sustainable development.
- **Economic Incentives**
  Carbon pricing mechanisms, such as carbon taxes and cap-and-trade systems, encourage industries to adopt cleaner technologies and reduce emissions.

Public Health Interventions

Community-driven and health-centered approaches are critical in reducing exposure to air pollution and enhancing resilience:

- **Urban Greening**
  Planting trees and establishing green belts in urban areas can act as natural air filters, reducing PM and improving air quality. Rooftop gardens and vertical forests further optimize limited urban spaces for greening.
- **Awareness Campaigns**
  Educating communities about the health risks of air pollution and promoting behavioral changes, such as using public transport and reducing open burning, can reduce pollution exposure.
- **Protective Measures**
  Encouraging the use of masks, especially during high pollution episodes, and improving indoor ventilation systems can minimize personal exposure to pollutants.
- **Vulnerable Population Support**
  Special interventions targeting children, the elderly, and individuals with pre-existing health conditions, such as air-conditioned shelters during smog events, can reduce health risks.
- **Healthcare System Strengthening**
  Expanding healthcare access and monitoring air pollution-related diseases can enable early diagnosis and treatment, reducing the long-term impact on public health.

Integrated Approach

While these strategies are effective individually, an integrated approach yields the best outcomes. Combining technological advancements with stringent policies and community-driven initiatives ensures comprehensive mitigation. Investments in research and development, cross-sector collaboration, and global cooperation are essential for sustainably addressing air pollution and its effects on human health.

CONCLUSION

This chapter reflects air pollution's multifaceted and profound impacts on human health and the environment, emphasizing its global relevance as a critical public health challenge. Through an extensive analysis of major air pollutants, including PM, nitrogen oxides, sulfur oxides, ozone, carbon monoxide, and lead, this chapter underscores the interconnected mechanisms through which these pollutants adversely affect respiratory, cardiovascular, neurological, reproductive, and overall systemic health. Vulnerable populations, such as children, the elderly, and those with pre-existing conditions, bear a disproportionate burden of these health risks, with air pollution exacerbating chronic conditions like asthma, cardiovascular diseases, and cancers, as well as contributing to cognitive decline and mental health issues. Additionally, the environmental ramifications, including contributions to global warming, acid rain, and ecosystem degradation, highlight the far-reaching consequences of unchecked pollution. This chapter stresses the urgent need for targeted interventions—ranging from stringent regulatory measures and innovative pollution control technologies to public awareness campaigns—to mitigate health risks and safeguard the environment. By integrating global research findings and proposing actionable strategies, this work calls for a collaborative and sustained effort to combat the escalating threats posed by air pollution, thereby advancing both human well-being and ecological resilience.

REFERENCES

Ahmad, S., Zeb, B., Ditta, A., Alam, K., Shahid, U., Shah, A. U., & Alqurashi, M. (2023). Morphological, mineralogical, and biochemical characteristics of particulate matter in three size fractions (PM10, PM2.5, and PM1) in the urban environment. *ACS Omega*, *8*(35), 31661–31674.

Ali, A., Khan, H., Bahadar, R., Riaz, A., & Asad, M. H. H. B. (2020). The impact of airborne pollution and exposure to solar ultraviolet radiation on skin: mechanistic and physiological insight. *Environmental Science and Pollution Research*, *27*, 28730–28736.

Austin, K. F., & Mejia, M. T. (2017). Household air pollution as a silent killer: women's status and solid fuel use in developing nations. *Population and Environment*, *39*, 1–25.

Bartaria, V., Jangid, A., & Kumar, R. (2023). Pyrotechnic impact of diwali 2023 on black carbon and aerosol levels in Taj City, India: relative endangerments and meteorological influences. *Urban Climate*, *61*, 102386.

Bhui, K., Newbury, J. B., Latham, R. M., Ucci, M., Nasir, Z. A., Turner, B., & Coulon, F. (2023). Air quality and mental health: evidence, challenges and future directions. *BJPsych Open*, *9*(4), e120.

Chauhan, B. V., Corada, K., Young, C., Smallbone, K. L., & Wyche, K. P. (2024). Review on sampling methods and health impacts of fine ($PM_{2.5}$, ≤2.5 μm) and ultrafine (UFP, $PM_{0.1}$, ≤0.1 μm) particles. *Atmosphere*, *15*(5), 572.

Chelani, A. B., Vyawahare, R. & Gautam, S. (2023). Study of ground ozone and precursors along with particulate matter at residential sites in the vicinity of power plant. *Waste Disposal & Sustainable Energy, 5*, 535–549. https://doi.org/10.1007/s42768-023-00163-1

Clark, C., Sbihi, H., Tamburic, L., Brauer, M., Frank, L. D., & Davies, H. W. (2017). Association of long-term exposure to transportation noise and traffic-related air pollution with the incidence of diabetes: a prospective cohort study. *Environmental Health Perspectives, 125*(8), 087025.

Collin, M. S., Venkatraman, S. K., Vijayakumar, N., Kanimozhi, V., Arbaaz, S. M., Stacey, R. S., & Swamiappan, S. (2022). Bioaccumulation of lead (Pb) and its effects on human: a review. *Journal of Hazardous Materials Advances, 7*, 100094.

ESCAP. (2024). *Compendium of Technologies for Air Pollution Control*. The Economic and Social Commission for Asia and the Pacific, Bangkok, Thailand.

García-Ortiz, A., & Serrador, J. M. (2018). Nitric oxide signaling in T cell-mediated immunity. *Trends in Molecular Medicine, 24*(4), 412–427.

Gautam, S., Pillarisetti, A., Yadav, A. et al. (2019). Daily average exposures to carbon monoxide from combustion of biomass fuels in rural households of Haryana, India. *Environment, Development and Sustainability* 21, 2567–2575. https://doi.org/10.1007/s10668-018-0131-1

Gupta, M. K. (2024). Status of air quality and its impact on human health in India: a review. *International Journal of Research and Analytical Reviews, 11*, 248–255.

Guttikunda, S. K., Goel, R., & Pant, P. (2014). Nature of air pollution, emission sources, and management in the Indian cities. *Atmospheric Environment, 95*, 501–510.

Iqubal, A., Ahmed, M., Ahmad, S., Sahoo, C. R., Iqubal, M. K., & Haque, S. E. (2020). Environmental neurotoxic pollutants. *Environmental Science and Pollution Research, 27*, 41175–41198.

Izah, S. C., Ogwu, M. C., Etim, N. G., Shahsavani, A., & Namvar, Z. (2024). Short-term health effects of air pollution. In: Izah, S. C., Ogwu, M. C., & Shahsavani, A. (eds) *Air Pollutants in the Context of One Health. The Handbook of Environmental Chemistry* (vol. 134). Springer, Cham.

Khajeamiri, Y., Sharifi, S., Moradpour, N., & Khajeamiri, A. (2021). A review on the effect of air pollution and exposure to PM, NO_2, O_3, SO_2, CO and heavy metals on viral respiratory infections. *Journal of Air Pollution and Health, 5*(4), 243–258.

Kim, J. B., Prunicki, M., Haddad, F., Dant, C., Sampath, V., Patel, R., & Nadeau, K. C. (2020). Cumulative lifetime burden of cardiovascular disease from early exposure to air pollution. *Journal of the American Heart Association, 9*(6), e014944.

Knibbs, L. D., Van Donkelaar, A., Martin, R. V., Bechle, M. J., Brauer, M., Cohen, D. D., & Barnett, A. G. (2018). Satellite-based land-use regression for continental-scale long-term ambient $PM_{2.5}$ exposure assessment in Australia. *Environmental Science & Technology, 52*(21), 12445–12455.

Kumar, A., Kumar, S., & Kumari, A. (2023, June). Carbon monoxide concentration in atmosphere—a review. In *IRC-SET 2022: Proceedings of the 8th IRC Conference on Science, Engineering and Technology* (pp. 97–109). Springer Nature, Singapore.

Levin, R., Vieira, C. L. Z., Rosenbaum, M. H., Bischoff, K., Mordarski, D. C., & Brown, M. J. (2021). The urban lead (Pb) burden in humans, animals and the natural environment. *Environmental Research, 193*, 110377.

Li, J., Lu, X., Liu, F., Liang, F., Huang, K., Yang, X., & Gu, D. (2020). Chronic effects of high fine particulate matter exposure on lung cancer in China. *American Journal of Respiratory and Critical Care Medicine, 202*(11), 1551–1559.

Li, W. W. (2020). Air pollution, air quality, vehicle emissions, and environmental regulations. In: Haneen, K., Mark, N., Josias, Z., & Tara, R. (eds) *Traffic-Related Air Pollution* (pp. 23–49). Elsevier, Amsterdam, Netherlands.

Lin, C. C., Chiu, C. C., Lee, P. Y., Chen, K. J., He, C. X., Hsu, S. K., & Cheng, K. C. (2022). The adverse effects of air pollution on the eye: a review. *International Journal of Environmental Research and Public Health, 169*(1), e240194.

Luderer, U. (2024). Adverse impacts of particulate matter air pollution on female and male reproductive function. *Reproduction, 72*(17), 2054–2070.

Ma, F. (2022). The current status of cigarette smoking and air pollution associated lung cancer. *Preprints, 2022*, 2022060056. https://doi.org/10.20944/preprints202206.0056.v1.

Mallah, M. A., Changxing, L., Mallah, M. A., Noreen, S., Liu, Y., Saeed, M., & Zhang, Q. (2022). Polycyclic aromatic hydrocarbon and its effects on human health: an overeview. *Chemosphere, 296*, 133948.

Manisalidis, I., Stavropoulou, E., Stavropoulos, A., & Bezirtzoglou, E. (2020). Environmental and health impacts of air pollution: a review. *Frontiers in Public Health, 8*, 14.

Maung, T. Z., Bishop, J. E., Holt, E., Turner, A. M., & Pfrang, C. (2022). Indoor air pollution and the health of vulnerable groups: a systematic review focused on particulate matter (PM), volatile organic compounds (VOCs) and their effects on children and people with pre-existing lung disease. *International Journal of Environmental Research and Public Health, 19*(14), 8752.

Mohania, D., Chandel, S., Kumar, P., Verma, V., Digvijay, K., Tripathi, D., & Shah, D. (2017). Ultraviolet radiations: skin defense-damage mechanism. *Ultraviolet Light in Human Health, Diseases and Environment, 996*, 71–87.

Paul, I. (2023). *Air Pollution in India.* Università degli Studi di Torino, Italy.

Perumpully, S. J., Gautam, S., John, J. P., & Sreenath, M. (2024). Evaluating the impact of personal exposure to emissions from sustainable commercial heating and cooking fuels on women in rural southern India and their alignment with sustainable development goals. *Water, Air, & Soil Pollution, 235*, 54. https://doi.org/10.1007/s11270-023-06854-z

Piracha, A., & Chaudhary, M. T. (2022). Urban air pollution, urban heat island and human health: a review of the literature. *Sustainability, 14*(15), 9234.

Popov, O., Taraduda, D., Sobyna, V., Sokolov, D., Dement, M., & Pomaza-Ponomarenko, A. (2020). Emergencies at potentially dangerous objects causing atmosphere pollution: peculiarities of chemically hazardous substances migration. In: Babak, V., Isaienko, V., Zaporozhets, A. (eds) *Systems, Decision and Control in Energy I* (pp. 151–163). Springer International Publishing, Cham.

Rajagopalan, S., Al-Kindi, S. G., & Brook, R. D. (2018). Air pollution and cardiovascular disease: JACC state-of-the-art review. *Journal of the American College of Cardiology, 72*(17), 2054–2070.

Schraufnagel, D. E., Balmes, J. R., Cowl, C. T., De Matteis, S., Jung, S. H., Mortimer, K., & Wuebbles, D. J. (2019). Air pollution and noncommunicable diseases: a review by the Forum of International Respiratory Societies' Environmental Committee, Part 2: air pollution and organ systems. *Chest, 155*(2), 417–426.

Sellami, F., Baati, H., Gautam, S., Azri, C., & Elaksher, A. F. (2025). Impact assessment of particulate deposits bound heavy metals on superficial soils along a street canyon at Sfax City (Tunisia): Pollution degree and toxicity risks. *Earth Systems and Environment.* Advance online publication. https://doi.org/10.1007/s41748-025-00719-7

Shabani, S. (2021). A mechanistic view on the neurotoxic effects of air pollution on central nervous system: risk for autism and neurodegenerative diseases. *Environmental Science and Pollution Research, 28*, 6349–6373.

Singh, P., & Yadav, D. (2021). Link between air pollution and global climate change. In: Singh, S., Singh, P., Rangabhashiyam, S., & Srivastava, K. K. (eds) *Global Climate Change* (pp. 79–108). Elsevier, Amsterdam, Netherlands.

Thanvisitthpon, K., Akbar, A. P., Chao, W. H. J., & Nawhath, K. (2024). Respiratory diseases caused by air pollutants. In: Mohammad, H. D., Rama, R. K., Teresa, V., & Salwa Kamal, M. H. (eds) *Diseases and Health Consequences of Air Pollution: Volume 3: Air Pollution, Human Health, and the Environment* (pp. 27–53). Academic Press, New York.

Theodorakidou, M., & Lambrou, G. I. (2017). Public health issues from the exposure to nitrogen oxides: a brief review. *ARC Journal of Public Health and Community Medicine*, *2*, 44–56.

Thiele, J. J., Traber, M. G., Tsang, K., Cross, C. E., & Packer, L. (1997). In vivo exposure to ozone depletes vitamins C and E and induces lipid peroxidation in epidermal layers of murine skin. *Free Radical Biology and Medicine*, *23*(3), 385–391.

Ukaogo, P. O., Ewuzie, U., & Onwuka, C. V. (2020). Environmental pollution: causes, effects, and the remedies. In: Pankaj, C., Abhay, R., Digvijay, V., & Yusuf, A. (eds) *Microorganisms for Sustainable Environment and Health* (pp. 419–429). Elsevier, Amsterdam, Netherlands.

Warburton, D. E., Bredin, S. S., Shellington, E. M., Cole, C., de Faye, A., Harris, J., & Abelsohn, A. (2019). A systematic review of the short-term health effects of air pollution in persons living with coronary heart disease. *Journal of Clinical Medicine*, *8*(2), 274.

Wei, J., Li, H., & Liu, J. (2021). Fate of dioxins in a municipal solid waste incinerator with state-of-the-art air pollution control devices in China. *Environmental Pollution*, *289*, 117798.

World Health Organization. (2016). *Burning Opportunity: Clean Household Energy for Health, Sustainable Development, and Wellbeing of Women and Children.* World Health Organization, Geneva, Switzerland.

Zhang, L., Ou, C., Magana-Arachchi, D., Vithanage, M., Vanka, K. S., Palanisami, T., & Kirkham, M. B. (2021). Indoor particulate matter in urban households: sources, pathways, characteristics, health effects, and exposure mitigation. *International Journal of Environmental Research and Public Health*, *18*(21), 11055.

Global Health Threats of Air Pollution

Risk Research and Future Directions

2

Hamid T. Al-Saad, Nader A. Salman, Maha K. Al-Mishrey, and Ahmed Rushdi

AIR POLLUTION AND GLOBAL HEALTH

People are becoming more aware of the positive impact that human activities, such as technology and medicine, have had on our lives. However, there is growing concern about the detrimental effects that the depletion of the Earth's natural resources will have on the environment and society. Air pollution is a major issue that is recognized as a significant threat to public health, causing illnesses and disabilities. It can stem from both natural and human activities, containing harmful substances from various sources. As a result, understanding the impact of air pollution on human health is crucial for developing effective environmental policies to reduce the burden of disease (Popov et al., 2020; Manisalidis et al., 2020).

 DOI: 10.1201/9781003613183-2

Air pollution has become the fourth leading risk factor for premature death. The 2020 State of Global Air report, which presents the most recent scientific insights on global air quality, estimated that in 2019, outdoor air pollution was associated with 4.5 million deaths, while indoor pollution contributed to an additional 2.2 million fatalities (HEI, 2020). Furthermore, World Health Organization (WHO) announced that ambient and household air pollution together account for 7 million premature deaths globally every year, making it more deadly than a combination of malaria, tuberculosis, and AIDS (WHO, 2021a). Specifically, ambient air pollution is estimated to cause the premature deaths of approximately 4.2 million people globally, primarily due to lung cancer, stroke, chronic obstructive pulmonary disease (COPD), heart disease, and acute respiratory infections (ARI) in children (WHO, 2021b). It disproportionately affects women, children, and the elderly in developing nations, who are frequently exposed to ambient and indoor air pollution (Pillarisetti et al., 2022; Dida et al., 2022). Air pollution is a global problem, and it is estimated to contribute to one in every 16 deaths worldwide (Diener and Mudu, 2021). Identifying the most affected population groups is essential for implementing strategies to minimize the effects of pollution.

GLOBAL AIR POLLUTION EXPOSURE LEVEL

The exposure to air pollution is particularly high for people living in Asia and Africa compared to people residing in America and Europe (Fisher et al., 2021; Rentschler and Leonova, 2023; Sicard et al., 2023). For instance, the exposure level of $PM_{2.5}$ annual average in East Asia and Pacific regions is 43 g/m^3 and Sub-Saharan Africa is 67 μg g/m^3. On the other hand, the annual average $PM_{2.5}$ exposure in North America is 9 μg g/m^3 and that in the European Union is 14 μg/m^3. Therefore, the exposure level can be seven orders of magnitude higher for the population in Africa as compared to people in North America. Moreover, 95% of the world population is currently exposed to $PM_{2.5}$ higher than 10 m^3, which is the recommended standard by the WHO to reduce deaths attributable to air pollution by 15%. Fatimah et al. (2024) mentioned that Africa exhibits the highest number of death per 100,000 people from ambient and household air pollution, combined with 180.9 deaths annually, followed by Southeast Asia with 165.8 deaths. Europe and the Americas experience much fewer deaths from air pollution with 29.7 and 36.3 deaths per 100,000 people, respectively (Ozkaynak and Thurston, 1987).

The increase in pollution due to industrialization and intensified agriculture in developing countries has further worsened the global epidemic of air pollution. Numerous studies have been conducted to understand the risks associated with air pollution, and these findings are vital for identifying gaps in research and shaping future policies (Tainio et al., 2021). This comprehensive review of air pollution and its impact on global health will help in identifying priorities for future research and policy formation. It is essential to address the worsening issue of atmospheric pollution to prevent further negative effects on human health.

DEFINITION OF AIR POLLUTION

Air pollution is the release of harmful pollutants and particulate matter into the atmosphere, which can have negative effects on humans, plants, animals, and the environment (Alharbi et al., 2024; Rushdi et al., 2023). There are two types of air pollution: primary pollutants, which are emitted directly into the atmosphere from sources like heavy metals and sulfur dioxide, and secondary pollutants, which are formed from chemical reactions between primary pollutants and natural elements such as water vapor and sunlight, resulting in substances like smog and ground-level ozone (Liao et al., 2021; Sridharan et al., 2021; Schraufnagel, 2020). Air pollution refers to a deterioration in air quality, identified through the presence of chemical, biological, or physical contaminants in the atmosphere. It signifies either the unwanted introduction of harmful substances or an abnormal increase in certain atmospheric components. More broadly, air pollution involves the emission of pollutants that pose serious risk to both human health and the environment. According to the WHO, indoor and outdoor air pollution together contribute to nearly seven million deaths worldwide each year (Turner et al., 2020; Tran et al., 2020).

SOURCES AND TYPES OF AIR POLLUTION

According to Moreno-Ríos et al. (2022), the main sources of air pollution include industrial complexes, power plants, waste incinerators, car exhaust, the petrochemical industry, oil refineries, fires, and natural sources such as volcanoes. Urban areas are particularly susceptible to air quality deterioration due to industrial discharges, vehicular emissions, and rapid population growth after industrialization. The complexities of air pollution sources make it challenging to compare emissions and their health impacts, underscoring the need for strategic planning for pollution control and a better understanding of local pollution and its impacts (Abaje et al., 2020). Types of air pollution as mentioned by Fatimah et al. (2024) can be classified into two categories: visible and invisible air pollution. Or local for the quality of ambient air within a radius of a few kilometers and regional for pollution like acid rain, photochemical reactions, and degradation of water quality at distances of a few kilometers to a thousand kilometers. The global category refers to the depletion of the ozone layer and global warming caused by the emission of greenhouse gases, mainly carbon dioxide (CO_2).

CAUSES OF AIR POLLUTION

Air pollution is generally caused by gas-phase pollutants (e.g., ozone, nitrogen dioxide, sulfur dioxide, and carbon monoxide) and particulate matter, especially particles less than 2.5 µm in diameter ($PM_{2.5}$). Air pollution results from the presence of toxic substances in the atmosphere, primarily generated by human activities. However, it can also stem from natural events such as dust storms, wildfires, and volcanic eruptions, all of which contribute to the deterioration of air quality (Inauen et al., 2021; Awewomom et al., 2024).

According to Solarimpulse (2020), major sources of anthropogenic air pollution include:

1. The combustion of fossil fuels such as oil and coal for transportation and electricity generation, which releases pollutants like nitrogen oxides and sulfur dioxide.
2. Industries and factory emissions, which discharge significant amounts of hydrocarbons, organic compounds, carbon monoxide, and chemicals, into the atmosphere.
3. Agricultural practices, particularly the use of fertilizers, insecticides, and pesticides, which emit hazardous chemicals.
4. Waste production, primarily from methane emissions generated in landfills.
5. Most atmospheric contamination comes from energy use and production. Driving a car on gasoline, heating a home with oil, running a power plant on fracked gas: In each case, a fossil fuel is burned, and harmful chemicals and gases are released into the air.

GAS EMISSION RISKS

According to the emissions gap report 2023, issued by UNDP (2023), there is an escalating challenge due to greenhouse gas emissions, rising temperatures, and increasingly intense climate impacts (Gautam et al., 2021; Stepanov and Makarov, 2024; Skillington et al., 2022). In 2022, global greenhouse gas emissions reached a record high of 57.4 $GtCO_2e$, with fossil fuel combustion and industrial processes accounting for two-thirds of the total emissions. The rapid increase in methane (CH_4), nitrous oxide (N_2O), and fluorinated gases (F-gases) further exacerbates these environmental challenges (Inauen et al., 2021). The emissions of gases ultimately lead to increased air pollution. According to the latest WHO report, 99% of the global population is exposed to air pollution levels that elevate the risk of diseases such as heart disease, stroke, COPD, cancer, and pneumonia (WHO, 2021a,b). WHO statistics indicate that air pollution contributes to

6.7 million deaths annually, with household air pollution remaining a significant concern in 2021 as 2.3 billion people relied on polluting fuels and technologies for cooking (Awewomom et al., 2024).

IMPACT OF AIR POLLUTION

It is impossible to describe the whole extent of potential and actual damage caused by all forms of air pollution. But here are the main consequences (Solarimpulse, 2020):

1. **Environmental Impact**: Air pollution has a major impact on the process of plant evolution by preventing photosynthesis in many cases, with serious consequences for the purification of the air we breathe. It also contributes to the formation of acid rain, atmospheric precipitations in the form of rain, frost, snow, or fog. These substances are emitted during the combustion of fossil fuels and undergo transformation upon interaction with water vapor and steam in the atmosphere.
2. **Impact on Global Warming**: Moreover, air pollution plays a significant role in global warming and climate change. One key factor is the high concentration of carbon dioxide, which contributes to the greenhouse effect. Under normal conditions, greenhouse gases are beneficial, as they help to trap infrared radiation emitted by the Earth's surface, maintaining a stable climate. However, the excessive accumulation of these gases in the atmosphere is driving the recent changes in the global climate.
3. **Impact on Human Health**: Continuous exposure to air pollutants contributes significantly to the decline of human health. Air pollution is a major risk factor, linked to respiratory and cardiovascular diseases, allergies, and damage to lung tissue (Sarla, 2020; Fatimah et al., 2024).

IMPACT ON HUMAN HEALTH

Air pollution has tremendous effects on human health in the form of respiratory diseases and aggravations in the form of asthma and lung cancer, cardiovascular dysfunctions, and malignant growth (Sarla, 2020; Fatimah et al., 2024). An affiliation has been found to exist between male infertility and air pollution, and a relationship has been established between air contamination and greater danger of immune dysfunction, neuroinflammation, neurobehavioral hyperactivity, crime, age-unseemly behaviors, and Alzheimer's and Parkinson's disease (Guo et al., 2022). Traffic-related air pollutants have been found to affect skin aging and cause pigmented spots on the face (Dominici et al., 2022). An association exists between air pollution and irritation of the eyes, dry eye syndrome, risk of retinopathy, and adverse ocular outcomes. Chronic exposure to air pollutants during pregnancy has been associated with adverse effects on the developing fetus in the form of low birth weight and stillbirth. Air contamination has been seen as a significant supporter of the expanded predominance of allergic diseases in children. Air pollution has a wide range of negative effects on human health. Short-term exposure to particulate matter and ground-level ozone has been linked to acute respiratory diseases, including wheezing, coughs, phlegm, and acute bronchitis (Perumpully et al., 2024; Dominici et al., 2022).

Living in heavily polluted areas can also have psychosocial effects such as an increased risk of developing mental health conditions like anxiety and depression. Exposure to stressors like noise, odor, dust, and air pollution can lead to stress, worries, restlessness, and helplessness. Additionally, high levels of particulate matter are associated with more doctor visits for mental health disorders. These impacts come with significant societal costs in terms of health, healthcare, and quality of life (Comunian et al., 2020; Kim et al., 2020). The effects of air pollution on the human body vary, depending on the type of pollutant, the length and level of exposure, and other factors, including a person's health risks and the cumulative impacts of multiple pollutants or stressors (Solomon et al., 2012).

IMPACT ON RESPIRATION AND CARDIOVASCULAR ACTIVITIES

Research shows that these pollutants are involved in various molecular pathways that contribute to respiratory inflammation. Long-term exposure to air pollution also increases cardiovascular problems and is associated with other serious health effects. Studies have found higher rates of asthma, COPD, lung and heart cancer, cardiovascular mortality, low birth weights, premature death, and reduced life expectancy in areas with high levels of pollution (Hayes et al., 2020). Wheezing, dry mouth, cough, and reduced ability to perform activities due to breathing difficulties are the most common clinical symptoms of respiratory illnesses caused by air pollution (Gao et al., 2014; Bentayeb et al., 2013). Air contamination is likewise considered the major natural hazard factor for some respiratory illnesses, such as asthma and lung cancer, as pointed out by Weisel (2002) and Brunekreef et al. (2009). Epidemiologic examinations have demonstrated the immediate relationship between air toxin presentation and heart-related illnesses as mentioned by Nogueira (2009). Health effects and long-haul impacts of air contamination on the beginning of maladies, such as respiratory diseases and aggravations, cardiovascular dysfunctions, and malignant growth, are generally accepted (Zhang et al., 2014; Gautam et al., 2020).

IMPACT ON NEUROPSYCHIATRY MANIFESTATIONS

Studies, such as those of Haynes et al. (2011) and Newman et al. (2013), have shown the connection between air contamination and neurobehavioral hyperactivity, crime, and age-unseemly behaviors. Some other studies have likewise uncovered the relationship between air contamination and higher danger of neuroinflammation (Calderón-Garcidueñas et al., 2008a), Alzheimer's and Parkinson's disease (Calderón-Garcidueñas, 2008b). A few investigations demonstrated that hostility and tension in megacities are in close association with the significant level of air pollutants (Rotton et al., 1979).

SKIN CHANGES

The study of Vierkötter et al. (2010) revealed that traffic-related air pollutants contribute to skin aging and cause pigmented spots on the face.

IMMUNE SYSTEM

Poor air quality can cause serious complications in the immune system such as an abnormal increase in the serum levels of immunoglobulin (Ig); IgA, IgM, and the complement component C3 in humans. It can also contribute to chronic inflammatory diseases of the respiratory system (Hadnagy et al., 1996). Exposure to these immunotoxicants may also cause immune dysfunction at different stages, increasing the risks of numerous diseases such as neuroinflammation (Calderón-Garcidueñas, 2008b; Rider and Carlsten, 2019).

EFFECTS ON EYES

According to Rozanova et al. (2009), chronic exposure to air pollutants increases the risk of retinopathy and adverse ocular outcomes. In addition, there is now evidence suggesting the association between air pollution and irritation of the eyes, dry eye syndrome, and some of the major blinding (Lin et al., 2022; Trpevska-Shekerinov et al., 2022).

IMPACT ON FERTILITY AND PREGNANCY

Recent study, such as that of Backes et al. (2013), has shown that air pollution can affect the developing fetus through maternal exposure, leading to preterm birth, low birth weight, growth restriction, and potentially adverse cardiovascular and respiratory outcomes. Several epidemiological and clinical studies conducted in low-income countries found an association between exposure to indoor air pollution during pregnancy and low birth weight and stillbirth (Lakshmi et al., 2013). Allergy air contamination has been identified as a significant contributor to the increasing predominance of allergic diseases in children (Leung et al., 2012). An affiliation has been found to exist between male infertility and air pollution in a study by Zhou et al. (2014).

EPIDEMIOLOGICAL STUDIES ON AIR POLLUTION AND HEALTH RISKS

Environmental risk, although it is exposed to the residential environment and may be affected by interfering factors, has received more attention worldwide since the development of various monitoring and individual-level exposure measurements and risk evaluations (Abaje et al., 2020; Blessy et al., 2023). In epidemiological studies, the longitudinal study method is classified into two types: cohort and case-control studies, which are epidemiologically expected to establish causation (Checkoway et al., 2004, 2007; Morgenstern and Thomas, 1993; Silman et al., 2018). Many studies have shown that the morbidity and mortality of the population with better air quality at the research site are significantly lower than those at the reference site with worse air quality (Kim et al., 2020; Kumar et al., 2023). Recently, the value of exposure and risk attributable to the entire country and other populations as a result of considering an analysis of accrued exposures to exposure assessment and follow-ups with a common cohort has also been evaluated by other updates. At present, there are only a few investigations of the general population and community-based epidemiological follow-up studies on air pollution and health risks (Mukherjee et al., 2021).

EPIDEMIOLOGICAL KEY FINDINGS

Health research on the spread of diseases has played a vital role in providing evidence about the potential dangers of being exposed to air pollution. Numerous studies have uncovered connections between exposure to air pollution, particularly $PM_{2.5}$ and NO_2, and both short-term and long-term health effects (Karimi et al., 2019; Kloog et al., 2013; Mainka and Żak, 2022). These connections are often expressed as a percentage increase in health problems for each unit of increase in air pollution. This percentage increase, known as the risk metric, varies based on the population and the specific health issue being studied (Nicolaou and Checkley, 2021). These studies typically report their findings based on statistical significance and confidence intervals. When a relationship is found to be statistically significant, it indicates that at a certain level of exposure, the selected population is at risk from the pollution. This creates a scale of acceptable exposure levels, with each level linked to a quantified increase in the risk of harm (Ma et al., 2020).

EPIDEMIOLOGICAL METHODS AND ASSESSMENT APPROACHES

Recent evaluations offer in-depth information on various epidemiological methods and assessment approaches for exposure-response (Arroyave et al., 2021; Checkoway et al., 2004; Forastiere et al., 2024; Viet et al., 2021). These include time series, panels, and cohort studies, each with specific strengths and limitations for different research questions. The emphasis on exposure assessment techniques for population exposure to air pollutants at micro-level spatial resolutions while considering confounding factors has been a major focus for decades. There is growing concern about the impact of methodological issues on public health evidence, especially in more recent and broader studies. Landmark studies continue to significantly influence air pollution health risk policy debates and outcomes (Peng et al., 2022). The value of an interdisciplinary, collaborative approach to considering air pollution exposure and health is well-documented, with emphasis on understanding long-term cumulative exposure-response, non-linear concentration response, and individual risks across sub-populations. Epidemiological findings have been critical in identifying public health awareness needs and prioritizing targets for public intervention and policy initiatives, including setting internationally accepted air quality and health standards. While most evidence is based on short-term or annual average mean concentration exposures, there is increasing interest in more robust, comprehensive, and longitudinal health studies (Lippmann et al., 2013).

WAYS TO REDUCE AIR POLLUTION

Reducing pollution must involve the adoption of cleaner production technologies, enforcement of regulations, promotion of sustainable transportation, and implementation of proper waste disposal systems (Inauen et al., 2021). Preventive environmental management approaches are essential, yet the implementation of environmental policies needs careful attention to tackle global challenges (Oloruntobi et al., 2023). Recent studies have explored strategies such as computable general equilibrium models and the enforcement of environmental policies, which demonstrate the potential for effective pollution control (Inauen et al., 2021). Strengthening institutions and global environmental governance is crucial to mitigate environmental impacts and ensure sustainable development. Addressing global pollution requires a holistic approach, proactive management strategies, and international cooperation to transition toward sustainability and reduce the negative impacts of pollution on the planet (Oloruntobi et al., 2023).

WAYS FOR PREVENTION OF AIR POLLUTION

In recent decades, global death attributable to air pollution has shown a decreasing trend, mainly because of the reduced indoor air pollution, while outdoor particulate matter and ozone pollution are exhibiting minor changes.

There are ways to prevent, control, and eventually reduce air pollution as recorded by Solarimpulse (2020):

1. **Renewable Fuel and Clean Energy Production**: One of the most fundamental solution to air pollution is transitioning away fossil fuels and adopting alternative energy sources such as wind, solar, and geothermal power.
2. **Energy Conservation and Efficiency**: While generating clean energy is essential, it is equally important to minimize energy consumption by adopting sustainable practices and using energy-efficient technologies.
3. **Eco-Friendly Transportation**: Reducing air pollution can be achieved by promoting electric vehicles and hydrogen-powered vehicles, as well as encouraging shared mobility options like public transport and carpooling.
4. **Green Building**: From design to demolition, green buildings focus on developing environmentally sustainable and resource-efficient structures to lower their overall carbon footprint.

Preventive measures for legitimate cooking stoves might be utilized rather than wood-consuming cooking stoves to decrease air contamination. Romieu et al. (2009) showed that in rural Mexico, a randomized trial of properly vented wood-burning cooking stoves versus open fires showed reductions in the longitudinal decline in forced expiratory volume in one second and improvements in respiratory symptom when proper cooking stoves were used. The use of improved cooking stoves has also been found to halve the exposure to carbon monoxide and resulted in a lower rate of diagnosis of pneumonia (Smith-Sivertsen et al., 2009). Urban forests and green roofs should be promoted as effective strategies for mitigating pollution in city environment (Escobedo et al., 2011). According to Smith et al. (2011), vegetation eliminates pollutants through various mechanisms, including the absorption of gaseous pollutants, the interception of particulate matter by leaves, and the degradation of organic compounds such as polycyclic aromatic hydrocarbons. Upgrades in the ventilation of homes in which biomass fuels are used will reduce the health effects of air pollution. Empowering walking and cycling as opposed to utilizing vehicles and bikes while voyaging short distances will help to check air contamination.

TECHNOLOGICAL SOLUTIONS FOR AIR POLLUTION REDUCTION

In recent years, new monitoring technologies and networks have emerged, providing researchers and policymakers with a more comprehensive understanding of air quality, including gaseous emissions and particulate matter. Research has identified and tested various technological solutions to air pollution, showing that emissions from industrial and vehicular sources can be effectively reduced using different technologies. Additionally, a variety of emissions control devices have been created to trap or break down air pollutants, and it has been found that energy sources such as gas produce lower polluting emissions compared to other energy sources (Rahman and Thurston, 2022).

Further attention should be given to the research and development of new and emerging technological innovations. Investments in technology can have additional benefits, particularly in terms of energy saving and reducing short-lived climate pollutants. There are various examples of technology-driven programs being used to address air pollution (Alvarado et al., 2019). The European Union has utilized technological advances to develop initiatives, including installing air quality monitoring stations at fixed and mobile sites. Other technologies being studied in different cities include reducing diesel truck emissions, such as deploying hybrid electric trucks and using engines with diesel particulate filters and catalysts. While these solutions may be more costly, further research is necessary to bring down costs and make them applicable in a wider range of situations. Ongoing public–private partnerships are crucial to address these technological frontiers (Popov et al., 2020; Manisalidis et al., 2020).

Overall, variations in air pollution levels in different regions emphasize the need for customized air pollution control programs. It is crucial to embrace advancements in clean energy technology and employ a focused strategy for monitoring various pollutants. It also indicates the potential for science to guide in addressing both indoor and outdoor air pollution. Organizations and government agencies should pay attention to these insights, as they are well-positioned to lead innovative future research in this area (Thurston et al., 2013).

CURRENT TECHNOLOGIES AND INNOVATIONS

Modern advancements and creative solutions are crucial in the battle against air pollution. Indoor and outdoor air purifiers are now essential tools in many countries, while high-performance air filtration and emission-reducing technologies, like diesel particulate filters for vehicles and urban transportation systems, are also making an impact (Varughese et al., 2023). In addition, new innovative technologies and solutions are being developed to address air pollution in cities, such as integrating smart sensors into urban systems for real-time air quality monitoring (Rahman and Thurston, 2022; Blaga and Gautam, 2024). The increasing use of data-driven technologies has also sparked interest in AI-driven approaches to recommend solutions to local governments and authorities. The next frontier is the development of AI-driven technologies that use predictive analytics to proactively reduce air pollution levels based on weather forecasting and construction patterns. Other technological initiatives are also in progress to combat urban air pollution (Molina et al., 2020).

Despite the success of many technologies in preventing air pollution, there are still challenges in controlling and monitoring it. For example, air quality monitoring technology in cities using real-time data faces issues with data integration and setting standards for external systems (Alvarado et al., 2019). Collaboration between stakeholders is crucial to test and validate technologies, and ongoing research into current and potential future technologies is essential. It is important to consider how to best use innovative technologies for indoor and outdoor air quality monitoring and to explore the potential of emerging technologies such as AI. Investigators must work together to identify and address obstacles that may become increasingly challenging over time (Hajat et al., 2015).

POLICY INTERVENTIONS AND GOVERNANCE FRAMEWORKS

Air pollution is a multifaceted societal issue that necessitates cooperation to tackle. Effective solutions must focus on both reducing climate change and transitioning transportation to electric power to effectively address these interrelated problems. There are established national and international guidelines that dictate the standards for air quality and control the levels of air pollution caused by transportation. Individual countries have the authority to create their air quality standards, usually following recommended guidelines. Rules on emissions have been slowly put into place, encompassing various aspects of transportation and engine technology. Additionally, more stringent regulations are also being introduced to address the requirements after the use of vehicles.

Research into air pollution's effects on public health has influenced the creation of policies. Some countries have implemented effective social policies that resulted in a significant decrease in emissions and improved air quality and public health. Additionally, an international organization has established regulations for international shipping emissions. One potential approach to address the health effects of air pollution and greenhouse gas emissions is to create a national policy that focuses on improving air quality in the energy and transport sectors. This policy could also help to meet climate change targets by reducing emission sources of $PM_{2.5}$ and ground-level ozone (Maciejczyk et al., 2021). Previous policies have aligned with air quality standards to regulate greenhouse gas emissions from light-duty vehicles. It is widely recognized that emissions from transportation contribute to harmful air pollution and have significant public health impacts. Therefore, it is crucial for air pollution policies to prioritize public health, reduce traffic-related pollution, and address environmental and climate change injustices as key objectives (GBDRF, 2019; US-EPA, 2022).

International regulations for controlling air pollutants go beyond regulating vehicle emissions to also encompass managing emissions from transportation, waste, domestic, and industrial activities. These regulations cover a range of areas such as land planning, waste management, energy production, and transportation and aim to address vehicle emissions and governance gaps. The Paris Agreement, ratified by 196 member countries within the United Nations Framework Convention on Climate Change, focuses on mitigating greenhouse gas emissions, adapting to climate change, providing financial support, and responding to loss and damage. It is expected to become effective in 2021 (Manisalidis et al., 2020; Abaje et al., 2020).

NATIONAL AND INTERNATIONAL POLICIES

The harmful effects of particulate matter ($PM_{2.5}$) on human health, such as COPD, cardiovascular disease (CVD), ARI, and lung cancer, have become a major concern, especially in urban areas with high levels of smog. It is crucial to quantitatively assess the amount of $PM_{2.5}$ that people inhale to better understand the risks and develop effective policies (Maciejczyk et al., 2021). The scientific community, as well as industries like petroleum, coal, and chemicals, needs to work together to raise public awareness and takes action to address this urgent issue. It is essential to research to evaluate policy measures, law enforcement, future development plans, and the economic impact of premature deaths caused by air pollution (Marlier et al., 2016; Comunian et al., 2020).

While numerous national and international policies have led to improvements in emissions and air quality, as well as beneficial effects on health and the environment, it is evident that there is a gap between these advanced policies and the estimated reductions in health, climate, and vegetation co-benefits resulting from air pollution and precursor emission reductions. Laws and technical feasibility point toward more sustainable directions for the future. It is crucial for policymakers, public health researchers, and national and international emission regulators to remain vigilant about $PM_{2.5}$ emissions from emerging processing activities such as electronics, diesel, and synthesis gas manufacturing, as well as incineration soon. Furthermore, they must continuously monitor formulated policies, adapt them promptly, and work to improve them continuously (Yang et al., 2020).

FUTURE DIRECTIONS IN AIR POLLUTION RESEARCH

An increasingly important area of interest is the group of pollutants that are currently not the focus of research. Examples include the harmful effects of heating systems and cooking fires and the long-term health impacts of military burn pits. Studying these alternative pollutants provides valuable insights into potential sources of exposure that may be contributing to disease levels, as mentioned by Pinder et al. (2019). The impact of emerging pollutants is a topic that will require further investigation in the future (Rathikannu et al., 2025). It is becoming increasingly clear that studying different types of pollutants is crucial in understanding air quality within a broader context. This will necessitate greater interdisciplinary collaboration in future research (Wild, 2005; Harker, 2022).

In addition to environmental science, epidemiology, and public health research, other fields such as atmospheric science, biostatistics, economic geography, architecture, and urban planning may also play a significant role. The sociology of risk research has shown that taking an interdisciplinary approach can change the way risk is viewed, incorporating socioeconomic factors along with environmental evidence. There is a need to conduct more surveillance and monitoring of exposure, which is a challenging area of research in air pollution, particularly in prioritizing research for vulnerable populations (Zhang et al., 2021). Van Donkelaar et al. (2019) studied the regional estimates of the chemical composition of fine particulate matter using a combined geoscience-statistical method with information from satellites, models, and monitors.

Collaborating with policymakers and other stakeholders is crucial to emphasize the importance of conducting research in this area. It is essential to understand the connections between the views of policy developers and the public, especially about environmental policy. It has been recognized that evidence of the impact on health is necessary to motivate policymakers to prioritize reducing air pollution across different sectors. Finally, studying future projections related to climate change will be an emerging area of importance. Addressing climate change and improving air quality can not only benefit local communities but can also significantly reduce the global burden of disease (Perumpully and Gautam, 2025).

EMERGING RISKS AND VULNERABLE POPULATIONS

Air pollution impacts populations differently, with low-income communities, children, the elderly, asthmatics, diabetics, and communities of color being especially vulnerable. Even in developed countries, low-income populations are often more heavily impacted due to their proximity to pollution sources. The risk of disease and adverse health effects is higher with sustained exposure to elevated concentrations of air pollutants. This suggests that short-term studies of the general public may not accurately represent the most vulnerable populations. Children, as mentioned by Korsiak et al. (2022), are more susceptible due to their higher rate of breathing relative to body weight. From a public health perspective, interventions and regulatory guidelines need to consider these groups to better protect public health. It's important to also consider ethical factors when conducting studies on vulnerable populations (Korsiak et al., 2022; Molina et al., 2020).

From an environmental justice standpoint, it is crucial to address the vulnerability of children and evaluate the risks associated with establishing National Ambient Air Quality Standards. The focus should be on populations rather than individuals to inform public policy decisions, applying the same safety principles used for chemicals and the environment. Climate change presents an additional challenge, with the impact on air quality and public health being an emerging area of concern. Many studies are being conducted to assess the potential impact on air pollution emissions and quality, especially concerning human health (Arias-Pérez et al., 2020). The intricate link between air pollution, climate change, and human health necessitates a comprehensive research agenda that emphasizes collaboration and active involvement from vulnerable communities (Ainali et al., 2022). This will help to develop pollution reduction policy options that are closely linked with research and policy development, promoting the use of mitigation technologies.

CONCLUSION AND CALL TO ACTION

Air pollution continues to be a major global issue, with developing regions like Africa and Asia experiencing significantly higher exposure levels than developed areas such as North America and Europe. Elevated concentrations of find atmospheric particulate matter (e.g., $PM_{2.5}$) and other pollutants in industrialized and urban areas pose serious environmental, health, and socioeconomic risks. The impact of pollution-related diseases, including respiratory and cardiovascular ailments, as well as neuropsychiatric disorders, is particularly severe among vulnerable groups, especially those in low-income and densely populated communities.

Despite numerous global and national policy efforts, substantial gaps persist in pollution control, public awareness, and technological advancements. Studies emphasize the importance of a multidisciplinary approach that combines environmental science, public health, policy development, and innovation to tackle air pollution effectively. The rise of emerging pollutants, the impacts of climate change, and social inequalities in pollution exposure further underscore the urgency for ongoing research and adaptive policy measures.

The following actions are necessary to enhance the air quality:

1. **Policy and Governance Enhancements**:
    - Strengthen and enforce air quality regulations at the national and international levels.
    - Align policies with WHO air quality standards and enhance monitoring systems for real-time pollution tracking.
    - Introduce stricter emission limits for industrial, transportation, and energy sectors.
    - Promote the adoption of clean energy policies, including incentives for renewable energy and sustainable transportation.

2. **Technological Innovations and Solutions**:
   - Invest in research and development of air pollution control technologies such as carbon capture, improved air filtration, and cleaner combustion processes.
   - Implement smart city solutions for real-time air quality monitoring and AI-driven pollution management.
   - Encourage the widespread use of electric and hydrogen-powered vehicles and expand public transport infrastructure.
3. **Public Awareness and Community Engagement**:
   - Launch large-scale educational campaigns to inform citizens about the health risks of air pollution and preventive measures.
   - Encourage community-driven initiatives such as urban greening projects, waste management improvements, and local air quality monitoring networks.
   - Develop school and workplace programs to reduce indoor air pollution through better ventilation and cleaner fuel use.
4. **Research and Data Collection**:
   - Expand long-term epidemiological studies to assess the cumulative health impacts of air pollution.
   - Focus research efforts on underexplored pollutants, including emerging contaminants from industrial waste and new chemical compounds.
   - Enhance cross-disciplinary collaboration among scientists, urban planners, and policymakers to develop holistic solutions.
5. **Climate Change and Sustainable Development**:
   - Integrate air pollution mitigation strategies with global climate change policies, emphasizing reductions in greenhouse gas emissions.
   - Encourage urban planning that prioritizes green infrastructure, sustainable construction practices, and low-emission zones.
   - Foster international cooperation to address transboundary air pollution and its broader ecological impacts.

By taking these measures, we can significantly reduce air pollution, enhance public health, and promote sustainable economic growth and environmental preservation on a global scale.

REFERENCES

Abaje, I. B., Bello, Y., & Ahmad, S. A. (2020). A review of air quality and concentrations of air pollutants in Nigeria. *Journal of Applied Sciences and Environmental Management*, 24(2), 373–379.

Ainali, N. M., Kalaronis, D., Evgenidou, E., Kyzas, G. Z., Bobori, D. C., Kaloyianni, M., & Lambropoulou, D. A. (2022). Do poly (lactic acid) microplastics instigate a threat? A perception for their dynamic towards environmental pollution and toxicity. *Science of the Total Environment*, 832, 155014.

Alharbi, H. A., Rushdi, A. I., Bazeyad, A., & Al-Mutlaq, K. F. (2024). Temporal variations, air quality, heavy metal concentrations, and environmental and health impacts of atmospheric $PM_{2.5}$ and PM_{10} in Riyadh City, Saudi Arabia. *Atmosphere*, 15(12), 1448.

Alvarado, M. J., McVey, A. E., Hegarty, J. D. et al. (2019). Evaluating the use of satellite observations to supplement ground-level air quality data in selected cities in low- and middle-income countries. *Atmospheric Environment*, 218, 117016.

Arias-Pérez, R. D., Taborda, N. A., Gómez, D. M., Narvaez, J. F., Porras, J., & Hernandez, J. C. (2020). Inflammatory effects of particulate matter air pollution. *Environmental Science and Pollution Research*, 27(34), 42390–42404.

Arroyave, W. D., Mehta, S. S., Guha, N., Schwingl, P., Taylor, K. W., Glenn, B., & Lunn, R. M. (2021). Challenges and recommendations on the conduct of systematic reviews of observational epidemiologic studies in environmental and occupational health. *Journal of Exposure Science & Environmental Epidemiology*, 31(1), 21–30.

Awewomom, J., Benyade, B. K., Osei, F. E., Azanu, D., Opoku, F., Sackey, L. N., & Akoto, O. (2024). A review of health hazards associated with exposure to galamsey-related pollutants. *Health Sciences Investigations Journal*, 5(2), 726–734.

Backes, C. H., Nelin, T., Gorr, M. W., & Wold, L. E. (2013). Early life exposure to air pollution: How bad is it? *Toxicology Letters*, 216, 47–53. doi: 10.1016/j.toxlet.2012.11.007.

Bentayeb, M., Simoni, M., Norback, D., et al. (2013). Indoor air pollution and respiratory health in the elderly. *Journal of Environmental Science and Health, Part A: Toxic/Hazardous Substances*, 48, 1783–1789.

Blaga, R., Gautam, S. (2024). Improving PM10 sensor accuracy in urban areas through calibration in Timişoara. *Nature Partner Journals Climate and Atmospheric Science,* **7**, 26. https://doi.org/10.1038/s41612-024-00812-0

Blessy, A., John Paul, J., Gautam, S., Shany, V. J., & Sreenath, M. (2023). IoT-based air quality monitoring in hair salons: Screening of hazardous air pollutants based on personal exposure and health risk assessment. *Water, Air, & Soil Pollution*, 234, 336. https://doi.org/10.1007/s11270-023-06350-4

Brunekreef, B., Beelen, R., Hoek, G., et al. (2009). Effects of longterm exposure to traffic-related air pollution on respiratory and cardiovascular mortality in the Netherlands: The NLCS-AIR study. *Research Reports Health Effects Institute*, 139, 5–71.

Calderón-Garcidueñas, L., Mora-Tiscareño, A., Ontiveros, E., et al. (2008a). Air pollution, cognitive deficits and brain abnormalities: A pilot study with children and dogs. *Brain and Cognition*, 68, 117–127.

Calderón-Garcidueñas, L., Solt, A. C., Henríquez-Roldán, C., et al. (2008b). Long-term air pollution exposure is associated with neuroinflammation, an altered innate immune response, disruption of the blood-brain barrier, ultrafine particulate deposition, and accumulation of amyloid beta-42 and alpha-synuclein in children and young adults. *Toxicologic Pathology*, 36, 289–310.

Checkoway, H., Pearce, N., & Kriebel, D. (2007). Selecting appropriate study designs to address specific research questions in occupational epidemiology. *Occupational and Environmental Medicine*, 64(9), 633–638.

Checkoway, H., Pearce, N., & Kriebel, D. (2004). *Research Methods in Occupational Epidemiology* (Vol. 34). Oxford University Press, New York.

Comunian, S., Dongo, D., Milani, C., & Palestini, P. (2020). Air pollution and COVID-19: The role of particulate matter in the spread and increase of COVID-19's morbidity and mortality. *International Journal of Environmental Research and Public Health*, 17(12), 4487.

Dida, G. O., Lutta, P. O., Abuom, P. O., Mestrovic, T., & Anyona, D. N. (2022). Factors predisposing women and children to indoor air pollution in rural villages, Western Kenya. *Archives of Public Health*, 80(1), 46.

Diener, A., & Mudu, P. (2021). How can vegetation protect us from air pollution? A critical review on green spaces' mitigation abilities for air-borne particles from a public health perspective-with implications for urban planning. *Science of the Total Environment*, 796, 148605.

Dominici, F., Zanobetti, A., Schwartz, J., Braun, D., Sabath, B., & Wu, X. (2022). Assessing adverse health effects of long-term exposure to low levels of ambient air pollution: Implementation of causal inference methods. *Research Reports from the Health Effects Institute*, 211, 1–56

Escobedo, F. J., Kroeger, T., & Wagner, J. E. (2011). Urban forests and pollution mitigation: Analyzing ecosystem services and disservices. *Environmental Pollution*, 159(8–9), 2078–2087.

Fisher, S., Bellinger, D. C., Cropper, M. L., Kumar, P., Binagwaho, A., Koudenoukpo, J. B., & Landrigan, P. J. (2021). Air pollution and development in Africa: Impacts on health, the economy, and human capital. *The Lancet Planetary Health*, 5(10), e681–e688.

Forastiere, F., Spadaro, J. V., Ancona, C., Andersen, Z. J., Cozzi, I., Gumy, S., & Krzyzanowski, M. (2024). Choices of morbidity outcomes and concentration–response functions for health risk assessment of long-term exposure to air pollution. *Environmental Epidemiology*, 8(4), e314.

Gao, Y., Chan, E. Y., Li, L., Lau, P. W., & Wong, T. W. (2014). Chronic effects of ambient air pollution on respiratory morbidities among Chinese children: A cross-sectional study in Hong Kong. *BMC Public Health*, 14, 105.

Gautam, S., Talatiya, A., Patel, M., Chabhadiya, K., & Pathak, P. (2020). Personal exposure to air pollutants from winter season bonfires in rural areas of Gujarat, India. *Exposure and Health*, 12(1), 89–97. https://doi.org/10.1007/s12403-018-0287-9.

Gautam, Y. K., Sharma, K., Tyagi, S., Ambedkar, A. K., Chaudhary, M., & Pal Singh, B. (2021). Nanostructured metal oxide semiconductor-based sensors for greenhouse gas detection: Progress and challenges. *Royal Society Open Science*, 8(3), 201324.

GBDRF. (2019). Global burden of 87 risk factors in 204 countries and territories, 1990–2019: A systematic analysis for the Global Burden of Disease Study 2019. *Lancet*, 396(10258), 1223–1249.

Guo, H., Li, X., Wei, J., Li, W., Wu, J., & Zhang, Y. (2022). Smaller particular matter, larger risk of female lung cancer incidence? Evidence from 436 Chinese counties. *BMC Public Health*, 22(1), 344.

Hadnagy, W., Stiller Winkler, R., & Idel, H. (1996). Immunological alterations in sera of persons living in areas with different air pollution. *Toxicology Letters*, 88, 147–153.

Hajat, A., Hsia, C., & O'Neill, M. S. (2015). Socioeconomic disparities and air pollution exposure: A global review. *Current Environmental Health Reports*, 2(4), 440–450

Harker, J. (2022). Future research priorities: Precision health and predictive tox. https://factor.niehs.nih.gov/2022/6/feature/2-feature-future-of-environmental-health-sciences (2022), Accessed 28th Oct 2022.

Hayes, R. B. Lim, C., Zhang, Y., et al. (2020). $PM_{2.5}$ air pollution and cause-specific cardiovascular disease mortality. *International Journal of Epidemiology*, 49(1), 25–35.

Haynes, E. N., Chen, A., Ryan, P., Succop, P., Wright, J., & Dietrich, K. N. (2011). Exposure to airborne metals and particulate matter and risk for youth adjudicated for criminal activity. *Environmental Research*, 111, 1243–1248.0

HEI (Health Effect Institute) (2020). *State of Global Air 2020.* Special Report. Health Effects Institute, Boston, MA.

Inauen, J., Contzen, N., Frick, V., Kadel, P., Keller, J., Kollmann, J., & van Valkengoed, A. M. (2021). Environmental issues are health issues. *European Psychologist*, 26(3), 219–229.

Karimi, A., Shirmardi, M., Hadei, M., Birgani, Y. T., Neisi, A., Takdastan, A., & Goudarzi, G. (2019). Concentrations and health effects of short-and long-term exposure to $PM_{2.5}$, NO_2, and O_3 in ambient air of Ahvaz city, Iran (2014–2017). *Ecotoxicology and Environmental Safety*, 180, 542–548.

Kim, H., Kim, W. H., Kim, Y. Y., & Park, H. Y. (2020). Air pollution and central nervous system disease: A review of the impact of fine particulate matter on neurological disorders. *Frontiers in Public Health*, 8, 575330.

Kloog, I., Ridgway, B., Koutrakis, P., Coull, B. A., & Schwartz, J. D. (2013). Long-and short-term exposure to $PM_{2.5}$ and mortality: Using novel exposure models. *Epidemiology*, 24(4), 555–561.

Korsiak, J., Lavigne, E., You, H., et al. (2022). Air pollution and pediatric respiratory hospitalizations: Effect modification by particle constituents and oxidative potential. *American Journal of Respiratory and Critical Care Medicine*, 206(11), 1370–1378.

Kumar, R. P., Perumpully, S. J., Samuel, C., & Gautam, S. (2023). Exposure and health: A progress update by evaluation and scientometric analysis. *Stochastic Environmental Research and Risk Assessment*, 37, 453–465. https://doi.org/10.1007/s00477-022-02313-z

Lakshmi, P. V. M., Virdi, N. K., Sharma, A., Tripathy, J. P., Smith, K. R., Bates, M. N., & Kumar, R. (2013). Household air pollution and stillbirth in India: Analysis of the DLHS-II national survey. *Environmental Research*, 121, 17–22.

Leung, T. F., Ko, F. W., & Wong, G. W. (2012). Roles of pollution in the prevalence and exacerbations of allergic diseases in Asia. *Journal of Allergy and Clinical Immunology*, 129, 42–47.

Liao, K., Huang, X., Dang, H., Ren, Y., Zuo, S., & Duan, C. (2021). Statistical approaches for forecasting primary air pollutants: A review. *Atmosphere*, 12(6), 686.

Lin, C. C., Chiu, C. C., Lee, P. Y., Chen, K. J., He, C. X., Hsu, S. K., & Cheng, K. C. (2022). The adverse effects of air pollution on the eye: A review. *International Journal of Environmental Research and Public Health*, 19(3), 1186.

Lippmann, M., Chen, L.C., Gordon, T., Ito, K., & Thurston, G. D. (2013). National particle component toxicity (NPACT) Initiative: Integrated epidemiologic and toxicologic studies of the health effects of particulate matter components. *Research Reports from the Health Effects Institute*, 177, 5–13.

Ma, Q., Qi, Y., Shan, Q., Liu, S., & He, H. (2020). Understanding the knowledge gaps between air pollution controls and health impacts including pathogen epidemic. *Environmental Research*, 189, 109949.

Maciejczyk, P. Chen, L.-C., & Thurston, G. (2021). The role of fossil fuel combustion metals in $PM_{2.5}$ air pollution health associations. *Atmosphere*, 12(9), 1086.

Marlier, M. E., Jina, A. S., Kinney, P. L., & De Fries, R. S. (2016). Extreme air pollution in global megacities. *Current Climate Change Reports*, 2(1), 15–27.

Mainka, A., & Żak, M. (2022). Synergistic or antagonistic health effects of long-and short-term exposure to ambient NO_2 and $PM_{2.5}$: A review. *International Journal of Environmental Research and Public Health*, 19(21), 14079.

Manisalidis, I., Stavropoulou, E., Stavropoulos, A., & Bezirtzoglou, E. (2020). Environmental and health impacts of air pollution: A review. *Frontiers in Public Health*, 8, 14.

Mukherjee, S., Dasgupta, S., Mishra, P. K., & Chaudhury, K. (2021). Air pollution-induced epigenetic changes: Disease development and a possible link with hypersensitivity pneumonitis. *Environmental Science and Pollution Research International*, 28(40), 55981–56002

Molina, L. T., Zhu, T., Wan, W., & Gurjar, B. R. (2020). Impacts of megacities on air quality: Challenges and opportunities. *Oxford Research Encyclopedia of Environmental Science*. Retrieved 9 Aug. 2025, from https://oxfordre.com/environmentalscience/view/10.1093/acrefore/9780199389414.001.0001/acrefore-9780199389414-e-5.

Moreno-Ríos, A. L., Tejeda-Benítez, L. P., & Bustillo-Lecompte, C. F. (2022). Sources, characteristics, toxicity, and control of ultrafine particles: An overview. *Geoscience Frontiers*, 13(1), 101147.

Morgenstern, H., & Thomas, D. (1993). Principles of study design in environmental epidemiology. *Environmental Health Perspectives*, 101(Suppl 4), 23–38.

Fatimah, M., Sunarsih, E., Sitorus, R. J., & Nurhayati, N. (2024). Environmental health issues and the solution in Indonesia: A narrative review. *Science Midwifery*, 12(2), 722–735.

Nicolaou, L., & Checkley, W. (2021). Inequities in air pollution exposure and gaps in air quality monitoring. *Journal of Allergy and Clinical Immunology*, 148(1), 64–66.

Nogueira, J. B. (2009). Air pollution and cardiovascular disease. *Revista Portuguesa de Cardiologia*, 28, 715–733.

Newman, N. C., Ryan, P., Lemasters, G., et al. (2013). Traffic-related air pollution exposure in the first year of life and behavioral scores at 7 years of age. *Environmental Health Perspectives*, 121, 731–736.

Oloruntobi, O., Mokhtar, K., Gohari, A., Asif, S., & Chuah, L.F. (2023). Sustainable transition towards greener and cleaner seaborne shipping industry: Challenges and opportunities. *Cleaner Engineering and Technology*, 13(2), 100628. doi: 10.1016/j.clet.2023.100628.

Ozkaynak, H., & Thurston, G. D. (1987). Associations between 1980U.S. mortality rates and alternative measures of airborne particle concentration. *Risk Analysis*, 7(4), 449–461.

Peng, R. D., Liu, J. C., McCormack, M. C., Mickley, L. J., & Bell, M. L. (2022). Estimating the health effects of environmental mixtures using principal stratification. *Statistics in Medicine*, 41(10), 1815–1828.

Perumpully, S. J., & Gautam, S. (2025). Impact of aerosols on atmospheric processes and climate variability: A synthesis of recent research findings. *Geosystems and Geoenvironment*, *4*(1), 100317. https://doi.org/10.1016/j.geogeo.2024.100317.

Perumpully, S. J., Gautam, S., John, J. P., & Sreenath, M. (2024). Evaluating the impact of personal exposure to emissions from sustainable commercial heating and cooking fuels on women in rural southern India and their alignment with sustainable development goals. *Water, Air, & Soil Pollution*, *235*, 54. https://doi.org/10.1007/s11270-023-06854-z

Pillarisetti, A., Ye, W., & Chowdhury, S. (2022). Indoor air pollution and health: Bridging perspectives from developing and developed countries. *Annual Review of Environment and Resources*, 47(1), 197–229.

Pinder, R. W., Klopp, J. M., Kleiman, G., Hagler, G. S. W., Awe, Y., & Terry, S. (2019). Opportunities and challenges for filling the air quality data gap in low- and middle-income countries. *Atmospheric Environment*, 215, 116794.

Popov, O., Taraduda, D., Sobyna, V., Sokolov, D., Dement, M., & Pomaza-Ponomarenko, A. (2020). Emergencies at potentially dangerous objects causing atmosphere pollution: Peculiarities of chemically hazardous substances migration. *Systems, Decision and Control in Energy I*, 2020, 151–163.

Rahman, M. M., & Thurston, G. (2022). A hybrid satellite and land use regression model of source-specific $PM_{2.5}$ and $PM_{2.5}$ constituents. *Environment International*, 163, 107233.

Rathikannu, S., Gautam, S., Joshi, S. K., Katharine, P., Mithra, K. E., Banusaranya, P., Amudhavarshini, V. M., Gayatri, R., & Ho, C.-H. (2025). FTIR-based assessment of microplastic contamination in soil, water, and insect ecosystems reveals environmental and ecological risks. *Scientific Reports*, *15*, 28615. https://doi.org/10.1038/s41598-025-14507-w

Rentschler, J., & Leonova, N. (2023). Global air pollution exposure and poverty. *Nature Communications*, 14(1), 4432.

Rider, C. F., & Carlsten C. (2019). Air pollution and DNA methylation: Effects of exposure in humans. *Clinical Epigenetics*, 11(1), 131

Romieu, I., Riojas-Rodríguez, H., Marrón-Mares, A. T., et al. (2009). Improved biomass stove intervention in rural Mexico: Impact on the respiratory health of women. *American Journal of Respiratory and Critical Care Medicine*, 180, 649–656.

Rotton, J., Frey, J., Barry, T., Milligan, M., & Fitzpatrick, M. (1979). The air pollution experience and physical aggression1. *Journal of Applied Social Psychology*, 9, 397–412.

Rozanova, E., Heilig, P., & Godnic-Cvar, J. (2009). The eye – A neglected organ in environmental and occupational medicine: An overview of known environmental and occupational non-traumatic effects on the eyes. *Archives of Industrial Hygiene and Toxicology*, 60, 205–215.

Rushdi, A. I., Simoneit, B. R., Lijotra, L., Bazeyad, A. Y., Dumenden, R., El-Mubarak, A. H., & Al-Mutlaq, K. F. (2023). Phthalates, non-phthalates, polychlorinated biphenyls, and phenyl phosphates in atmospheric suspended particulate matter of Dhahran City, Saudi Arabia: Levels and seasonal variation. *International Journal of Environmental Science and Technology*, 20(4), 3561–3576.

Sarla, G. S. (2020). Air pollution: Health effects. *Revistamedicina Legel De Costarics*, 37(1): 33–38.

Schraufnagel, D. E. (2020). The health effects of ultrafine particles. *Experimental & Molecular Medicine*, 52(3), 311–317.

Sicard, P., Agathokleous, E., Anenberg, S. C., De Marco, A., Paoletti, E., & Calatayud, V. (2023). Trends in urban air pollution over the last two decades: A global perspective. *Science of the Total Environment*, 858, 160064.

Silman, A. J., Macfarlane, G. J., & Macfarlane, T. (2018). *Epidemiological Studies: A Practical Guide*. Oxford University Press, Oxford.

Skillington, K., Crawford, R. H., Warren-Myers, G., & Davidson, K. (2022). A review of existing policy for reducing embodied energy and greenhouse gas emissions of buildings. *Energy Policy*, 168, 112920.

Smith-Sivertsen, T., Díaz, E., Pope, D., et al. (2009). Effect of reducing indoor air pollution on women's respiratory symptoms and lung function: The respire randomized trial, Guatemala. *American Journal of Epidemiology*, 170, 211–220.

Smith, K. R., McCracken, J. P., Weber, M. W., et al. (2011). Effect of reduction in household air pollution on childhood pneumonia in Guatemala (RESPIRE): A randomized controlled trial. *Lancet*, 378, 1717–1726.

Solarimpulse (2020). *New Energy Opportunities (NEO) Network. Connecting Solutions on Renewable Energy and Clean Technology*. The Solar Impulse Foundation. https://www.solarimpulse.com

Solomon, P. A., Costantini, M., Grahame, T. J., et al. (2012). Air pollution and health: Bridging the gap from sources to health outcomes: Conference summary. *Air Quality, Atmosphere and Health*, 5(1), 9–62.

Sridharan, S., Kumar, M., Singh, L., Bolan, N. S., & Saha, M. (2021). Microplastics as an emerging source of particulate air pollution: A critical review. *Journal of Hazardous Materials*, 418, 126245.

Stepanov, I. A., & Makarov, I. A. (2024). Greenhouse gas emissions regulation in fossil fuels exporting countries: Opportunities and challenges for Russia. *Strategies and Challenges of Sustainable Development in Eurasia*, 34, 82–109.

Tainio, M., Andersen, Z. J., Nieuwenhuijsen, M. J., Hu, L., De Nazelle, A., An, R., & de Sá, T. H. (2021). Air pollution, physical activity and health: A mapping review of the evidence. *Environment International*, 147, 105954.

Thurston, G. D., Ito, K., Lall, R., Burnett, R. T., Turner, M. C., Krewski, D., & Pope III, C. A. (2013). NPACT study 4. Mortality and long-term exposure to $PM_{2.5}$ and its components in the American cancer society's cancer prevention study II cohort. In: Lippmann, M., Chen, LC.,

Gordon, T., Ito, K., & Thurston, GD. (eds), *National Particle Component Toxicity (NPACT) Initiative: Integrated Epidemiologic and Toxicologic Studies of the Health Effects of Particulate Matter Components* (pp. 127–166), Health Effects Institute, Boston, MA.

Tran, V. V., Park, D., & Lee, Y. C. (2020). Indoor air pollution related human diseases, and recent trends in the control and improvement of indoor air quality. *International Journal of Environmental Research and Public Health*, 17(8), 2927.

Trpevska-Shekerinov, N., Petrushevska, A., Gjoshevska-Dashtevska, E., Shekerinov, T., & Nivichka-Kjaeva, J. (2022). Dry eye disease and risks of urban air pollution–literature review. *Archives of Public Health*, 14(2), 113–119.

Turner, M. C., Andersen, Z. J., Baccarelli, A., Diver, W. R., Gapstur, S. M., & Cohen, A. (2020). Outdoor air pollution and cancer: An overview of the current evidence and public health recommendations. *CA: A Cancer Journal for Clinicians*, 70(6), 460–479.

UNDP (2023). United Nations Development Program Annual Report. New York, p. 21. www.undp.org.

United States Environmental Protection Agency (US-EPA) (2022). *Criteria Air Pollutants.* https://www.epa.gov/criteria-air-pollutants#self Accessed 30th Sep 2022.

Van Donkelaar, A., Martin, R. V., Li, C., & Burnett, R. T. (2019). Regional estimates of chemical composition of fine particulate matter using a combined geoscience-statistical method with information from satellites, models, and monitors. *Environmental Science & Technology*, 53(5), 2595–2611.

Varughese, S. P., Raj, S. M. G., Joel, T. J., Gautam, S. (2023). Detecting airborne pathogens: A computational approach utilizing surface acoustic wave sensors for microorganism detection. Technologies, 11, 135. https://doi.org/10.3390/technologies11050135

Vierkötter, A., Schikowski, T., Ranft, U., et al. (2010). Airborne particle exposure and extrinsic skin aging. *Journal of Investigative Dermatology*, 130, 2719–2726.

Viet, S. M., Falman, J. C., Merrill, L. S., Faustman, E. M., Savitz, D. A., Mervish, N., & O'Brien, B. (2021). Human health exposure analysis resource (HHEAR): A model for incorporating the exposome into health studies. *International Journal of Hygiene and Environmental Health*, 235, 113768.

Weisel, C. P. (2002). Assessing exposure to air toxics relative to asthma. *Environmental Health Perspectives*, 110(Suppl 4), 527–537.

Wild, C. P. (2005). Complementing the genome with an "exposome": The outstanding challenge of environmental exposure measurement in molecular epidemiology cancer. *Cancer Epidemiology, Biomarkers & Prevention*, 14(8), 1847–1850.

World Health Organization. (2021a). *What are the WHO Air quality guidelines?* https://www.who.int/about/contact-us (2021), Accessed 1st Oct 2022.

World Health Organization. (2021b). *WHO Global Air Quality Guidelines: Particulate Matter ($PM_{2.5}$ and PM_{10}), Ozone, Nitrogen Dioxide, Sulfur Dioxide and Carbon Monoxide*. World Health Organization, Geneva, Switzerland.

Yang, M., Guo, Y. M., Bloom, M. S., et al. (2020). Is $PM_{(1)}$ similar to $PM_{(2.5)}$? A new insight into the association of $PM_{(1)}$ and $PM_{(2.5)}$ with children's lung function. *Environment International*, 145, 106092.

Zhang, W., Qian, C. N., & Zeng, Y. X. (2014). Air pollution: A smoking gun for cancer. *Chinese Journal of Cancer Research*, 33, 173–175.

Zhang, P., Carlsten, C., Chaleckis, R., et al. (2021). Defining the scope of exposome studies and research needs from a multidisciplinary perspective. *Environmental Science & Technology Letters*, 8(10), 839–852.

Zhou, N., Cui, Z., Yang, S., et al. (2014). Air pollution and decreased semen quality: A comparative study of Chongqing urban and rural areas. *Environmental Pollution*, 187, 145–152.

3 The Silent Killer *Uncovering the Health Risks of Air Pollution*

Murali Dadi, Stanford Siachoono, and Kenneth Maseka

INTRODUCTION

Air pollution refers to the presence of an excess amount of unwanted and undesirable substances in the atmosphere, which can change the nature or quality of the air that leads to negative impacts on the health of humans (Sellami et al., 2025). There are two main sources of air pollution: natural and anthropogenic. These sources include volcanic eruptions, forest fires, dusty storms, power plants and automobile emissions, burning of wreckages, including plastics, and chemicals that contaminate the outdoor and indoor air through physical, chemical, or biological means (Shetty et al., 2023).

The major pollutants present in these sources are particulate matter (PM), oxides of nitrogen and sulfur, carbon monoxides (CO), polycyclic aromatic hydrocarbons (PAHs), ozone, dioxins, and volatile organic compounds (VOCs), which are harmful and cause serious health diseases in humans (Setu et al., 2024).

These pollutants contribute to creating serious environmental and health consequences. These include climate change, the formation of smog and acid rains, and health issues including respiratory, cardiovascular, neurological disorders, cancer, and reproductive problems (Shetty et al., 2023). Moreover, air pollution affects vulnerable groups, such as children, the elderly, those with pre-existing health issues, and those in low-income regions.

DOI: 10.1201/9781003613183-3

According to 2021 WHO statistics, 7.3 billion people worldwide are affected by direct exposure to $PM_{2.5}$, annually out of which 80% live in middle- and low-income regions. In Sub-Saharan Africa, 716 million people live where unsafe environmental conditions are endured.

Contextual data presentation of the WHO (2021) statistics on the global burden of air pollution:

As per WHO reports due to air pollution annually, approximately 7 million deaths happen, with 91% occurring in low- and middle-income countries (LMICs). Southeast Asia and Africa are the most affected, primarily due to biomass fuel use and industrial emissions, while Europe and China have achieved $PM_{2.5}$ reductions of 30% and 28%, respectively. Vulnerable groups, such as children and the elderly, face heightened risks, with LMIC $PM_{2.5}$ levels exceeding WHO guidelines by 5–10 times (Figure 3.1).

Globally, urban areas experience ample air pollution due to industrialization and dense populations. Mainly, the urban populations suffer from respiratory problems, cardiovascular issues, asthma, lung cancer, infant mortality, and skin allergies (Kumar et al., 2023).

The amount of air pollution is becoming one of the main concerns in developing countries, particularly in Asia and Africa, and it is noticed that many disparities exist among these countries in handling or controlling the pollution due to these issues often related to socioeconomic status and the availability of resources needed (Bherwani et al., 2020).

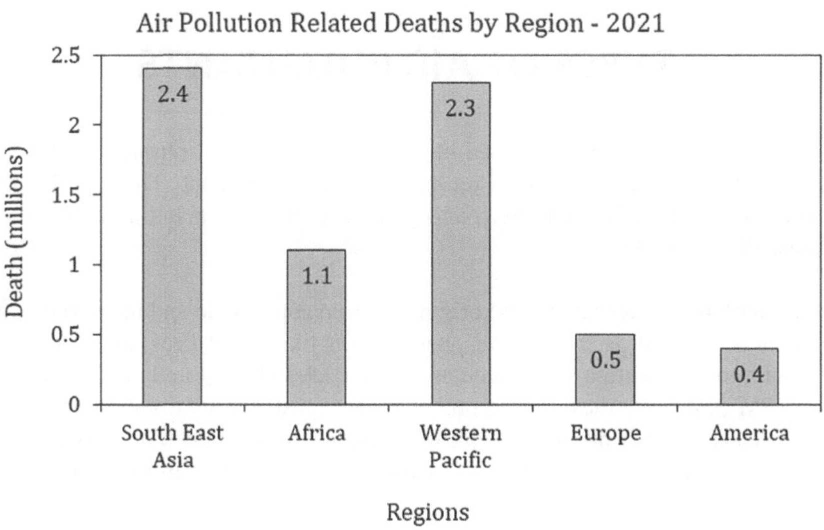

FIGURE 3.1 Comparative analysis of mortalities due to air pollution as per WHO (2021) statistics across the regions.

Control measures for air pollution can vary from country to country. Some countries are taking measures to minimize the pollution, such as implementing strict emission guidelines, encouraging green technology methods, and conducting public awareness campaigns. On the other hand, some countries are struggling to enforce these measures due to a lack of sufficient facilities, infrastructure, socioeconomic issues, and inadequate public awareness, leading to an increase in air pollution, which further impacts the living conditions and health of people (Chen et al., 2024).

Moreover, in recent past, it has been noticed that a bidirectional relationship existing between climate change and air pollution underlining how each aggravates the other. For instance, air pollutants such as methane, carbon emissions, and tropospheric ozone are directly contributing to global warming by absorbing heat and altering the chemical nature of air (Mar et al., 2022).

Escalating global temperatures and altering weather patterns can intensify the incidence of wildfires, increase ground-level ozone formation, and reduce the dispersion of pollutants, leading to higher pollutant concentrations and deteriorating air quality (Afifa et al., 2024).

In view of the effects of air pollution on human health and the environment, this chapter describes issues such as types of air pollutants and their mechanisms of action in relation to health complications in humans and other living organisms, including a venerable population such as children, the elderly, and people with pre-existing health complications of air pollution impacts. It also discusses various measures and practices to control and mitigate air pollution at the local and international levels.

TYPES OF AIR POLLUTANTS

Air pollutants are categorized based on the sources, namely, physical and chemical nature and their significant impact on human and environment. These pollutants are classified into PM ($PM_{2.5}$ and PM_{10}) and gaseous pollutants including NO_x, SO_2, CO, aerosols, VOCs, and O_3.

Particulate Matter (PM): PM originates from automobile and industrial emissions, as well as the natural phenomenon like forest fires, and it consists of two different particles such as fine particles ($PM_{2.5}$) and coarse particles (PM_{10}). Among these, $PM_{2.5}$ pollutants are most common in the polluted air because of their minute size, allowing them to easily penetrate into the lungs and bloodstream, resulting in serious respiratory and cardiovascular problems in affected people.

Gaseous Pollutants: The gaseous pollutants include oxides of nitrogen (NO_x), oxides of sulfur mainly sulfur dioxide (SO_2), and carbon monoxide (CO), which are mainly impacting the health of the living organisms (Li et al., 2017).

Oxides of Nitrogen and Sulfur (NO_x and SO_2): These pollutants are the result of the burning of fossil fuels, combustion of automobiles, and industrial emissions. These oxides together adversely affect terrestrial ecosystem and aquatic ecosystems, as well as human health, by polluting the air, water, and soil. High concentrations of these pollutants cause severe respiratory illness in humans and other living organisms.

Carbon Monoxide (CO): This pollutant is mainly produced from the incomplete combustion of carbon elements. Majorly, automobile vehicle emissions are responsible for high concentration of this pollutant in the atmosphere. Intoxication from (Carbon monoxide) CO in the organisms is highly dependent on oxygen because CO has 250 times high affinity for binding to hemoglobin (Hb) than that of oxygen. The symptoms of CO intoxication are fatigue, dizziness, headache, etc.

Ozone (O_3): It is a harmful secondary pollutant at ground level formed by reactions between oxides and nitrogen and VOCs in the presence of sunlight. It mainly affects the respiratory system and causes respiratory illness, bronchitis, and asthma. Furthermore, it also affects the cardiovascular, neurological, and reproductive systems.

Secondary pollutants, such as ozone and PM, are formed in the atmosphere through complex chemical reactions involving precursor compounds such as nitrogen oxides (NO_x) and VOCs. These reactions take place in the presence of sunlight. The following are the steps for formation processes of O_3, PM, H_2SO_4, and HNO_3:

Ozone Formation

Ozone (O_3) is a secondary pollutant formed in the troposphere by photochemical reactions involving NO_x and VOCs under sunlight (Nguyen et al., 2022).

Steps:

NO_2 Photolysis

Sunlight converts nitrogen dioxide (NO_2) into nitrogen oxide (NO) and a free oxygen atom (O):

$$NO_2 + h\nu \rightarrow NO + O$$

Oxygen Atom Reaction

The free oxygen atom reacts with molecular oxygen (O_2) to form ozone (O_3):

$$O + O_2 \rightarrow O_3$$

NO_x and VOCs Contribution

VOCs react with hydroxyl radicals (OH) to form peroxy radicals (RO_2). These peroxy radicals oxidize NO to NO_2, which can again undergo photolysis, maintaining the ozone formation cycle:

$$RO_2 + NO \rightarrow RO + NO_2$$

This cycle can result in high concentrations of ozone, particularly in urban areas with high NO_x and VOC emissions.

Particulate Matter (PM) Formation

PM, especially fine particles ($PM_{2.5}$), is formed by chemical reactions involving precursor gases like sulfur dioxide (SO_2), NO_x, ammonia (NH_3), and VOCs (Hodan and Barnard, 2023).

Key Processes:

Sulfuric Acid Formation

SO_2 is oxidized in the atmosphere to form sulfuric acid (H_2SO_4), which can condense to form sulfate aerosols (Yang et al., 2023):

$$SO_2 + OH \rightarrow HOSO_2$$

$$HOSO_2 + O_2 \rightarrow HO_2 + SO_3$$

$$SO_3 + H_2O \rightarrow H_2SO_4$$

Nitrate Formation

NO_x reacts to form nitric acid (HNO_3), which can condense or react with ammonia to form nitrate aerosols (Ren et al., 2021):

$$NO_2 + OH \rightarrow HNO_3$$

Organic Aerosol Formation

VOCs undergo oxidation to form secondary organic aerosols (SOAs). These reactions occur via:

Oxidation by OH, O_3, or NO_3 radicals to form products that condense into fine particles (Sasso et al., 2024).

Volatile Organic Pollutants (VOCs): These are synthetic gaseous chemicals used in the manufacturing of domestic and commercial products. Vehicles and various industries like petroleum, coal, and dry cleaning contribute to these pollutants' emissions and are responsible for indoor and outdoor chemical pollution. VOCs are precursors for the formation of ground-level ozone. High concentrations of VOCs affect plant metabolic processes and cause skin and respiration-related ailments in humans.

Heavy Metals: The heavy metals such as lead, mercury, and cadmium are common air pollutants. These are disposed into the atmosphere as a result of anthropogenic activities. Even low accumulated concentrations of these pollutants in the atmosphere enter into organisms through respiration and the food chain, causing various health issues in living organisms including neurological damage, lung cancer, cardiovascular problems, and bone and kidney damage.

Heavy metals such as arsenic (As), lead (Pb), cadmium (Cd), and mercury (Hg) are toxic pollutants brought into the atmosphere through fossil fuel combustion, industrial processes, mining, and waste incineration. These metals can remain airborne as PM or deposit onto soil and water, posing significant risks to human health and ecosystems (Afzal & Mahreen, 2024).

ATMOSPHERIC PREVALENCE OF HEAVY METALS AND HEALTH EFFECTS

Lead (Pb): Lead is commonly emitted from leaded gasoline, metal smelting, and battery recycling. For instance, in urban areas of India, airborne lead levels ranged between 0.2 and 2.2 $\mu g/m^3$, exceeding WHO guidelines. Lead pollution causes neurological disorders, especially in children, leading to cognitive impairments. WHO estimates 0.9 million premature deaths annually linked to lead exposure (Levin et al., 2021; Mandal et al., 2016).

Mercury (Hg): It is mainly introduced into the atmosphere through industrial emissions, coal combustion, and artisanal gold mining. Mercury pollution is responsible for neurological problems, nausea, gingivostomatitis, and inflammation in the affected people (Kung et al., 2024). Annually in the USA, approximately 50 tons of mercury emissions are recorded, which have a significant impact on the aquatic food chains of polluted areas (Brocza et al., 2024).

Cadmium (Cd): Cd is produced from waste incineration and zinc smelting. The industrial region of China has cadmium contamination of about 0.02–0.1 $\mu g/m^3$ in ambient air (Khan et al., 2022). Cd pollution damages kidneys and increases the risk of cancer. Chronic exposure in industrial workers leads to diseases like itai-itai (Rasin et al., 2025).

Arsenic (As): Arsenic is discharged from coal combustion and metal ore processing industries. For example, in Bangladesh, areas near coal power plants reported arsenic levels of 0.01–0.05 $\mu g/m^3$, contributing to contaminated water supplies (Hasan et al., 2019). Long-term exposure of Arsenic leads to lung, skin, and bladder cancers and an increased morality rate from respiratory diseases (López et al., 2020) (Table 3.1).

TABLE 3.1 Types of air pollutants and their characteristics

S.NO.	*TYPES OF AIR POLLUTANTS*	*SOURCES*	*HEALTH EFFECTS*	*REFERENCES*
1	Particulate Matter (PM)	Fine ($PM_{2.5}$) and coarse (PM_{10}) particles from automobile emissions, industries, and natural events like forest fires.	$PM_{2.5}$ penetrates deep into lungs and bloodstream, causing respiratory and cardiovascular diseases.	Undavalli and Khandelwal (2021), Basith et al. (2022)
2	Oxides of Nitrogen (NO_x)	Fossil fuel, vehicular and industrial emissions.	Causes respiratory illnesses and contributes to acid rain, impacting terrestrial and aquatic ecosystems.	Perera (2017)
3	Sulfur Dioxide (SO_2)	Released from burning fossil fuels and industrial emissions.	Causes severe respiratory diseases and acid rain, harming human health and ecosystems.	Chen et al. (2007), de Vries (2021)
4	Carbon Monoxide (CO)	Produced by incomplete combustion of carbon materials, especially from vehicles.	Causes intoxication by binding to hemoglobin, leading to symptoms like fatigue, dizziness, and headaches.	Kashfi and Patel (2022), Savioli et al. (2024)
5	Ozone (O_3)	A secondary pollutant formed by reactions between NO_x and VOCs in sunlight.	Impacts respiratory, cardiovascular, neurological, and reproductive systems; exacerbates asthma and bronchitis.	Soni et al. (2021)
6	Volatile Organic Compounds (VOCs)	Synthetic chemicals from industrial and household sources, including petroleum and coal-based industries.	Cause respiratory and skin ailments in humans; disrupt metabolic processes in plants; precursors to ground-level ozone formation.	Zhou et al. (2023)
7	Heavy Metals (e.g., Lead, Mercury, Cadmium)	Released through anthropogenic activities like industrial processes and fossil fuel combustion.	Lead to neurological damage, lung cancer, cardiovascular problems, and bone and kidney disorders through respiration and bioaccumulation in the food chain.	Briffa et al. (2020), WHO (2007)

MECHANISM OF IMPACTS OF AIR POLLUTION

Air pollutants affect human health through complex mechanisms. They enter the body through inhalation and penetrate into the alveoli and blood vessel. These penetrating toxic particles can cause oxidative stress and inflammation, eventually damaging lung and heart tissues and affecting normal cellular functions of the tissues. Moreover, recent studies indicated that the $PM_{2.5}$ is responsible for the occurrence of cardiopulmonary diseases including chronic lung infections, respiratory tract infections, heart disease, lung cancer, and neurological disorders, which lead to death of 4.5 million people worldwide.

Gaseous pollutants such as oxides of nitrogen and sulfur dioxide and ozone trigger acute events such as stroke, myocardial infraction, heart and respiratory failures, asthma, and lung cancers in human and other animals. These pollutants prompt these events through mechanisms such as damaging the endothelial tissues of organs like lungs, air pathways, and heart by their chemical nature leading to alterations in endothelial functions of those organs, which eventually affect the cardiovascular and respiratory functions, ultimately causing disorders such as diabetes, hypertension, atherosclerosis, lung cancer, asthma, and other respiratory-related problems.

On the other hand, carbon monoxide (CO) forms carboxyhemoglobin by strongly binding to hemoglobin, which affects the binding of oxygen to hemoglobin leading to reduction in the supply of oxygen to tissues such as brain, heart, and other essential organs, which ultimately generate hypoxia-ischemia conditions and cause cellular damage, finally leading to death of tissues of those organs. Furthermore, the air pollutants also have an impact on pregnant women, increasing the risk of fetus development, premature birth, and low birth weight of newborns (Rani & Dhok, 2023). Most of the air pollutants discussed here systemically trigger inflammatory responses that affect the respiratory, cardiovascular, and neurological systems that lead to the death of the organism. The mechanism of action of these air pollutants emphasizes the systemic impacts, imposing urgent interventions to lessen revelation and safeguard public health.

SYSTEMIC EFFECTS OF AIR POLLUTION

Respiratory System Effects

Air pollution adversely affects the respiratory system of humans and other organisms, leading to both acute and chronic respiratory health effects. Long-term exposure to various air pollutants including nitrogen oxides, ozone and PM results in many respiratory-related diseases such as allergic rhinitis, chronic obstructive pulmonary disease (COPD), lung infections, lung cancer, and asthma.

Individuals suffering from various respiratory conditions like COPD are susceptible to detrimental effects of air pollution, which can intensify conditions like asthma and lead to hospitalization due to acute episodes. Moreover, air pollutants induce inflammation in the lungs and respiratory tract. They produce oxidative stress and impair functions of immune system of lungs.

According to the WHO, air pollution alone caused about 6.7 million deaths globally in 2019. This is due to the progression of non-communicable diseases (NCDs) in affected individuals such as stroke, asthma, COPD, lung cancer, diabetes, and ischemic heart disease, which makes air pollution the second largest cause of NCDs after tobacco globally (WHO, 2021).

Continued exposure to toxic air pollutants such as $PM_{2.5}$ and gases may affect the mucosal cilia function of airways by interfering with the cilia movement, thereby decreasing their ability to clear mucus and dust particles. These may lead to an increase in the number of pollutants and particles in the airways, making individuals more vulnerable to various respiratory infections (Krismanuel & Hairunisa, 2024).

Air pollutants mainly $PM_{2.5}$, nitrogen dioxide (NO_2), O_3, and SO_2 are responsible for inflammation, irritation, and narrowing of airways, leading to the development of chronic respiratory condition known as asthma with symptoms like coughing, wheezing, tightness of chest, difficulty breathing, and decreased lung functions. It further creates challenges like breathlessness and difficulty in removing mucus and other particles from airways, which may finally lead to COPD. This can cause severe respiratory infections, higher mucus secretion, and coughing, and these exacerbations lead to life threatening situations for COPD patients.

Many individuals with severe COPD problems can further develop other diseases like cardiovascular diseases, hypertension, and diabetes, which increases the cascade of health issues and complications.

The air polluted environment can also develop certain complication in patients such as pneumonia, bronchitis, and other upper respiratory tract infections due to weakening immune system functions in the respiratory tract (Wibowo & Saputra, 2022). Prolonged exposure to air pollutants such as $PM_{2.5}$, VOCs, and PAHs leads to lung cancer is due to the penetration of these substances into the cells of lungs, and the buildup of genetic mutations leads to the occurrence of cancer.

Air pollutants, ozone, and PM induce chronic inflammation in the respiratory system, which may damage cells and enhance mutations that increase the risk of lung cancer. Moreover, these pollutants can create oxidative stress, which leads to damage of cellular structures and DNA, ultimately promoting the growth of cancer cells in lungs. Moreover, air pollution causes the development of allergies and sensitization to allergens driving the occurrence of allergic respiratory diseases like allergic rhinitis.

Cardiovascular System Effects

Air pollution has detrimental effects on cardiovascular illness and death. Particularly, the air pollutant, PM, causes a wide range of cardiovascular illness such as elevated blood pressure, blood clotting disorders, vascular functions, stroke, alterations in cardiac rhythm, and coronary artery disease.

Exposure to 1%–2% of 10 μg/m³ increase in concentration of $PM_{2.5}$ can increase the risk of stroke, arrhythmia, heart failure, myocardial infarctions, and sudden death, and it will increase to 5%–10% risk if a person is exposed long term to this pollutant. This is due to fact that the $PM_{2.5}$ is primarily in contact with cardiopulmonary organs through penetration, which can provoke direct oxidative stress and subsequently cause chronic inflammatory responses, leading to effects of the circulatory system and its vasculatures (Ji, 2022).

Air pollution emerges the cardio metabolic disease by disturbing the endothelial function of the heart muscle, inflammation, interrupting pro-thrombotic pathways, creating autonomic imbalance, distracting the link between cardiovascular system and CNS, and causing epigenomic changes (Rajagopalan et al., 2018).

Additionally, fine particles ($PM_{2.5}$) exposure causes chronic heart diseases with the symptoms such as fatigue, breathing problem, palpitation of heart, lightheadedness, and chest pain. $PM_{2.5}$ pollution also linked to thrombosis, raised blood pressure, and insulin resistance. It also enhances the injuries in the blood vessels, which leads to thermogenesis. Long-term inhalation of fine particles leads to premature death due to ischemic heart disease, stroke, and heart failure (Bhatnagar, 2022).

Central Nervous System (CNS) Effects

Like respiratory and cardiovascular systems, CNS is also significantly impacted by air pollution. Recent research studies indicated that air pollution has been closely associated with CNS allied diseases such as Parkinson's disease, Alzheimer's disease, neurodevelopmental disorders, and stroke. The fine particles in the air penetrate into the CNS and activate innate immune responses, leading to the neuroinflammation. Moreover, air pollution induces microglial activation, changes the blood–brain barrier, creates oxidative stress, and causes cerebrovascular dysfunction, which influence the mental and nervous system health (Genc et al., 2012).

The brain and lungs directly absorb fine PM particles through the nasal olfactory mucosa (Ajmani et al., 2016). Fine PM ($PM_{2.5}$) and oxides of nitrogen affect the nervous system and cause various nervous system abnormalities such as oxidative stress, neuroinflammation, and systemic inflammation. $PM_{2.5}$ is mainly responsible for increased incidents of Alzheimer's disease and associated cognitive functions. Similarly, oxides of nitrogen and ozone are causative agents for vascular dementia, which may lead to hospitalization of patients if not treated (Kim et al., 2020).

Depression is a kind of psychological problem that may occur in the population when men are continuously exposed to $PM_{2.5}$ and short-term exposed to PM_{10}, CO, SO_2, and NO_2, whereas women are exposed to ozone along with these pollutants (Nuyts et al., 2019). Outdoor air pollution is responsible for many kinds of mental illnesses such as Multiple sclerosis (MS), Schizophrenia, and attention-deficit/hyperactivity disorder (ADHD).

MS is a chronic disease of CNS, which is caused by neurodegeneration due to autoimmune diseases. The results have shown that PM is responsible for these autoimmune diseases as these particles are drawn into the lungs through the respiratory tract, causing pulmonary inflammation, which triggers the activation of T cells and in turn secretes pro-inflammatory cytokines in the brain (Zhao et al., 2019).

On the other hand, Schizophrenia is an enduring and unadorned mental disorder affecting more than 21 million people globally. This disease is more prevalent in urban areas than rural areas. The incidents of this disease are due to short-term exposure to air pollutants such as PM_{10}, NO_2, and SO_2 (Liang et al., 2019).

As per the study conducted, about 60% of higher mental illness has been noticed in teenagers who are exposed to $PM_{2.5}$, NO_2, and NO_x than the unexposed ones (Qiu et al., 2019). In addition to this, studies on human and animals indicated that the air pollution promotes neurotoxicity and can take part in certain neurodevelopmental disorders, including autism spectrum disorder (ASD) (Costa et al., 2020).

Research studies revealed the links between pre-natal and early-life exposure to pollutants such as $PM_{2.5}$, NO_2, and PAHs with developmental disorders, including ASD, ADHD, and cognitive delays in children (Volk et al., 2021).

Pollutant $PM_{2.5}$ damages the brain and peripheral tissue by generating reactive oxygen species (ROS), which causes cellular damage and dysregulation of neural growth and synaptic development leading to the loss of memory and difficulty in learning. Air pollutants trigger the generation of cytokines such as IL-6 and TNF-α, which cause chronic inflammation and interfere the neural circuit functions in the prefrontal cortex and hippocampus of the brain, leading to memory loss in children and the elderly people (Jaiswal & Singh, 2024).

Other Systemic Health Effects

Air pollution along with respiratory, cardiovascular, and neural systems also affects multiple body systems including reproductive, endocrine, and immune systems.

Air pollutants including PM and VOCs significantly impact the reproductive health of people, leading to infertility, premature births, low birth weights, and developmental abnormalities in fetuses. Predominantly, inhaled Endocrine-disrupting chemicals (EDCs) interfere with the neuroendocrine and endocrine systems and disrupt the production, regulation, and signaling of natural hormones. These chemicals mainly effect the functions of the hypothalamic–pituitary–gonadal (HPG) axis, which in turn have an impact on the fertility of both men and women by affecting the sperm production, morphology, and menstrual cycle (Plunk & Richards, 2020).

In recent past, the incidences of diabetes have been gradually increasing in both developed and developing countries. Moreover, recent studies suggest a potential link between air pollution and disturbances in insulin resistance (IR), glucose metabolism, and the development of type 2 diabetes mellitus (T_2DM). The key air pollutants that cause diabetes are $PM_{2.5}$, PM_{10}, NO_x, NO_2, CO, and SO_2 (Li et al., 2019).

Ambient air pollutants stimulate adaptive immunity, thereby enhancing the activity of T helper lymphocyte type 2 (Th2) and T helper lymphocyte type 17 (Th17) cells, resulting in various health issues such as asthma and allergies clinically, leading to a life-threating disease, i.e., COPD.

Inter-system Impacts

Air pollution impacts the CVD, respiratory, and other systems through interconnected mechanism. For instance, $PM_{2.5}$, nitrogen oxides, and ozone enter the lungs through inhalation, causing inflammation due to irritation in the air track and lungs, and particularly, $PM_{2.5}$ enters the bloodstream by penetration into the alveoli causing inflammation in the blood vessels and endothelial dysfunction, which leads to atherosclerosis, arrhythmias and stroke (Konduracka & Rostoff, 2022).

Similarly, through the circulation, these pollutants reach the cellular level, thereby disturbing the neural, immunological, and metabolic systems leading to acute and chronic diseases such as COPD, diabetes, neural disorders, hormonal imbalances, organ stress, immunological disorders, and hypertension (Shetty et al., 2023) (Table 3.2).

Figure 3.2 illustrates the systemic impacts of air pollution, including respiratory diseases (asthma, COPD), cardiovascular disorders (stroke, heart failure), neurological

TABLE 3.2 Systemic effects of air pollution

S.NO	*SYSTEM AFFECTED*	*DESCRIPTION*	*HEALTH IMPACTS*	*REFERENCES*
1	Respiratory System	Acute and chronic impacts, including allergic rhinitis, asthma, COPD, and lung cancer.	Inflammation, oxidative stress, impaired mucus clearance, asthma and COPD.	Kurt et al. (2016), Tran et al. (2023)
2	Cardiovascular System	Oxidative stress, inflammation, and disrupt vascular functions, hypertension, stroke, and heart attacks.	Fatigue, breathing difficulties, chest pain; atherosclerosis, and ischemic heart diseases.	Miller (2022), Bourdrel et al. (2017), Ji (2022)
3	Central Nervous System	Oxidative stress, inflammation, effects the blood-brain barrier, neurodevelopmental and neurodegenerative diseases.	Alzheimer's, Parkinson's, multiple sclerosis, schizophrenia, and depression;	Genc et al. (2012), Kim et al. (2020)
4	Reproductive System	Interfere with hormone signaling and reproductive health.	Infertility, preterm births, low birth weights, and fetal developmental abnormalities; disrupted HPG axis affecting fertility in men and women.	Afridah et al. (2024), Aguilera et al. (2023)
5	Immune System	Enhances the activity of inflammatory pathways and immune cell dysregulation.	Increased susceptibility to infections, asthma, allergies, and autoimmune disorders.	Glencross et al. (2020)

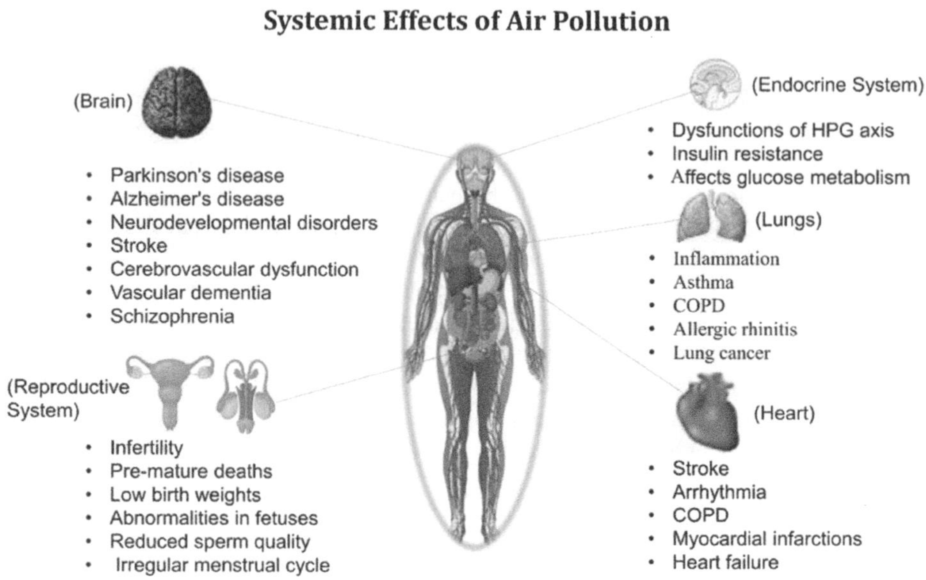

FIGURE 3.2 Systemic effects of air pollution.

issues (Alzheimer's, neurodevelopmental disorders), reproductive health effects (infertility, low birth weight), and metabolic dysfunctions (insulin resistance)

EFFECTS OF AIR POLLUTION ON VULNERABLE POPULATION

Vulnerable populations, including children and pregnant women, and people with pre-existing health issues elevate intensified risks, emphasizing the systemic reach of air pollution's health impacts. Children are recognized as high-risk group among vulnerable populations to air pollution impacts, though it is varying from various stages of childhood, while pregnant women, despite increasing evidence of reproductive health risks, are not consistently classified as a vulnerable group.

The factors include genetic disorders, socioeconomic issues, age group, lifestyle, pre-existing health issues like cystic fibrosis, asthma, lung transplantation, and COPD, which contribute to enhancing the effects of air pollution on the population.

The elderly people are more prone to air pollution due to weak immune system and age-associated health issues like cardiovascular and respiratory dysfunctions, which may lead to experiencing of life-threating conditions such as COPD, asthma, diabetes, lung cancer, emphysema, atherosclerosis, and bronchitis. These conditions or symptoms can lead to hospitalization due to exposure to higher levels of air pollution. In addition, socioeconomic issues also contribute to severe effects of air pollution. People who live with low income and inadequate facilities are more susceptible to air pollution effects.

Air pollution pretences intensified risks for specific groups, including industrial workers and populations facing urban–rural disparities:

Occupational Exposure: Workers in various industries like mining, manufacturing, and construction are frequently exposed to higher levels of airborne pollutants such as PM, heavy metals, and VOCs. Prolonged exposure to these pollutants causes respiratory problems like asthma, bronchitis, COPD, and increased risks of cardiovascular diseases, lung cancer, and neurological effects due to oxidative stress and systemic inflammation (Chen et al., 2024).

URBAN–RURAL DISPARITIES

Urban Areas: Urban populations are due to traffic, industries, and waste burning, as well as high exposure to pollutants like $PM_{2.5}$, NO_2, and ozone. These pollutants extremely affect residents in densely populated, low-income neighborhoods due to proximity to pollution sources (Nakhjiri & Kakroodi, 2024).

Rural Areas: While rural regions have better air quality than urban, the sources like agricultural practices, biomass burning, and dust can lead to localized pollution. Furthermore, rural residents often lack access to healthcare facilities that can treat population affected by air pollution (Hilly et al., 2024).

Vulnerable Subpopulations: Vulnerable population such as children and the elderly individuals are highly susceptible to air pollution due to prolonged exposure to polluted environment, which leads to respiratory and cardiovascular health issues, particularly in urban areas. Rural populations may face challenges in managing pollution-related illnesses due to limited resources and health facilities.

Pregnancy-Related Risks: Air pollution significantly has an impact on pregnancy-related issues. Long-term exposure to air pollutants like $PM_{2.5}$ and VOCs can have an impact on the fetal development. For instance, the pollutant $PM_{2.5}$ can cross the placental barrier through penetration and causes oxidative stress, leading to a reduction or restriction of the flow of nutrients and oxygen to the fetus, which will have effect on growth, development, and birth. On the other hand, VOCs can create endocrine disruption, which also have an impact on the development of the fetus, leading to low weight, premature birth, and mortality. Prolonged exposure to these pollutants further leads to neurological disorders and congenital defects in children (Rani & Dhok, 2023).

GLOBAL AND LOCAL PROSPECTIVE OF AIR POLLUTION EFFECTS

Annually at global level, about 6 million mortality cases have been happening due to air pollution exposure. Over 200 countries in the world are emitting toxic air pollutants such as PM, O_3, NO_x, SO_2, CO, and VOCs, out of which about 171 countries discharge air pollutants above the threshold level, causing serious health issues in people of those countries. Furthermore, according to the Ella's tragic and preventable death statistics, each year about 6.7 million deaths are occurring due to air pollution, and at the same time, about 213 million disabilities have been registered due to the same problem, indicating that the air pollution has become the fourth highest health hazard globally.

The *Lancet* Commission health report stated that the pollution was accountable for around 9 million premature deaths worldwide in 2015 due to extreme poverty. The recent data from the Global Burden of Diseases, Injuries, and Risk Factors Study 2019 indicated that the rate of premature death has been reduced due to the improvement in economic conditions. Nevertheless, at the same time, the increased rate of death from toxic air pollutants is caused by industrialization and urbanization from year 2000 (7%) to 2015 (66%).

Climate change also contributes to the rise of air pollution incidents locally, which is due to burning of fossil fuels and the decomposition of waste matter that releases gases like CO_2, oxides of nitrogen and sulfur, and methane, which are responsible for global warming and in turn accountable for the formation of smog, fog, and ground-level ozone; these factors result in the rise of air pollution incidents, which directly affects the health of local communities and causes premature mortality; these also directly have an impact on the low- and middle-income nation's economy like Africa, Asia, and Europe (Afifa et al., 2024).

CASE STUDIES OF AIR POLLUTION EFFECTS

Air Pollution Incidents in Kyrgyzstan

In Kyrgyzstan, air pollution is the most hazardous environmental issue, contributing to a high rate of premature mortalities, mainly young children. The UNICEF study in Bishkek, Kyrgyzstan revealed that the people of Bishkek are exposed to air pollutants extremely above the WHO air quality guidelines due to the burning of raw coal for cooking and heating in winter throughout the year, resulting in higher exposure to $PM_{2.5}$ (44 $\mu g/m^3$) pollution. In summer, residents of this city are exposed to air pollutants of 7.3 ± 1.7 μg/m^3, leading to serious health impacts on residents, particularly the younger population.

Cross-Regional Comparisons

Air pollution significantly impacts health across various regions, with distinct socioeconomic and environmental contexts influencing its severity.

Kyrgyzstan: In recent winters in Bishkek, average daily $PM_{2.5}$ concentrations are reported to be more than 200 μg/m^3, which is far than the WHO guideline of 15 μg/m^3. Between 2010 and 2019, 12%–13% of annual deaths in Kyrgyzstan were attributed to air pollution, which is equal to approximately 4,100–5,000 deaths per year (Dzushupov et al., 2022).

South Asia: This region remains the world's most polluted, accounting for 45% of global pollution exposure. Despite some improvements, air quality continues to pose severe health risks. Between 1990 and 2020, greenhouse gas (GHG) emissions in South Asia increased significantly. In India, CO_2 emissions rose from 578.5 to 2,441.8 million tons, with similar increases in Bangladesh, Nepal, and Pakistan. Between 1990 and 2018, GHG emissions nearly tripled in India (1,990.4–3,346.6 MT CO_3-equivalents) and Nepal (20.6–54.6 MT), while they doubled in Bangladesh. Methane emissions increased the most in Pakistan (70.4–151 MT), and nitrous oxide emissions almost doubled in all four countries.

$PM_{2.5}$ exposures have also worsened, with India's levels rising from 81.3 μg/m^3 (1990) to 90.9 μg/m^3 (2017) and Nepal experiencing the largest increase (87.6–99.7 μg/m^3). During the COVID-19 pandemic (2018–2021), Bangladesh had the highest $PM_{2.5}$ concentrations (76.9 μg/m^3), followed by Pakistan (66.8 μg/m^3), India (58.1 μg/m^3), and Nepal (46 μg/m^3). These trends highlight the urgent need to improve air quality in South Asia (Abdul Jabbar et al., 2022).

Sub-Saharan Africa: Air pollution has a heavy impact on health and livelihoods in Africa, with significant differences across countries and regions. In 2019, $PM_{2.5}$ pollution, including outdoor and household air pollution, was the leading cause of deaths across the continent. Northern Africa had the highest disease burden from $PM_{2.5}$, while Eastern Africa had the lowest.

From 2010 to 2019, $PM_{2.5}$ exposure levels in Africa remained largely unchanged. Country-level data shows disparities, with $PM_{2.5}$ levels ranging from 17.1 μg/m^3 in Comoros to 80.1 μg/m^3 in Niger. The air pollution-related death rate in Africa (155 deaths per 100,000 people) is nearly double the global average of 85.6 deaths per 100,000. Egypt recorded the highest number of deaths from $PM_{2.5}$ (90,600 in 2019), while São Tomé and Príncipe and Comoros had fewer than 100 deaths each (Safo-Adu et al., 2023).

THE EFFECT OF AIR POLLUTION ON COVID-19 INCIDENCES IN EUROPE AND NORTH AMERICA

One of the factors contributing to the worsening of COVID-19 cases in Europe and North America was air pollution. The COVID-19 estimate reports of both countries stated that the there is a significant connection between air pollution and COVID-19 cases, which rise up to about 52.7%, out of which about 48.1% were morality cases and 41.2% were confirmed as non-fatal severity cases.

Prolonged exposure to air pollutants increased COVID-19 cases by 68.3%. Air pollutants such as PM ($PM_{2.5}$ and PM_{10}), CO, NO_2, and O_3 were positively allied with COVID-19 events, with NO_2 and $PM_{2.5}$ mainly contributing to the COVID-19 deaths in both countries.

Biological Mechanisms Linking to Air Pollution and COVID-19: Air pollution worsens COVID-19 severity and mortality through its effects on inflammation and immune responses. Prolonged exposure to air pollutants increased production of cytokines like IL-6, IL-1β and TNF-α, causing systemic inflammation and amplifying cytokine levels in severe COVID-19 (Bayram et al., 2024).

Moreover, pollutants weaken macrophages activity and T-cell responses, leading to weakening of the immune system to fight against COVID-19. Furthermore, pollutants produce ROS, which damages lung tissue and increases viral entry and replication through upregulated ACE2 receptors (Sanduzzi Zamparelli et al., 2025).

MITIGATION STRATEGIES FOR REDUCTION OF AIR POLLUTION EFFECTS

From the experiences of events of air pollution impacts globally, it is understood that the air pollution is one of the most dangerous environmental issues in the world, since it has claimed millions of deaths to date. Hence, it is understood that there is a need to address the global air pollution issue for the safety of public health (Sofia et al., 2020).

Various mitigation strategies include the involvement of different stakeholders to get health benefits by the reduction of air pollution by providing specific guidelines to different sectors including household, transport, agriculture, industry, shipping, and energy generation. These guidelines can be treated as a foundation for governments for the execution of various strategic plans for reducing all kinds 'of air pollutant emission and controlling overall air pollution-related risk.

Moreover, it is necessary to explore the role of environmental policies, preventive environmental management strategies, and their implications setup by the government in the interest of protection of health of environment and living beings. Apart from this, individuals and public play an important role in controlling air pollution, which is done by creating public awareness by the government and environmental protective agencies.

Furthermore, adopting advanced green technology like air filtration systems namely high efficiency particulate air (HEPA) filters, which remove 99.97% of dust, $PM_{2.5}$, bacteria, pollens, and mold, improves indoor and outdoor air quality (Chen et al., 2022). Renewable energies (Solar, wind, hydro power, and biomass) are used to reduce emissions of fossil fuels while also lowering GHGs and other pollutants in the atmosphere (Paraschiv & Paraschiv, 2023). Advances in vehicular emissions like introducing electrical vehicles, catalytic converters, and hybrid engines effectively reduce the exhaust pollutants like PM, CO, CO_2, hydrocarbons, and NO_x (Milku et al., 2024). Emerging technologies, such as carbon capture and storage (CCS), further mitigate industrial emissions.

Additionally, policy matters including enforcing strict emissions standards, implementing solid environmental regulations, and adopting new green techniques like solar-assisted cleaning systems, congestion charges, and photocatalytic materials can reduce air pollution impacts (Afifa et al., 2024).

Air Quality Policy Success Stories: Efficient air quality policies can considerably reduce pollution and mitigate its health impacts.

U.S. Clean Air Act (CAA): It was approved by EPA in 1970 to reduce the emissions of air pollutants such as nitrogen oxides, sulfur oxides, and PM to improve public health and control the accidental release of toxic air pollutants.

India's National Clean Air Programme (NCAP): NCAP was introduced in 2019, and the main objective of this programme is to reduce PM ($PM_{2.5}$ and PM_{10}) by 20%–30% by 2024 in 132 cities. Later, it was stretched to 2026 with a revised target of a 40% reduction in PM_{10} levels compared to 2017 and also it sets up stricter air emission standards for industries, better monitoring, and public awareness campaigns to contest air pollution (CREA, 2025).

China's Air Pollution Action Plan: This plan instituted in 2013 aims to implement strict air emission standards for industries, monitor them, and introduce new clean energy technology. China has substantial reductions in $PM_{2.5}$ levels since this policy introduced (Cheng et al., 2023).

CONCLUSION

Air pollution remains a critical global challenge, with intense implications for public health and environmental sustainability. Major pollutants present in air include PM, gaseous pollutants, and VOCs, originating from both natural and human activities.

The mechanisms by which these pollutants impact health are complex, involving oxidative stress, inflammation, and systemic absorption, leading to severe outcomes across organ systems.

The respiratory and cardiovascular systems are susceptible to air pollution contributing to conditions like asthma, COPD, hypertension, and stroke. Moreover, its effects extend to the nervous system, reproductive health, and immune system, exacerbating the global burden of diseases like diabetes, dementia, and cancer. Vulnerable populations, including children, the elderly, and low-income communities, face health risks due to air pollution.

Global and regional studies highlight the impacts of air pollution and its health effects and emphasizes mitigation strategies for controlling episodes of air pollution. This can be done by public awareness and partnership, strict policy measures and implementation, and the adoption of clean and green technologies.

Policymakers can reduce harmful effects of air pollutants by implementing and close monitoring strict air emission standards for industries and vehicles, capitalizing in public transport, boosting cleaner, efficient mobility, and reducing traffic emissions. Administering green building **codes** encourages energy-efficient designs and cuts pressure on fossil fuels consumption. Investing more on renewable energy and air quality monitoring systems brings long-term improvements. Public awareness campaigns can further drive collective action for cleaner air.

Filling research gaps such as toxic effects of air-borne microplastics on human health, the long-term effects of prolonged exposure of low levels of pollutants (NO_x, PM), and the role of indoor pollutants such as VOCs causing diseases like respiratory and neurological disorders. Moreover, further investigations are needed to determine how climate change impacts the air pollution effects on health of ecosystems and the efficacy and scalability of emerging green technologies in combating the air pollution impacts on human and environment.

By reducing exposure and improving air quality, we can protect public health, especially for the most vulnerable, and pave the way for a healthier and more sustainable future.

REFERENCES

Abdul Jabbar, S., Tul Qadar, L., Ghafoor, S., Rasheed, L., Sarfraz, Z., Sarfraz, A., Sarfraz, M., Felix, M., & Cherrez-Ojeda, I. (2022). Air quality, pollution and sustainability trends in south Asia: a population-based study. *International Journal of Environmental Research and Public Health*, *19*(12), 7534. https://doi.org/10.3390/ijerph19127534

Afifa, A. K., Hussain, N., Ashraf, M. H., & Saleem, M. Z. (2024). Air pollution and climate change as grand challenges to sustainability. *Science of The Total Environment*, *928*, 172370. https://doi.org/10.1016/j.scitotenv.2024.172370

Afzal, A., & Mahreen, N. (2024). Emerging insights into the impacts of heavy metals exposure on health, reproductive and productive performance of livestock. *Frontiers in Pharmacology*, *15*, 1375137. https://doi.org/10.3389/fphar.2024.1375137

Bayram, H., Konyalilar, N., Elci, M. A., Rajabi, H., Aksoy, G. T., Mortazavi, D., Kayalar, Ö., Dikensoy, Ö., Taborda-Barata, L., & Viegi, G. (2024). Issue 4- Impact of air pollution on COVID-19 mortality and morbidity: an epidemiological and mechanistic review. *Pulmonology*, *31*(1), 2426829. https://doi.org/10.1016/j.pulmoe.2024.04.005

Bherwani, H., Nair, M., Musugu, K., Gautam, S., Gupta, A., Kapley, A., & Kumar, R. (2020). Valuation of air pollution externalities: Comparative assessment of economic damage and emission reduction under COVID-19 lockdown. *Air Quality, Atmosphere & Health*, *13*, 683–694. https://doi.org/10.1007/s11869-020-00845-3

Brocza, F. M., Rafaj, P., Sander, R., Wagner, F., & Jones, J. M. (2024). Global scenarios of anthropogenic mercury emissions. *Atmospheric Chemistry and Physics*, *24*(12), 7385–7404. https://doi.org/10.5194/egusphere-2024-41

Chen, C.-F., Hsu, C.-H., Chang, Y.-J., Lee, C.-H., & Lee, D. L. (2022). Efficacy of HEPA air cleaner on improving indoor particulate matter 2.5 concentration. *International Journal of Environmental Research and Public Health*, *19*(18), 11517. https://doi.org/10.3390/ijerph191811517

Chen, F., Zhang, W., Mfarrej, M. F. B., Saleem, M. H., Khan, K. A., Ma, J., Raposo, A., & Han, H. (2024). Breathing in danger: understanding the multifaceted impact of air pollution on health impacts. *Ecotoxicology and Environmental Safety*, *280*, 116532. https://doi.org/10.1016/j.ecoenv.2024.116532

Cheng, Y., Du, K., & Yao, X. (2023). Stringent environmental regulation and inconsistent green innovation behavior: evidence from air pollution prevention and control action plan in China. *Energy Economics*, *120*, 106571. https://doi.org/10.1016/j.eneco.2023.106571

CREA. (2025). *Only 31% of the 131 NCAP Cities Meet Targets Set by Initial 2024 Deadline*. CREA, New Delhi, India.

Dzushupov, K. O., Buban, J. M. A., Aidaraliev, A. A., Ahmadi, A., Chahal, P., Ibrahim, M., Lin, X., & Kouwenhoven, M. B. N. (2022). Air pollution in Bishkek, Kyrgyzstan: driving factors and state response. *Public Health Challenges*, *1*(4), 22. https://doi.org/10.1002/puh2.22

Hasan, M. K., Shahriar, A., & Jim, K. U. (2019). Water pollution in Bangladesh and its impact on public health. *Heliyon*, *5*(8), e02145. https://doi.org/10.1016/J.HELIYON.2019.E02145

Hilly, J. J., Singh, K. R., Jagals, P., Mani, F. S., Turagabeci, A., Ashworth, M., Mataki, M., Morawska, L., Knibbs, L. D., Stuetz, R. M., & Dansie, A. P. (2024). Review of scientific research on air quality and environmental health risk and impact for PICTS. *Science of The Total Environment*, *942*, 173628. https://doi.org/10.1016/j.scitotenv.2024.173628

Hodan, W. M., & Barnard, W. R. (2023). Evaluating the contribution of $PM_{2.5}$ precursor gases and re-entrained road emissions to mobile source $PM_{2.5}$ particulate matter emissions. *MACTEC Federal Programs*, Research Triangle Park, NC, pp. 1–53.

Jaiswal, C., & Singh, A. K. (2024). Particulate matter exposure and its consequences on hippocampal neurogenesis and cognitive function in experimental models. *Environmental Pollution*, *363*, 125275. https://doi.org/10.1016/j.envpol.2024.125275

Khan, Z., Elahi, A., Bukhari, D. A., & Rehman, A. (2022). Cadmium sources, toxicity, resistance and removal by microorganisms-A potential strategy for cadmium eradication. *Journal of Saudi Chemical Society*, *26*(6), 101569. https://doi.org/10.1016/j.jscs.2022.101569

Konduracka, E., & Rostoff, P. (2022). Links between chronic exposure to outdoor air pollution and cardiovascular diseases: a review. *Environmental Chemistry Letters*, *20*(5), 2971–2988. https://doi.org/10.1007/s10311-022-01450-9

Kumar, R. P., Brema, J., Samuel, C., & Gautam, S. (2023). A bibliometric and scientometric analysis towards global patterns and trends related to aerosol and precipitation studies from 2002 to 2022. *Air Quality, Atmosphere & Health*, *16*, 613–628. https://doi.org/10.1007/s11869-022-01290-0

Kung, H.-C., Wu, C.-H., Huang, B.-W., Chang-Chien, G.-P., Mutuku, J. K., & Lin, W.-C. (2024). Mercury abatement in the environment: insights from industrial emissions and fates in the environment. *Heliyon*, *10*(7), e28253. https://doi.org/10.1016/j.heliyon.2024.e28253

Levin, R., Zilli Vieira, C. L., Rosenbaum, M. H., Bischoff, K., Mordarski, D. C., & Brown, M. J. (2021). The urban lead (Pb) burden in humans, animals and the natural environment. *Environmental Research*, *193*, 110377. https://doi.org/10.1016/j.envres.2020.110377

Li, Z., Liu, Y., Lin, Y., Gautam, S., Kuo, H.-C., Tsai, C.-J., Yeh, H., Huang, W., Li, S.-W., & Wu, G.-J. (2017). Development of an automated system (PPWD/PILS) for studying $PM_{2.5}$water-soluble ions and precursor gases: Field measurements in two cities, Taiwan. *Aerosol and Air Quality Research*, *17*, 426–443.

López, J. F., Fernández, M. I., & Coz, L. F. (2020). Arsenic exposure is associated with significant upper tract urothelial carcinoma health care needs and elevated mortality rates. *Urologic Oncology: Seminars and Original Investigations*, *38*(7), 638.e7–638.e13. https://doi.org/10.1016/j.urolonc.2020.01.014

Mandal, P., Sarkar, R., Mandal, A., Patel, P., & Kamal, N. (2016). Study on airborne heavy metals in industrialized urban area of Delhi, India. *Bulletin of Environmental Contamination and Toxicology*, *97*(6), 798–805. https://doi.org/10.1007/s00128-016-1944-y

Mar, K. A., Unger, C., Walderdorff, L., & Butler, T. (2022). Beyond CO_2 equivalence: the impacts of methane on climate, ecosystems, and health. *Environmental Science & Policy*, *134*, 127–136. https://doi.org/10.1016/j.envsci.2022.03.027

Milku, A. K., Attiogbe, F., Atombo, C., Derkyi, N. S. A., & Asuako, E. L. (2024). Evaluating the categorical effect of vehicle characteristics on exhaust emissions. *African Transport Studies*, *2*, 100008. https://doi.org/10.1016/j.aftran.2024.100008

Nakhjiri, A., & Kakroodi, A. A. (2024). Air pollution in industrial clusters: a comprehensive analysis and prediction using multi-source data. *Ecological Informatics*, *80*, 102504. https://doi.org/10.1016/J.ECOINF.2024.102504

Nguyen, D.-H., Lin, C., Vu, C.-T., Cheruiyot, N. K., Nguyen, M. K., Le, T. H., Lukkhasorn, W., Vo, T.-D.-H., & Bui, X.-T. (2022). Tropospheric ozone and NO_x: a review of worldwide variation and meteorological influences. *Environmental Technology & Innovation*, *28*, 102809. https://doi.org/10.1016/j.eti.2022.102809

Paraschiv, L. S., & Paraschiv, S. (2023). Contribution of renewable energy (hydro, wind, solar and biomass) to decarbonization and transformation of the electricity generation sector for sustainable development. *Energy Reports*, *9*, 535–544. https://doi.org/10.1016/j.egyr.2023.07.024

Rani, P., & Dhok, A. (2023). Effects of pollution on pregnancy and infants. *Cureus*, *15*(1), e33906. https://doi.org/10.7759/cureus.33906

Rasin, P., Ashwathi, A. V., Basheer, S. M., Haribabu, J., Santibanez, J. F., Garrote, C. A., Arulraj, A., & Mangalaraja, R. V. (2025). Exposure to cadmium and its impacts on human health: a short review. *Journal of Hazardous Materials Advances*, *17*, 100608. https://doi.org/10.1016/j.hazadv.2025.100608

Ren, C., Huang, X., Wang, Z., Sun, P., Chi, X., Ma, Y., Zhou, D., Huang, J., Xie, Y., Gao, J., & Ding, A. (2021). Nonlinear response of nitrate to NO_x reduction in China during the COVID-19 pandemic. *Atmospheric Environment*, *264*, 118715. https://doi.org/10.1016/j.atmosenv.2021.118715

Safo-Adu, G., Attiogbe, F., Emahi, I., & Ofosu, F. G. (2023). Outdoor and indoor particle air pollution and its health consequences in African cities: new evidence and an exhortation. *Sustainable Environment*, *9*(1), 729. https://doi.org/10.1080/27658511.2023.2265729

Sanduzzi Zamparelli, S., Sanduzzi Zamparelli, A., & Bocchino, M. (2025). Immune-boosting and antiviral effects of antioxidants in COVID-19 pneumonia: a therapeutic perspective. *Life*, *15*(1), 113. https://doi.org/10.3390/life15010113

Sasso, F., Picca, F., Pignatelli, A., Commodo, M., Minutolo, P., & D'Anna, A. (2024). A laboratory study of secondary organic aerosol formation in an oxidation flow reactor. *Fuel*, *367*, 131491. https://doi.org/10.1016/j.fuel.2024.131491

Sellami, F., Baati, H., Gautam, S., Azri, C., & Elaksher, A. F. (2025). Impact assessment of particulate deposits bound heavy metals on superficial soils along a street canyon at Sfax City (Tunisia): Pollution degree and toxicity risks. *Earth Systems and Environment*. Advance online publication. https://doi.org/10.1007/s41748-025-00719-7

Setu, S., Gautam, S., Khan, H. M. S., Baten, M. A., & Khan, M. B. (2024). Phytomonitoring of air pollution around brick kilns in urban area: Exploring the potential of plants for the remediation of pollutants. *Chemosphere*, *368*, 143721. https://doi.org/10.1016/j.chemosphere.2024.143721

Shetty, S. S., Deepthi, D., Harshitha, S., Sonkusare, S., Naik, P. B., Kumari, N. S., & Madhyastha, H. (2023). Environmental pollutants and their effects on human health. *Heliyon*, *9*(9), e19496. https://doi.org/10.1016/j.heliyon.2023.e19496

Volk, H. E., Perera, F., Braun, J. M., Kingsley, S. L., Gray, K., Buckley, J., Clougherty, J. E., Croen, L. A., Eskenazi, B., Herting, M., Just, A. C., Kloog, I., Margolis, A., McClure, L. A., Miller, R., Levine, S., & Wright, R. (2021). Prenatal air pollution exposure and neurodevelopment: a review and blueprint for a harmonized approach within ECHO. *Environmental Research*, *196*, 110320. https://doi.org/10.1016/j.envres.2020.110320

Yang, Z., Li, K., Tsona, N. T., Luo, X., & Du, L. (2023). SO_2 enhances aerosol formation from anthropogenic volatile organic compound ozonolysis by producing sulfur-containing compounds. *Atmospheric Chemistry and Physics*, *23*(1), 417–430. https://doi.org/10.5194/acp-23-417-2023

4

Exposure to Fine Particulate Matter among Women in Rural Households using Biomass Fuels

Analysis of Elemental Composition, Morphological Characteristics, and Health Risks

Yaparla Deepthi, Amit Passi, Chithra V. S., Smaranika Panda, and S. M. Shiva Nagendra

 DOI: 10.1201/9781003613183-4

INTRODUCTION

Air pollution is the second prominent health risk factor in India after malnutrition, accounting for 1.67 million deaths (17.8% of all deaths in India) (Yunusa et al., 2023; Yadav et al., 2021; Kaur and Pandey, 2021; Arora et al., 2020). According to a 2019 Global Burden Diseases (GBD) study, 6.1 lakh premature deaths in India were caused by household air pollution (HAP). Biomass fuels used for cooking and heating in rural Indian households contribute to HAP. Of the three billion people globally who rely on solid fuels, over 0.82 billion live in India, most of whom are from rural areas. The inefficient and incomplete burning of these fuels releases harmful pollutants, contributing to the loss of approximately 86 million healthy life years. (WHO, 2024). As per a 2019 GBD study, HAP in India accounts for 4.5% of disability-adjusted life years (DALYs). In addition, HAP is a leading cause of mortality related to biomass burning, contributing to conditions like stroke, lung cancer, chronic obstructive pulmonary disease, ischemic heart disease, and lower respiratory infections. These diseases account for 32%, 23%, 21%, 19%, and 6% of fatalities, respectively, as reported by the World Health Organization (WHO, 2024).

The biomass burnings emit several harmful pollutants, including nitrogen dioxide (NO_2), sulphur dioxide (SO_2), particulate matter (PM), carbon monoxide (CO), formaldehyde (HCHO), and various other gaseous pollutants (Yao et al., 2023; Jiang et al., 2024; Tripathi et al., 2024). Biomass combustion emits PM with a high concentration of volatile organic compounds (VOCs), including formaldehyde, acrolein, benzene, toluene, styrene, and 1,3-butadiene. In addition to VOCs, the smoke from biomass burning also includes polycyclic aromatic hydrocarbons (PAHs) like benzo(a)pyrene, fluoranthene, pyrene, benzo(ghi)fluoranthene, and chrysene. These compounds are harmful due to their carcinogenic properties. The PM further consists of carbonaceous materials, comprising organic carbon (OC) and elemental carbon (EC), along with heavy metals such as chromium, cadmium, lead, nickel, and manganese, which contribute to the overall toxic nature of biomass smoke (Samburova et al., 2016; Liu et al., 2022; Jiang et al., 2024). Many of these pollutants are recognized as highly hazardous and harmful to human health. Prolonged exposure to HAP, particularly fine particles, elevates the risk of various health issues, including cardiac events, respiratory infections, stroke, worsening of inflammatory lung conditions, eye diseases, tuberculosis, and cancer (Ahmed et al., 2019; Lee et al., 2020; Juginović et al., 2021; Dehghani et al., 2023). The harmful effects on health are worsened by the presence of toxic compounds such as EC, PAH, and heavy metals in fine and ultrafine particles (Liu et al., 2022; Thangavel et al., 2022; Li et al., 2023; Alves et al., 2023).

The PM released from biomass burning primarily consists of carbonaceous particles along with nitrate (NO_3^-), sulphate (SO_4^{2-}), ammonium (NH_4^+), organic compounds, water-soluble ions dominated by potassium, and various heavy metals. Additionally, chloride (Cl^-) is a significant inorganic species, accounting for 40%–45% of the soluble inorganic components produced during biomass burning (Satsangi et al., 2014). Trace metals such as chromium (Cr), cobalt (Co), copper (Cu), iron (Fe), magnesium (Mg), nickel (Ni), manganese (Mn), and zinc (Zn) are known to contribute to both short-term and long-term health problems. The PM from biomass burning contains carcinogenic heavy metals, including Cr, Cd, Pb, and Ni. These metals have biomagnification potential and pose significant health risks. Nickel (Ni) is among the heavy metals with a higher cancer risk, particularly linked to lung and nasal cancers. Satsangi et al. (2014) detected carcinogenic metals like Cr, Cd, and Ni in PM emitted from biomass combustion. Likewise, Massey et al. (2013) reported elevated concentrations of Fe, followed by Cr, Cd, Pb, Cu, Ni, and Mn, in PM at a rural Indian site. For a comprehensive toxicological assessment of PM and its associated elements, it is crucial to incorporate both qualitative and quantitative analyses, as well as particle size and morphology.

Personal monitoring is considered one of the most important approaches for assessing human exposure to vulnerable people in determining health risks. Multiple studies in rural India have reported indoor pollutant levels surpassing the National Ambient Air Quality Standards (NAAQS) and the WHO guidelines (Satsangi et al., 2014; Liu et al., 2022; Tripathi et al., 2024). Additionally, previous studies have indicated that personal exposure levels during cooking periods can reach as high as 2,000–6,800 $\mu g/m^3$ (Ansari et al., 2010; Nayek and Padhy, 2018; Yadav et al., 2023).

Many research studies have been made to investigate the levels of HAP in rural India. However, the majority of studies have primarily focused on villages situated in the Indo-Gangetic region of North India (Rajput et al., 2018) and a few studies in South India (Balakrishnan et al., 2004; Satsangi et al., 2014). Various factors, including climate, geography, population, and fuel usage, influence HAP. Compared to northern regions, the lack of sufficient studies in southern India provides a unique opportunity to focus on villages in the south for this research. Additionally, there is a lack of research linking personal exposure to respirable PM with resulting health outcomes. Limited studies have been reported on the chemical composition and morphological analysis of personal exposure samples from biomass smoke (Carabali et al., 2016; Ahmad et al., 2023; Alves et al., 2023). Thus, in this study, we investigated the personal exposure of women to fine PM ($PM_{2.5}$) during morning and evening cooking hours in households using biomass fuel for cooking, followed by an analysis of the elemental composition, morphological characteristics, and health risks associated with the women's exposure.

MATERIALS AND METHODS

Description of the Study Region

This study was conducted in Kishannagar village (17.3850°N, 78.4867°E) (Figure 4.1), in the Mahbubnagar district of southern Telangana, India. The study area experiences a typical tropical climate with semi-arid conditions, representing south India. According to the IMD, May means the summer season, with an average atmospheric temperature of 33°C. December represents the winter season, with an average atmospheric temperature of 22.1°C.

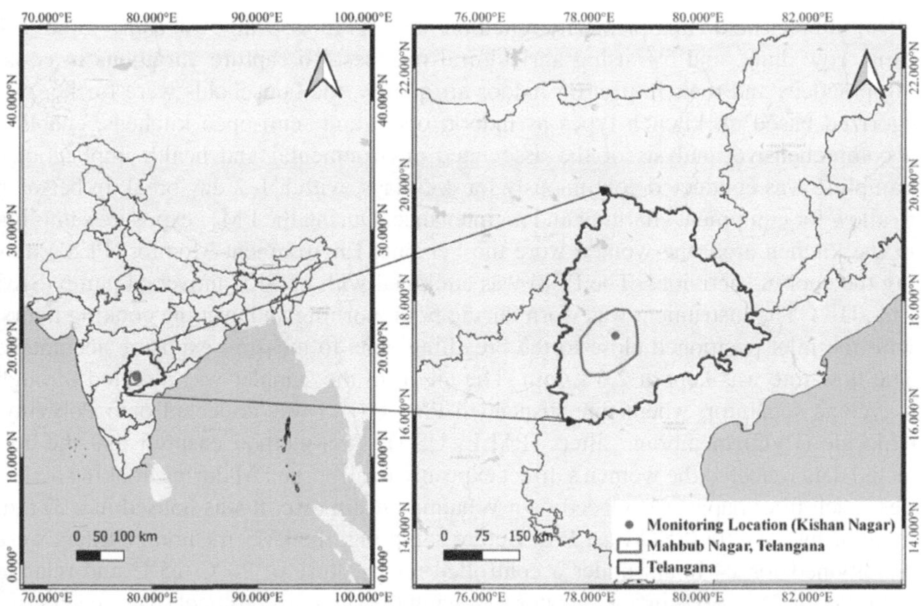

FIGURE 4.1 Study area map of Krishna Nagar village in Mahbub Nagar, Telangana.

Fuel usage, socioeconomic status, household structural characteristics, and climate were considered in the selection of the village. The village has a population of 3,639 residents residing in 737 households. It covers an area of 1,571 hectares. The total male population in the village was 51% and the female was 49%. Agriculture is the primary occupation, with 30% of the villagers engaged in cattle farming. The village cultivates a variety of crops, including cereal crops like paddy, jowar, ragi, and millets; leguminous crops such as pulses; oilseed crops like groundnut and castor; sugar crops like sugarcane; fruits such as mango, lime, guava, sapota, papaya, and sweet orange; and various vegetables. Kishannagar is known for its significant reliance on biomass for household energy, with 80.5% of households using it, followed by LPG at 17.7% (Census, 2011). To ensure the accuracy and relevance of the data, primary information on fuel consumption patterns was gathered through a questionnaire survey. The survey aimed to provide a detailed understanding of how households use different fuels. The findings from both the questionnaire and the data collected during the monitoring campaign indicated a decline in the number of households using firewood. This trend was consistent with the census data, which identified firewood as the predominant fuel source in these areas.

Monitoring Campaign

For the monitoring campaign, from a total of 737 households in the village, 40 households were selected using randomized sampling to ensure unbiased representation. Monitoring was conducted during the cold season (December 8, 2017–February 25,

2018) in households that primarily relied on biomass fuels, with 90% using wood, 5% using cow dung, and 5% using agricultural residues. To capture variations in cooking practices and their impact on indoor air quality, the households were further categorized based on kitchen types as indoor, open, and semi-open kitchens, enabling a comprehensive analysis of the associated environmental and health implications. Sampling was conducted continuously for 4–5 days, with a 1–2 day break in between to allow for equipment charging and maintenance. During the $PM_{2.5}$ exposure sampling in the kitchen area, the women wore the Personal Environment Monitor (PEM) during the cooking activities. The PEM was equipped with an SKC universal pump (SKC Ltd., UK). The instrument was worn during both morning and evening cooking hours, with the inlet positioned close to the breathing zone to measure exposure accurately. The flow rate was kept at 2.6 L/min. The air from the sampler was directed through a cyclone separator, where fine suspended PM ($PM_{2.5}$) was collected onto Polyvinyl Chloride (PVC) membrane filters (PALL, USA). This method ensured that the collected data reflected the women's direct exposure to airborne PM during cooking activities. Each filter paper was placed on a Whatman drain disc. It was housed in a 37 mm monitoring cassette (SKC Inc., PA). Before sampling, the PVC membrane filters were conditioned for 24 hours under a controlled temperature of 20°C–25°C and relative humidity of 25%–30% in a desiccator following the standard protocol. After each sampling, the pump was calibrated using a soap bubble flow corrector, adjusting the flow rates accordingly. Post-conditioning of $PM_{2.5}$ filter papers was done by placing them in the desiccator at a controlled temperature and relative humidity for 24 hours. The total mass of $PM_{2.5}$ deposited was quantified by weighing the filter paper before and after sampling. The weight was taken using a microbalance with an accuracy of ±1 μg (Sartorius ME5-F).

Chemical Characterization

The $PM_{2.5}$ samples were analysed for their elemental and morphological characteristics. The PVC filter paper was divided into two equal parts using a template and ceramic scissors. Elemental composition and morphological analysis were carried out by utilizing the filter portions.

Elemental Analysis

A total of 40 filter papers collected during the winter season were carefully extracted and analysed for their elemental composition as the season generally exhibits the highest concentrations of PM. This analysis adhered to the procedures outlined in the United States Environmental Protection Agency (USEPA). By following these standardized methods, the study aimed to ensure accuracy and consistency in measuring various elements present on the filter papers, providing reliable data for further interpretation. The filter papers were processed using a Microwave Digestion System Multiwave GO (Anton Paar). Half of the filter papers were placed in PTFE-TFM digestion vessels and subjected to acid digestion with a 7.5 mL hydrochloric acid mixture and 2.5 mL nitric acid. The digestion program was set to heat the samples to 180°C and maintain this temperature for 20 minutes. After digestion, the samples were diluted with Milli-Q water and filtered using syringe filters (Nylon, 0.45 µm pore size, Millipore, USA). The filtered samples were then analysed using Inductively Coupled Plasma Optical Emission Spectroscopy (ICP-OES, Perkin Elmer Optima 5300 DV, USA). The elemental composition was analysed for metals, including Aluminium (Al), Manganese (Mn), Barium (Ba), Potassium (K), Sodium (Na), Chromium (Cr), Strontium (Sr), Iron (Fe), Calcium (Ca), Cadmium (Cd), Cobalt (Co), Copper (Cu), Magnesium (Mg), Nickel (Ni), and Zinc (Zn).

The standards for all the elements analysed using ICP-OES were prepared from E. Merck multielement stock solutions. Standards for each element were introduced into the argon plasma in order of increasing concentration. Emission intensity against standard concentration was plotted to create the calibration graphs for each element. Subsequently, unknown samples were introduced into the argon plasma, and their emission intensity was measured. The corresponding concentration was determined using the calibration graph by interpolating the intensity on the concentration axis to obtain the precise sample concentration. Details on the wavelengths of the elements were analysed, as well as their detection limit (DL) and background equilibrium concentration (BEC). Their concentrations' relative standard deviation (RSD) was under 5% for all elements, with standards prepared from reference materials. Any instrument deviations were checked before and after each element's analysis.

Morphological Characteristics

The fine PM deposited on PVC membrane filters was examined for its morphology using a High-Resolution Scanning Electron Microscope (HRSEM) (FEI Quanta FEG 200) paired with Energy Dispersive X-ray Spectroscopy (EDX) (Genesis 2011 Emcrafts; Oxford Instruments). For the morphological analysis, the filter paper was precisely cut into 1 cm^2 sections, followed by a process of gold coating utilizing a sputter coater (Cresington sputter coater 108). The process enhances conductivity and maintains stability in a vacuum. The SEM stubs were subsequently introduced into the chamber to observe the morphological features of individual particles. The SEM analysis was conducted by scanning the sample with an electron beam. The accelerating voltage was set between 20 and 30 kV. During the scan, secondary and backscattered signals were detected. These signals were then amplified to produce images, with particles being identified when the signal surpassed the predetermined threshold. These signals provided valuable information about the sample's structure. The particle sizes were determined by outlining their dimensions with the electron beam. Further, the EDX method was used to analyse the characteristic X-rays emitted. The SEM-EDX analysis detected the presence of various elements, including sodium (Na), magnesium (Mg), aluminium (Al), silicon (Si), phosphorus (P), sulphur (S), fluorine (F), potassium (K), calcium (Ca), manganese (Mn), iron (Fe), copper (Cu), zinc (Zn), lead (Pb), titanium (Ti), chromium (Cr), vanadium (V), and nickel (Ni).

Quality Control and Quality Assurance

Instruments were calibrated according to the manufacturer's guidelines before use. Strict precautions were followed throughout the sampling and analysis processes to prevent any contamination. The sampler was regularly cleaned to ensure it remained free of dust. Great care was taken while handling the filters to avoid any contamination or damage. Only analytical reagent-grade chemicals were used. Deionized water was used to prepare reagents and solutions. The Milli-Q water purification system was used to collect deionized water. Repeated sample analysis of lab-defined quality control solutions of low concentrations was utilized to determine the minimum detection limit of elements. This process helped to ensure accurate measurement sensitivity. Field blanks were collected every ten days to evaluate any potential contamination that might occur during transportation and sampling. The field blanks were handled and analysed using the same procedure applied to the samples. This ensured consistency and accuracy in the results. In addition, laboratory blanks were also analysed to minimize potential errors arising from the laboratory procedures, further ensuring the reliability of the data.

Risk Assessment

The health hazards posed by exposure to heavy metals were assessed. The specific goal of the current investigation was to determine the level of heavy metal ingestion through breathing. The excess cancer risk (ECR) for carcinogenic elements such as Cadmium, Chromium, and Nickel elements was calculated using the measured concentrations of these metals in $PM_{2.5}$ as outlined in Equation 4.1. The elements' inhalation unit risk (IUR is available online at www.epa.gov/iris/ governed and maintained by USEPA)

$$\text{ECR (Inhalation)} = \text{heavy metal} \left(\frac{\mu g}{m^3}\right) \times \text{Inhalation unit risk} \left(\frac{\mu g}{m^3}\right)^{-1} \tag{4.1}$$

RESULTS AND DISCUSSION

Exposure to Fine Particulate

Women's exposure to $PM_{2.5}$ during morning and evening cooking activities was assessed in 40 rural households using biomass as the primary fuel (Figure 4.2a and b). The average $PM_{2.5}$ concentration (PM levels ± standard error) among the households was 2,274.53 ± 485.58 μg/m^3 ranging between 516 and 7,453 μg/m^3. The huge disparity in the exposures despite using the same fuel, i.e., biomass, was due to varied kitchen

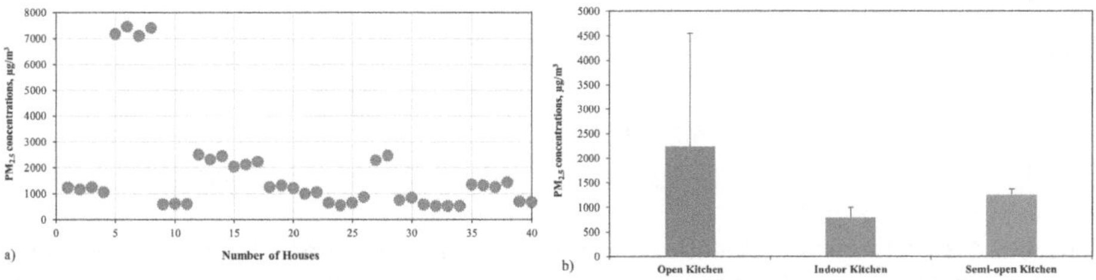

FIGURE 4.2 (a) $PM_{2.5}$ concentrations exposed to women during cooking in sampled rural households and (b) $PM_{2.5}$ concentrations variation among different kitchen types.

types such as indoor kitchens (784.17±209.34), open kitchens (2,232.0±2,312.15), or semi-open kitchens (1,249.99±115.35), their sizes and ventilation rates. Past studies reported similar attributing factors affecting PM levels, i.e., varied kitchen setups, sizes of kitchen, and air exchange (Nayek and Padhy, 2017; Sharma and Jain, 2019). Moreover, it is evident that women's exposure to $PM_{2.5}$ in all households significantly exceeded the WHO guideline of 25 μg/m^3. Exposure studies conducted in rural households of West Bengal, India, reported PM_4 concentrations ranging from 418 to 2,690 μg/m^3 and $PM_{2.5}$ concentrations ranging from 197 to 2,139 μg/m^3 (Nayek and Padhy, 2017; Nayek and Padhy, 2018). Furthermore, a study performed in a domestic fire site in southern West Africa reported that 5-hour $PM_{2.5}$ average personal exposure to open wood fires in households was 1,574 μg/m^3 (Xu et al., 2019). It is evident that the studies carried out in the past also reported similar concentrations (Balakrishnan et al., 2002; Nayek and Padhy, 2017, 2018; Sanchez et al., 2020; Ye et al., 2022). Given the evidence, personal exposure studies have increased in the past decade to capture the personal inhalation exposures of those performing cooking activities, specifically focusing on women. However, apart from women who are involved in cooking, children and the elderly who are accompanied or present in the kitchen during cooking activities are the most vulnerable due to these higher fine PM levels, which need more studies in the future.

Elemental Composition

The personal exposure samples were analysed for Barium (Ba), cobalt (Co), potassium (K), calcium (Ca), aluminium (Al), cadmium (Cd), chromium (Cr), iron (Fe), copper (Cu), magnesium (Mg), manganese (Mn), sodium (Na), nickel (Ni), strontium (Sr), and zinc (Zn) collected during cooking activities in rural households across different types of kitchen setups, namely, open kitchen, indoor kitchen, and semi-open kitchens, as presented in Figure 4.3. These elements are classified based on their primary sources as crustal or non-crustal. Crustal elements (Ca, Fe, Al, K, Ba, Mg, Sr, Mn) typically reflect geological inputs and are considered indicators of natural mineral sources. Non-crustal elements (Zn, Co, Cr, Cd, Ni, Cu) are indicative of external sources such as pollution and industrial activities, reflecting their significant anthropogenic

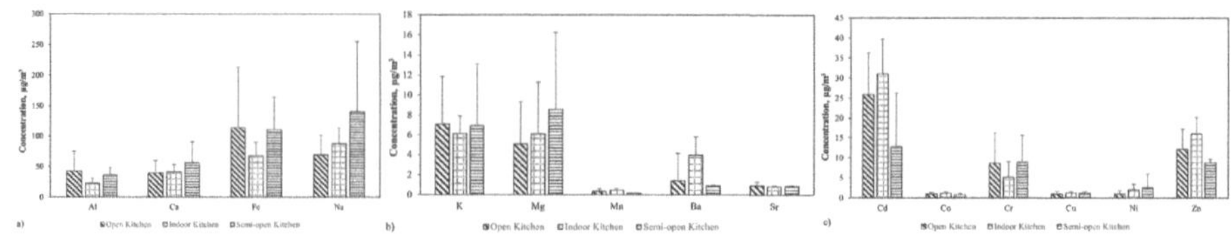

FIGURE 4.3 (a) Crustal elemental concentrations (range<20 μg/m³), (b) crustal elemental concentrations (range>20 μg/m³), and (c) non-crustal elements in different kitchen types.

TABLE 4.1 Elemental composition of $PM_{2.5}$ concentrations

	MEAN	*SD*	*MIN*	*MEDIAN*	*MAX*
Al (μg/m³)	39	27	16.4	27.8	117
Ba	1.9	2.5	0.1	0.8	7.5
Ca	46.2	24.3	20.8	38.8	99.7
Cd	25.1	10.7	3.2	23.7	43.8
Co	1.1	0.4	0.7	1.0	1.7
Cr	8.3	6.5	1.2	5.6	21.8
Cu	1.2	0.6	0.5	0.9	2.6
Fe	109.7	82.0	29.0	76.7	313.3
K	7.9	5.7	2.1	6.5	22.4
Mg	6.4	5.2	1.9	3.8	15.9
Mn	0.4	0.2	0.1	0.3	0.8
Na	93.3	60.1	42.7	65.3	229.6
Ni	1.6	1.6	0.1	1.0	5.1
Sr	0.9	0.3	0.5	0.9	1.7
Zn	12.8	4.8	6.4	11.5	20.0

contribution. However, certain elements, such as Cr, Co, Ni, and Cu, exhibit dual characteristics. These elements are naturally occurring but also have significant anthropogenic sources, especially in areas affected by industrial activities. The analysis revealed the concentration trend as Fe>Na>Ca>Al>Cd>Zn>K>Cr>Mg>Co>Ni>Cu, with corresponding mean values of 109, 93, 46, 39, 25, 12, 7, 8, 6, 1, 0.9, and 0.8 μg/m³, respectively (Table 4.1).

Past studies in rural households of North India recorded elevated levels of Cd, Cr, Cu, Fe, Mn, Ni, and Pb in PM compared to the present study. They also identified carcinogenic metals such as Cr, Cd, and Ni in PM generated from biomass combustion (Satsangi et al., 2014). However, a study conducted in Assam, a northeastern state of India, reported lower levels of Co, Mn, Cd, Zn, Cu, and Ni compared to the present study (Rabha et al., 2018). The composition difference primarily depends on PM levels, which are, in a way, dependent on the type of fuel used for cooking (Rabha et al., 2018). Various factors, such as the cooking method, fuel type, food ingredients, relative humidity, ventilation rate, cooking temperature and sampling height, influence the composition and properties of $PM_{2.5}$ generated from cooking. Among these factors, fuel type is crucial in determining PM composition. Rural households primarily rely on biomass fuels such as crop residue, firewood, and dung as their main energy sources in India. Pervez et al. (2019) conducted a study in ten Indian states to measure the emission factors (EFs) of elemental, ionic, and carbonaceous $PM_{2.5}$ components from burning of solid biomass fuels, including coal, dung cakes, and mixed fuels, in different stoves. The study found that regional variations in fuel usage patterns have a significant impact on emission rates and PM composition.

Additionally, Pearson's correlation coefficient was used to assess the relationship between heavy metals (Figure 4.4), providing an initial insight into the sources of elements in PM. Our results indicated that a positive correlation was observed for Ca-Mg, Ca-Na, Cd-Co, Cd-Mn, Co-Mn, Co-Sr, Cr-Fe, Cu-Mg, Cu-Na, Mg-Na, Mn-Sr, Mn-Zn, and Na-Ni. Ca has a good correlation with Na, K, and Mg as they are largely from crustal sources. Cd showed a significant positive correlation with Co, Mn, Ni, and Zn indicating emission during biomass fuel burning. It was observed that apart from the type of fuel, factors such as food, utensils, cooking oil, and flame characteristics also contributed to trace element emissions during various cooking activities (Padhi et al., 2022).

Further, the elements' enrichment factor (EF) was estimated to understand the relative contribution of crustal and non-crustal sources. The earth's crust is the primary

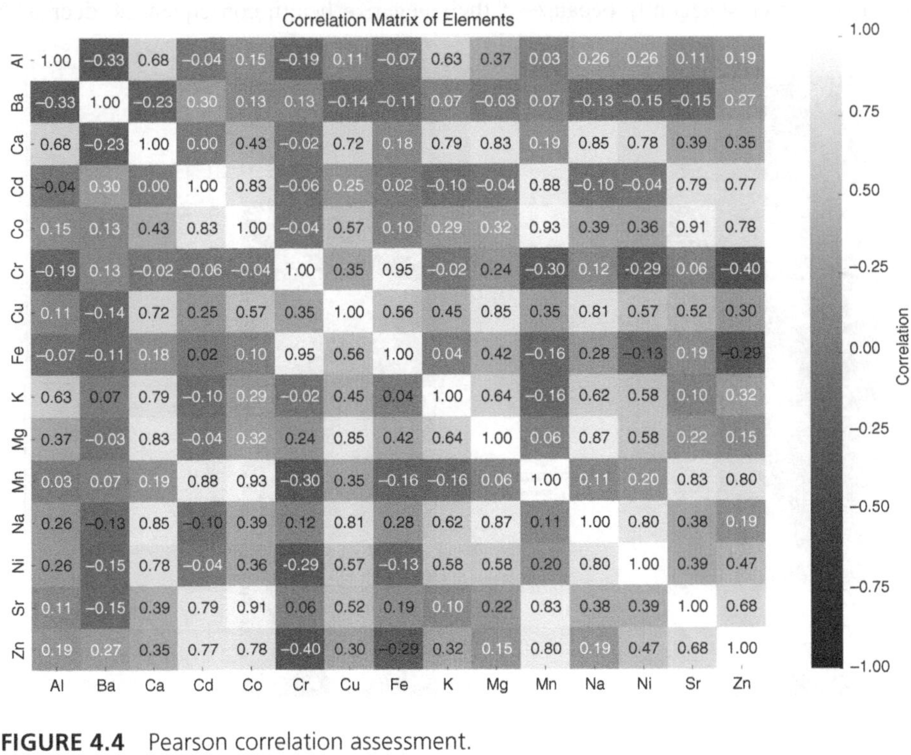

FIGURE 4.4 Pearson correlation assessment.

source when the EF value is close to unity. In contrast, a higher EF value indicates a greater contribution from non-crustal sources. The elements with the highest enrichment (EF>5) primarily from biomass combustion were Zn, Cu, Cr, Co, Ni, and Cd. In contrast, elements with lower EF values (<5), such as Ba, Sr, Mn, Mg, Fe, K, and Ca, suggest a crustal origin.

Morphological Characteristics

The morphological examination of $PM_{2.5}$ from rural biomass kitchens revealed that these particles are predominantly composed of soot or carbonaceous material, exhibiting chain-like aggregates of carbon-rich spheres generated through combustion processes (Singh et al., 2014; Carabali et al., 2016).

Analysing the size, morphology, and elemental composition of atmospheric particles provides valuable insights into the formation mechanisms, transport, and removal of chemical species in PM, as well as the identification of potential sources. Figure 4.5a–c depicts clustered and branched carbonaceous particles released from biomass combustion during cooking hours. These branched clusters form due to the interconnection of carbonaceous spherules, driven by partial coalescence and adhesive surface forces, typically happening at high temperatures during combustion (Pachauri et al., 2013). These particles, which are mainly released when biomass is burned, have garnered particular interest recently because of their negative health consequences, decreased

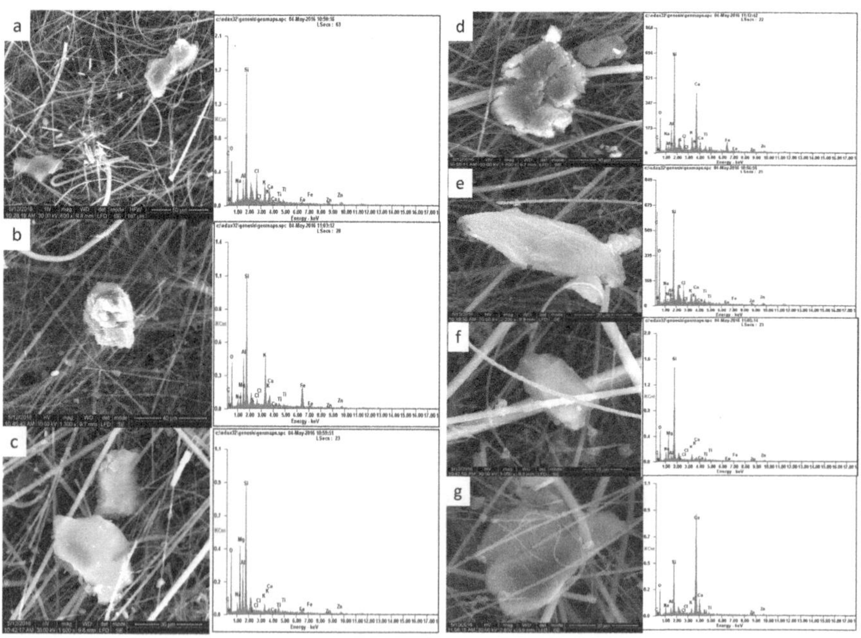

FIGURE 4.5 Structural and morphological traits of particulate matter.

visibility, and role in climate change (global warming). Figure 4.5d–g shows the presence of carbonaceous spherules containing traces of aluminium, chloride, calcium, magnesium, silicon, and potassium. Similar findings were reported by Pachauri et al. (2013) and Ahmed et al. (2023). Detecting trace potassium in these soot particles suggests firewood combustion, while carbon-rich particles with K and Cl traces further indicate wood burning emissions. Thus, the combustion of biomass fuels is a major source of chloride particles, particularly potassium chloride (KCl). Many spheroidal particles with a smooth texture are Cl-rich particles linked to trace Si and K in particles. These are typically the result of burning biomass. Additionally, Srivastava et al. (2009) detected chloride particles in aerosols across Delhi, India. Quijano et al. (2019) identified irregularly shaped particles with sharp edges, attributed to the presence of elements like Ca and K. These particles were associated with the formation of $CaSO_4$, resulting from crustal resuspension and the generation of secondary aerosols emitted during biomass or fossil fuel combustion. The analysis of certain particles (Figure 4.5e) revealed a predominant presence of C, with smaller traces of O, S, K, and Ca. This elemental composition indicates that the PM was generated from biomass combustion.

In coarse particles, irregularly shaped mineral particles were seen (Figure 4.5b); these could mostly result from soil dust resuspension. These mineral particles, most likely from geological origins, included Al, Ca, Si, Mg, O, C and K. The extremely prevalent particles in PM were found to be silica and aluminosilicates. It is distinguished by high Si and Al levels in PM together with variable Mg, K, Fe, and Ca, which can be attributed to the aeolian dispersion of soil particles.

Health Risk Assessment

Exposure to metals from human activities is linked to a higher risk of natural-cause mortality (Al-Harbi et al., 2021). Heavy metal contamination has gained significant attention due to its toxic effects, long-term persistence, and biomagnification potential (Ayaz et al., 2023). Since the harmful health effects vary depending on the toxicant's exposure routes, each heavy metal element causes a unique toxicity. The kidneys, skin, liver, heart, neurological system, reproduction, and immunological responses are among the various organs and systems in the human body that are affected by heavy metal toxicity. Furthermore, certain heavy metals cause genotoxicity and cancer. Table 4.2 highlights that the highest cancer risk from inhaling heavy metals was observed for chromium (67 cases per one million people), followed by cadmium (42 cases per one million people). These findings align with studies conducted in western and northern

TABLE 4.2 Cancer risk from heavy metal exposure

POLLUTANT	*RFC* ($\mu g/m^3)^{-1}$	*EXCESS CANCER RISK*
Cadmium	0.0018	42 (10^{-6} population)
Hexavalent chromium	0.012	67 (10^{-6} population)
Nickel	0.00024	1 (10^{-6} population)

India (Massey et al., 2013; Satsangi et al., 2014). Therefore, the elevated cancer risk associated with households relying on biomass combustion remains a critical concern, necessitating urgent mitigation measures.

The harmful consequences of inhaling heavy metals differ slightly from those of other exposure pathways (Mahmoud et al., 2023). For instance, hexavalent chromium is genotoxic and carcinogenic, especially to lung tissues. Chromium exposure may also damage the heart and lungs (Mitra et al., 2022). Since nickel exposure occurs mostly by inhalation, its carcinogenic effects extend beyond the development of lung and nasal malignancies to include several epigenetic consequences. Additionally, dermatitis, gastrointestinal symptoms, lung fibrosis, and cardiovascular disorders are among its non-carcinogenic consequences (Genchi et al., 2020). Long-term exposure to cadmium may result in cancer and cause damage to the kidneys and bones. Acute exposure, however, has a negative impact on the respiratory system. Asthma and other respiratory disorders are brought on by prolonged cobalt exposure, and cancer may result from long-term exposure to cobalt mixed with tungsten carbide (Godt et al., 2006).

Other than women, vulnerable groups, like children and the elderly, are also at greater risk. In their study, Yoo et al. (2016) identified that high levels of exposure to heavy metals increase the occurrence of Sarcopenia, a condition characterized by the gradual decline of muscle mass and strength with ageing in older adults. Additionally, research conducted by Cui et al. (2016) and Muller et al. (2007) revealed that oxidative DNA damage levels increase with age. Mizoue et al. (2006) found an association between the rise in oxidative DNA destruction and a drop in body mass index. Also, Hu et al. (2023) recognized chronic kidney disease as a major health concern; it was linked to exposure to multiple metals in the elderly.

Furthermore, in their research, Zeng et al. (2019) showed growth and development concerns among preschool children. They found that lead exposure hinders the growth and development of preschool children. The harmful impact of heavy metals on children's health includes neurocognitive disorders, developmental delays, behavioural issues, respiratory ailments, cancer, and cardiovascular diseases (Osman et al., 2019). Moreover, Zheng et al. (2023) noted that adverse health effects of heavy metal exposure are much higher in children than adults, and low immune system function is more prevalent.

CONCLUSIONS

The study examines the exposure of women in rural households of Kishannagar village, Telangana, to $PM_{2.5}$ during cooking in households using biomass fuels. The elemental composition, morphology, and associated health risks of $PM_{2.5}$, particularly from heavy metals, were also assessed. The average personal exposure to $PM_{2.5}$ was 2,274.53 $\mu g/m^3$, with significant variations based on kitchen type, size, and ventilation. Elemental analysis revealed the dominance of Fe, Na, Ca, Al, and Cd, with their respective concentrations being 109, 93, 46, 39, and 25 $\mu g/m^3$. Correlations between Ca and other elements like K, Mg, and Na were observed, suggesting crustal sources. The risk of cancer from heavy metal inhalation was highest for Chromium (Cr), with 67 cases per million, followed by Cadmium (Cd), with 42 cases per million. Long-term exposure to these metals, especially Cr and Cd, poses significant health risks, including cancer and damage to kidneys, bones, and lungs. Morphological analysis revealed the presence of carbonaceous particles, soot aggregates, and spherules containing traces of chlorine, calcium, and aluminium, indicating emissions from wood combustion.

It is evident from this study that IAP from biomass combustion in rural areas is a major concern. Tackling IAP in rural India demands a comprehensive approach to ensure the long-term adoption of clean cooking fuels and technologies. While the Pradhan Mantri Ujjwala Yojana (PMUY) has improved LPG access, barriers such as high refill costs, limited awareness, and traditional preferences hinder exclusive use. Behavioural factors, including cultural practices, perceptions of fuel efficiency, and gendered decision-making power, further perpetuate reliance on biomass. Policy measures must prioritize affordability and accessibility by subsidizing LPG refills and ensuring seamless supply chains. Enhanced community-level awareness campaigns highlighting the health benefits of cleaner fuels can address cultural resistance. Empowering women through financial inclusion and decision-making roles is crucial for sustained transitions. In parallel, investments in alternative solutions like improved cook stoves (ICS), including natural and forced draft technologies, should be scaled up. Lastly, bridging the affordability gap and ensuring robust policy mechanisms for continuous support are critical. Renewable energy alternatives like solar cookstoves must be developed as long-term solutions to complement existing efforts, ultimately guaranteeing widespread availability and affordability of clean cooking technologies.

REFERENCES

Ahmed, F., Hossain, S., Hossain, S., Fakhruddin, A. N. M., Abdullah, A. T. M., Chowdhury, M. A. Z., & Gan, S. H. (2019). Impact of household air pollution on human health: Source identification and systematic management approach. *SN Applied Sciences*, *1*, 1–19.

Ahmad, S., Bahadar, Z., Allah, D., Khan, A., Umer, S., Atta, U. S., Iftikhar, A., Abdulrahman, A., Mohamed, S., & Mohammed, A. (2023). Morphological, mineralogical, and biochemical characteristics of particulate matter in three size fractions (PM10, PM2.5, and PM1) in the urban environment. *ACS Omega*, *8*(35), 31661–31674. https://doi.org/10.1021/acsomega.3c01667.

Al-Harbi, M., Ibrahim, A., & Joann, K. W. (2021). Characteristics and health risk assessment of heavy metal contamination from dust collected on household HVAC air filters. *Chemosphere*, *277*, 130276. https://doi.org/10.1016/j.chemosphere.2021.130276.

Alves, C., Evtyugina, M., Vicente, E., Vicente, A., Rienda, I. C., de la Campa, A. S., & Duarte, I. (2023). $PM_{2.5}$ chemical composition and health risks by inhalation near a chemical complex. *Journal of Environmental Sciences*, *124*, 860–874.

Ansari, F. A., Khan, A. H., Patel, D. K., Siddiqui, H., Sharma, S., Ashquin, M., & Ahmad, I. (2010). Indoor exposure to respirable particulate matter and particulate-phase PAHs in rural homes in North India. *Environmental Monitoring and Assessment*, *170*, 491–497.

Arora, P., Sharma, D., Kumar, P., & Jain, S. (2020). Assessment of clean cooking technologies under different fuel use conditions in rural areas of Northern India. *Chemosphere, 257*, 127315. https://doi.org/10.1016/j.chemosphere.2020.127315.

Ayaz, H., Rab, N., Iqra, N., Muhammad, A. I., Ali, I., Iram, K., Mohammad, K. O., Gezahign, F. W., Zulkifl, A., & Mohammed, B. (2023). Comprehensive human health risk assessment of heavy metal contamination in urban soils: Insights from selected metropolitan zones. *Frontiers in Environmental Science, 11*, 1–17. https://doi.org/10.3389/fenvs.2023.1260317.

Balakrishnan, K., Sankar, S., Parikh, J., Padmavathi, R., Srividya, K., Venugopal, V., & Pandey, V. L. (2002). Daily average exposures to respirable particulate matter from combustion of biomass fuels in rural households of southern India. *Environmental Health Perspectives, 110*(11), 1069–1075.

Balakrishnan, K., Sambandam, S., Ramaswamy, P., Mehta, S., & Smith, K. R. (2004). Exposure assessment for respirable particulates associated with household fuel use in rural districts of Andhra Pradesh, India. *Journal of Exposure Science & Environmental Epidemiology, 14*(1), S14–S25.

Carabali, G., Castro, T., Cruz, W. De La., Peralta, O., Varela, A., Amelines, O., Rivera, M. et al. (2016). Morphological and chemical characterization of soot emitted during flaming combustion stage of native-wood species used for cooking process in Western Mexico. *Journal of Aerosol Science, 95*, 1–14. https://doi.org/10.1016/j.jaerosci.2016.01.008.

Cui, X., Ohtsu, M., Mise, N., Ikegami, A., Mizuno, A., Sakamoto, T., Ogawa, M., Machida, M., & Kayama, F. (2016). Heavy metal exposure, in combination with physical activity and aging, is related with oxidative stress in Japanese women from a rural agricultural community. *SpringerPlus, 5*(1), 885. https://doi.org/10.1186/s40064-016-2430-z.

Dehghani, S., Yousefi, S., Oskoei, V., Tazik, M., Moradi, M. S., Shaabani, M., & Vali, M. (2023). Ecological study on household air pollution exposure and prevalent chronic disease in the elderly. *Scientific Reports, 13*(1), 11763.

Genchi, G., Alessia, C., Graziantonio, L., & Maria, S. S. (2020). Nickel: Human health and environmental toxicology. *International Journal of Environmental Research and Public Health, 17*(3), 679–700.

Godt, J., Franziska, S., Christian, G-S., Vera, E., Paul, B., Andrea, R., & David, A. G. (2006). The toxicity of cadmium and resulting hazards for human health. *Journal of Occupational Medicine and Toxicology, 1*(1), 1–6. https://doi.org/10.1186/1745-6673-1-22.

Hu, W., Li, G., He, J., Zhao, H., Zhang, H., Lu, H., Liu, J., & Huang, F. (2023). Association of exposure to multiple serum metals with the risk of chronic kidney disease in the elderly: A population-based case–control study. *Environmental Science and Pollution Research, 30*(7), 17245–17256. https://doi.org/10.1007/s11356-022-23303-2.

Jiang, K., Xing, R., Luo, Z., Huang, W., Yi, F., Men, Y., & Shen, G. (2024). Pollutant emissions from biomass burning: A review on emission characteristics, environmental impacts, and research perspectives. *Particuology, 85*, 296–309.

Juginović, A., Vuković, M., Aranza, I., & Biloš, V. (2021). Health impacts of air pollution exposure from 1990 to 2019 in 43 European countries. *Scientific Reports, 11*(1), 22516.

Kaur, R., & Pandey, P. (2021). Air pollution, climate change, and human health in Indian Cities: A brief review. *Frontiers in Sustainable Cities, 3*, 705131. https://doi.org/10.3389/frsc.2021.705131.

Lee, K. K., Bing, R., Kiang, J., Bashir, S., Spath, N., Stelzle, D., & Shah, A. S. (2020). Adverse health effects associated with household air pollution: A systematic review, meta-analysis, and burden estimation study. *The Lancet Global Health, 8*(11), e1427–e1434.

Li, C., Li, Z., & Wang, H. (2023). Characterization and risk assessment of polycyclic aromatic hydrocarbons (PAHs) pollution in particulate matter in rural residential environments in China-A review. *Sustainable Cities and Society, 96*, 104690.

Liu, C., Cai, J., Chen, R., Sera, F., Guo, Y., Tong, S., & Kan, H. (2022). Coarse particulate air pollution and daily mortality: A global study in 205 cities. *American Journal of Respiratory and Critical Care Medicine*, *206*(8), 999–1007.

Mahmoud, N., Dana, Al -S., Hanadi, Al -T., & Rima, J. I. (2023). Risk assessment of the impact of heavy metals in urban traffic dust on human health. *Atmosphere*, *14*(6), 49. https://doi.org/10.3390/atmos14061049.

Massey, D. D., Kulshrestha, A., & Taneja, A. (2013). Particulate matter concentrations and their related metal toxicity in rural residential environment of semi-arid region of India. *Atmospheric Environment*, *67*, 278–286.

Mitra, S., Arka, J. C., Abu, M. T., Talha, B. E., Firzan, N., Ameer, K., Abubakr, M. I., et al. (2022). Impact of heavy metals on the environment and human health: Novel therapeutic insights to counter the toxicity. *Journal of King Saud University – Science*, *34*(3), 101865. https://doi.org/10.1016/j.jksus.2022.101865.

Mizoue, T., Kasai, H., Kubo, T., & Tokunaga, S. (2006). Leanness, smoking, and enhanced oxidative DNA damage. *Cancer Epidemiology Biomarkers and Prevention*, *15*(3), 582–585. https://doi.org/10.1158/1055-9965.EPI-05-0658.

Muller, F. L., Lustgarten, M. S., Jang, Y., Richardson, A., & Remmen, H. V. (2007). Trends in oxidative aging theories. *Free Radical Biology and Medicine*, *43*(4), 477–503. https://doi.org/10.1016/j.freeradbiomed.2007.03.034.

Nayek, S., & Padhy, P. K. (2018). Approximation of personal exposure to fine particulate matters ($PM_{2.5}$) during cooking using solid biomass fuels in the kitchens of rural West Bengal, India. *Environmental Science and Pollution Research*, *25*, 15925–15933.

Osman, M. Al., Yang, F., & Massey, I. Y. (2019). Exposure routes and health effects of heavy metals on children. *BioMetals*, *32*(4), 563–573. https://doi.org/10.1007/s10534-019-00193-5.

Pachauri, T., Aparna, S., Vyoma, S., Anita, L., & Kumari, K. M. (2013). Characteristics and sources of carbonaceous aerosols in $PM_{2.5}$ during winter time in Agra, India. *Aerosol and Air Quality Research*, *13*(3), 977–991. https://doi.org/10.4209/aaqr.2012.10.0263.

Padhi, A., Bansal, M., Habib, G., Samiksha, S., & Raman, R.S. (2022). Physical, chemical and optical properties of $PM_{2.5}$ and gaseous emissions from cooking with biomass fuel in the Indo-Gangetic Plain. *Science of the Total Environment*, *841*, 156730.

Pervez, S., Verma, M., Tiwari, S., Chakrabarty, R.K., Watson, J.G., Chow, J.C., Panicker, A.S., Deb, M.K., Siddiqui, M.N., & Pervez, Y.F. (2019). Household solid fuel burning emission characterization and activity levels in India. *Science of the Total Environment*, *654*, 493–504.

Quijano, M. F. C., Vinícius, L. M., Tatiana, D. S., Bott, I. S., & Adriana, G. (2019). Exploratory and comparative analysis of the morphology and chemical composition of $PM_{2.5}$ from regions with different socioeconomic characteristics. *Microchemical Journal*, *147*, 507–515. https://doi.org/10.1016/j.microc.2019.03.071.

Rabha, R., Ghosh, S., & Padhy, P. K. (2018). Indoor air pollution in rural north-east India: Elemental compositions, changes in haematological indices, oxidative stress and health risks. *Ecotoxicology and Environmental Safety*, *165*, 393–403.

Rajput, P., Singh, D. K., Singh, A. K., & Gupta, T. (2018). Chemical composition and source-apportionment of sub-micron particles during wintertime over Northern India: New insights on influence of fog-processing. *Environmental Pollution*, *233*, 81–91.

Samburova, V., Connolly, J., Gyawali, M., Yatavelli, R. L., Watts, A. C., Chakrabarty, R. K., & Khlystov, A. (2016). Polycyclic aromatic hydrocarbons in biomass-burning emissions and their contribution to light absorption and aerosol toxicity. *Science of the Total Environment*, *568*, 391–401.

Sanchez, M., Milà, C., Sreekanth, V., Balakrishnan, K., Sambandam, S., Nieuwenhuijsen, M., & Tonne, C. (2020). Personal exposure to particulate matter in peri-urban India: Predictors and association with ambient concentration at residence. *Journal of Exposure Science & Environmental Epidemiology*, *30*(4), 596–605.

Satsangi, P. G., Yadav, S., Pipal, A. S., & Kumbhar, N. (2014). Characteristics of trace metals in fine ($PM_{2.5}$) and inhalable (PM_{10}) particles and its health risk assessment along with in-silico approach in indoor environment of India. *Atmospheric Environment*, *92*, 384–393.

Sharma, D., & Jain, S. (2019). Impact of intervention of biomass cookstove technologies and kitchen characteristics on indoor air quality and human exposure in rural settings of India. *Environment International*, *123*, 240–255.

Singh, P., Renuka, S., & Ajay, T. (2014). Physicochemical characteristics of $PM_{2.5}$: Low, middle, and high-income group homes in Agra, India-A case study. *Atmospheric Pollution Research*, *5*(3), 352–60. https://doi.org/10.5094/APR.2014.041.

Srivastava, A., Sandeep, G., & Jain, V. K. (2009). Winter-time size distribution and source apportionment of total suspended particulate matter and associated metals in Delhi. *Atmospheric Research*, *92*(1), 88–99. https://doi.org/10.1016/j.atmosres.2008.09.005.

Thangavel, P., Park, D., & Lee, Y. C. (2022). Recent insights into particulate matter ($PM_{2.5}$)-mediated toxicity in humans: An overview. *International Journal of Environmental Research and Public Health*, *19*(12), 7511.

Tripathi, S., Yadav, S., & Sharma, K. (2024). Air pollution from biomass burning in India. *Environmental Research Letters*, *19*(7), 073007.

WHO (2024). Household air pollution. https://www.who.int/news-room/fact-sheets/detail/household-air-pollution-and-health#:~:text=Household%20air%20pollution%20accounted%20for,cataract%2C%20nasopharyngeal%20and%20laryngeal%20cancers.

Xu, H., Léon, J. F., Liousse, C., Guinot, B., Yoboué, V., Akpo, A. B., & Cao, J. (2019). Personal exposure to $PM_{2.5}$ emitted from typical anthropogenic sources in southern West Africa: Chemical characteristics and associated health risks. *Atmospheric Chemistry and Physics*, *19*(10), 6637–6657.

Yadav, R., Deora, S., & Yadav, G. (2021). Air pollution and its impact on cardiovascular health – it's time to act fast! *Indian Heart Journal*, *73*(1), 1–6. https://doi.org/10.1016/j.ihj.2021.01.021.

Yadav, S., Kapoor, T. S., Vernekar, P., & Phuleria, H. C. (2023). Examining the chemical and optical properties of biomass-burning aerosols and their impact on oxidative potential. *Aerosol and Air Quality Research*, *23*, 230102.

Yao, W., Zhao, Y., Chen, R., Wang, M., Song, W., & Yu, D. (2023). Emissions of toxic substances from biomass burning: A review of methods and technical influencing factors. *Processes*, *11*(3), 853.

Ye, W., Thangavel, G., Pillarisetti, A., Steenland, K., Peel, J. L., Balakrishnan, K., & Hapin Investigators. (2022). Association between personal exposure to household air pollution and gestational blood pressure among women using solid cooking fuels in rural Tamil Nadu, India. *Environmental Research*, *208*, 112756.

Yoo, J.-Il., Ha, Y-C., Lee, Y-K., & Koo, K-H. (2016). High levels of heavy metals increase the prevalence of sarcopenia in the elderly population. *Journal of Bone Metabolism*, *23*(2), 101. https://doi.org/10.11005/jbm.2016.23.2.101.

Yunusa, S. U., Mensah, E., Preko, K., Narra, S., Saleh, A., Sanfo, S., Isiaka, M., Dalha, I. B., & Abdulsalam, M. (2023). Biomass cookstoves: A review of technical aspects and recent advances. *Energy Nexus*, *11*(March), 100225. https://doi.org/10.1016/j.nexus.2023.100225.

Zeng, X., Xu, X., Qin, Q., Ye, K., Wu, W., & Huo, X. (2019). Heavy metal exposure has adverse effects on the growth and development of preschool children. *Environmental Geochemistry and Health*, *41*(1), 309–21. https://doi.org/10.1007/s10653-018-0114-z.

Zheng, K., Zeng, Z., Tian, Q., Huang, J., Zhong, Q., & Huo, X. (2023). Epidemiological evidence for the effect of environmental heavy metal exposure on the immune system in children. *Science of the Total Environment*, *868*, 161691. https://doi.org/https://doi.org/10.1016/j.scitotenv.2023.161691.

Brick Kilns

An Extreme Case of Personal Exposure to Particulate Matter in the South Asian Region

5

Parth Sarathi Mahapatra, Alpha Thapa, Bidya Banmali Pradhan, and Siva Praveen Puppala

ABBREVIATIONS

PM	Particulate Matter
$PM_{2.5}$	Particulate Matter with aerodynamic diameter < 2.5 μm
PM_{10}	Particulate Matter with aerodynamic diameter <10 μm
BC	Black Carbon
FCBTKs	Fixed Chimney Bull's Trench Kilns
IDZKs	Induced Draught Zig-Zag Kilns
NDZKs	Natural Draught Zig-Zag Kilns
mL/min	milliliter per minute
μm	micrometer
$μg/m^3$	microgram per meter cube
mg/m^3	milligram per meter cube

DOI: 10.1201/9781003613183-5

L/m	liter per minute
g/kg	gram per kilogram
NHRC	Nepal Health Research Council
TSPM	Total Suspended Particulate Matter

INTRODUCTION

Globally, air pollution is acclaimed as an important public health concern, ranking fourth on the list of global risk factors for early deaths due to all cause in both sexes in 2019 (Health Effects Institute, 2020). Various adverse health outcomes of air pollution exposure, i.e., declining cognitive function, diabetes, preterm births, respiratory and cardiovascular diseases and many more (Baumgartner et al., 2018; Kingsley et al., 2017; Pokhrel et al., 2005; Rajagopalan & Brook, 2012; Smith et al., 2000; Zhang et al., 2018), are well established. In occupational context, it is generally assumed that people are exposed to high concentration of different air pollutants (Toren et al., 2007) depending upon the nature of the occupation. Existing literature around this topic indicates a range of air pollution exposures (from few micrograms to milligrams in different sizes) in a variety of occupations. Pertaining to the limited number of studies, more sophisticated measurements are required at a finer scale at regional and global scales for a better understanding of exposure to air pollution based on different occupations. Some of these high-risk occupations include e-waste dismantling, cement factories, textile and mining industries, food delivery persons, roadside vendors, security personnel and so on (Bungadaeng et al., 2019; Pattinson et al., 2018; Shakya et al., 2019; Yang et al., 2020). Although the list is exhaustive (vary from region to region) and not just limited to the above-mentioned occupations, making it even more important to conduct such studies at a larger scale.

In this context considering the rapid growth of urban and industrial patches in the low- and middle-income countries, there is a high demand of building and construction industry (Maithel et al., 2014; Nepal et al., 2019; Weyant et al., 2014). Brick along with associated elements continues to be one of the most popular construction materials for structural integrity. To be more precise, clay bricks, due to cheaper cost of production and ease of handling, retain their spot to be the most popular among others (fly ash brick, concrete brick, cement stabilized soil blocks and autoclaved aerated concrete) (Valdes et al., 2020). In the recent past, the clay brick industry has seen a significant increase in their production capacities. Within South Asia, there are approximately 140,000–210,000 kilns operational in India, 18,000–20,000 kilns in Bangladesh, 7,000–7,200 kilns in Pakistan and 1,349 kilns in Nepal with approximately 70 workers employed in each kiln. The numbers would vary largely depending upon the size of the brick kiln and its production capacity, respectively. However, data on brick kilns in Bhutan or Sri Lanka are extremely patchy (Ball, 2024). In general, the few studies done on brick kilns in Sri Lank do indicate generalized aspects of environmental degradation (Herath et al., 2021; Jayasinghe, 2007), thus highlighting the dearth of data on brick

FIGURE 5.1 Brick kilns, an extreme case of air pollution exposure scenario.

kilns for these countries. Although it is an important industry in the South Asian context, employing many workers, little importance is bestowed upon their occupational health assessment. Those that did have largely been built on questionnaire data. Brick kiln operation being a seasonal work (in monsoon affected areas), workers and their families migrate from the nearby regions to earn their livelihoods. Hence, not only are the workers directly exposed to such high polluting environments, but their families also get exposed to these conditions by virtue of the nature of this job and associated social structure around it (Figure 5.1). Being a high polluting point source, some studies in the past have characterized emissions (define emission factor for regional modelling studies) from various designs of brick kilns operating in this region (Guttikunda et al., 2013; Nepal et al., 2019; Stockwell et al., 2016; Tibrewal et al., 2019; Weyant et al., 2014, 2016). Different studies indicate a varying contribution of emissions from the brick sector to different air pollutants in the South Asian region. One such study in Kathmandu, Nepal, indicates 15% contribution of brick sector to the PM_{10}. Another study in Dhaka, Bangladesh, suggests 22% contribution of this sector to $PM_{2.5}$ mass concentration. Few other studies have also revolved around the political economy, migration and child labor aspects associated with brick kilns (Ali et al., 2017; Daly et al., 2020; Sharma et al., n.d.). The study by Sharma et al. (n.d.) also suggests that exact data on units of brick manufactured is generally under reported, making it a non-economy reported sector (profitable in nature). Some studies also identify health vulnerabilities in brick kiln workers using questionnaires (Rupakheti et al., 2018; Thomas et al., 2014). Although some research has measured exposure to various air pollutants (Kamal et al., 2014; Khan et al., 2019), the findings are not easily generalizable at a more detailed spatial scale. Most of these studies were either performed qualitatively using a questionnaire or, if quantitatively measured, were either measured at a point location or had a small sample size or both. Overall, the brick sector is not well documented in terms of scientific literature, rendering it difficult to study and generalize. Considering the research gap, this chapter would discuss the findings from a field campaign that was designed to estimate the workers personal exposure to $PM_{2.5}$, PM_{10} (particulate matter with aerodynamic diameter <10 μm) and BC (black carbon) in representative brick kilns (n=21) operational in Nepal, followed by a comparison of occupations with high particulate matter exposure in their work environments and then a discussion on the social and health issues. The findings would not only provide a country statistic, but it can also be generalized for the entire South Asia.

GEOGRAPHIC FOCUS

This study was performed in Nepal, which roughly exists between 26°N and 31°N latitude and 80°E and 89°E longitude. Most brick kilns here are clustered around the low-lying plain regions (terai'), with a big cluster in and around the country capital, Kathmandu. This study was conducted in 21 brick kilns geographically distributed across the eastern, central and far-western parts (Figure 5.2). Three major kiln types, i.e., Fixed Chimney Bull's Trench Kilns (FCBTKs) that are straight line in nature, Induced Draught Zig-Zag Kilns (IDZKs) and Natural Draught Zig-Zag Kilns (NDZKs), were assessed in the present study. Personal exposure assessment was performed on 64 workers distributed across various working zones of the kilns. Considering the seasonal nature of this job, we conducted the study from February to July 2017.

MATERIAL AND METHODS

In this study, we discuss the workers personal exposure to BC, $PM_{2.5}$ and PM_{10}. These measurements are taken using portable equipment that are placed in a customized jacket worn by the workers with the inlet placed at the breathing zone of selected workers during real-time working conditions. The exposure monitoring for BC, $PM_{2.5}$ and PM_{10} was performed in four different work zones (loading, unloading, firing and crushing zones; Figure 5.3) of each brick kilns depending upon the availability and consent of the workers. As the name indicates, the unloading zone is the place from where baked bricks are transported elsewhere, whereas the loading zone is the place where raw bricks are stacked for the purpose of baking. The bricks are baked using fire in the firing zone. While the crushing zone is the place where coal (any other fuel mixture) is normally crushed before its use as fuel. The working hours in each zone were not fixed, and based on the workers' consent, the exposure monitoring was carried out. Hence, on average, at least 4 hours of personal exposure monitoring was performed for each worker.

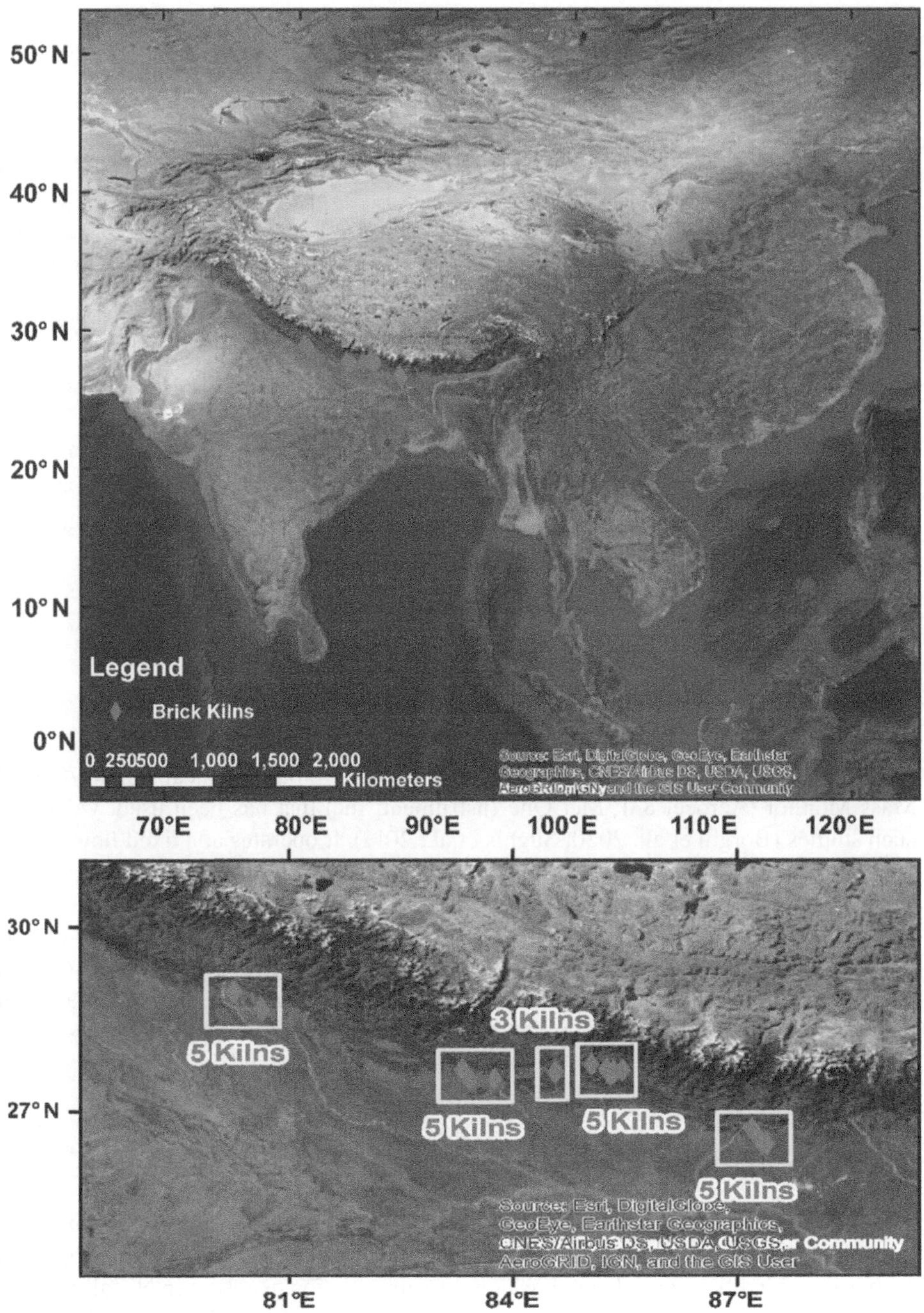

FIGURE 5.2 (Top panel) A regional map with the study area. (Bottom panel) A zoomed-in map of the study area along with location of brick kilns spread across different parts of Nepal.

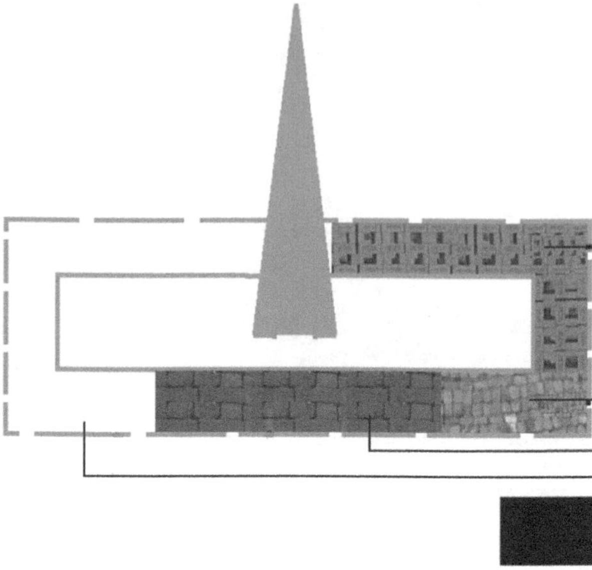

FIGURE 5.3 Different zones in a brick kiln where measurements were conducted.

Personal Exposure Assessment

Real-time $PM_{2.5}$ and PM_{10} personal exposures were monitored by means of an Aerosol Mass Monitor (Aerocet 831, Met One Instrument, Inc) that has been used widely for such studies (Borghi et al., 2020; Sughis et al., 2012). It operates at a fixed flow rate of 2.83 L/m and has a measurement time of 1 minute. Personal exposure to BC was measured using the microAeth, AE51 (AethLabs, CA). The device draws in air at a steady flow rate of 50 mL/min and is fitted with a $PM_{2.5}$ cyclone that blocks the entry of particles exceeding 2.5 μm into the instrument (Adhikari et al., 2020). Care was also taken to replace the filter strips used for operating the microAeth at regular intervals. All the equipment were factory calibrated and used in the field for conducting this study. Hence, no corrections were applied for these measurements.

Ethical Approval and Consent

The approval for conducting the field work and performing personal exposure monitoring was obtained from the Nepal Health Research Council (NHRC) under the approved proposal entitled 'Epidemiological response to air pollution exposure in Nepal' bearing registration no. 395/2016. Informed consents were obtained on field from each participant after completely explaining the details of the study, and participation was completely voluntary.

RESULTS AND DISCUSSIONS

Personal Exposure of $PM_{2.5}$, PM_{10} and BC

The brick kiln workers' average personal exposure to $PM_{2.5}$, PM_{10} and BC was observed to be approximately 80±48 μg/m³, 4,094±5,860 μg/m³ and 23±15 μg/m³, respectively. The measurement denoted a $PM_{2.5}/PM_{10}$ ratio of 0.02, indicating a strong dominance of coarse mode particle exposure. This infers that working in the brick kilns makes the worker vulnerable to dust exposure. Personal exposure in other studies has used stationery measurements with instruments placed in a fixed location within the breathing zone, whereas measurements in this study are mobile readings (moved along with the person's movement) of personal exposure. Hence, the scenario, although similar, is not directly comparable. Sanjel et al. (2016) in one of their studies conducted within the Kathmandu Valley indicated the TSPM, PM_{10}, $PM_{2.5}$ and PM_1 (in the breathing zone) to be 5.179 mg/m³ (5,179 μg/m³), 4.958 mg/m³ (4,958 μg/m³), 3.965 mg/m³ (3,965 μg/m³) and 3.954 mg/m³ (3,954 μg/m³). In another study, the same group indicated that the respirable silica (quartz) concentrations in the breathing zone were 71–331 μg/m³ (Sanjel et al., 2018). The difference in values compared to this study might be attributed to the location, measurement technique, design and number of measurements. The current research was undertaken in 21 different kilns and in four different locations of each kiln, whereas the study by Sanjel et al. (2016, 2018) was limited to the measurements conducted in kilns within the Kathmandu valley.

Now each kiln has four different zones of work, as shown in Figure 5.3. We thus compared the workers' personal exposure based on their work in different zones (Table 5.1). The average elevated personal exposure concentrations of $PM_{2.5}$ and PM_{10} were noted in the unloading zone while lowest was observed in the crushing zone. However, irrespective of the workers working in different zones, all had a similar kind of personal exposure to BC.

We also attempted to compare personal exposure to PM across various occupations with high air pollution exposure (Table 5.2). The comparison indicates that exposure to $PM_{2.5}$ and BC was found to be lower in brick kilns compared to many other occupations in other developing countries (Bungadaeng et al., 2019; Kc et al., 2020; Pokhrel et al.,

TABLE 5.1 Average personal exposure of pollutants in four zones of kilns. Comparison of occupation exposure in different work environments.

ZONES	*$PM_{2.5}$ (μG/M³)*	*PM_{10} (μG/M³)*	*BC (μG/M³)*	*$PM_{2.5}$:PM_{10}*
Unloading	131±42	9,734±7,508	23±13	0.01
Loading	67±31	3,292±2,939	23±16	0.02
Firing	56±32	845±516	23±17	0.07
Crushing	45±16	693±330	22±17	0.06

All numerical values were approximated to the nearest integer.

TABLE 5.2 Comparison of occupation with high particulate matter exposure

SL NO.	*OCCUPATION*	*$PM_{2.5}$ (μG/M³)*	*PM_{10}/ RESPIRABLE DUST (μG/M³)*	*BC/EC (μG/M³)*	*DUST EXPOSURE (μG/M³)*	*EXPOSURE TYPE*	*REFERENCES*
1.	Brick kiln workers	80±48	4,094±5,860	23±15		Personal exposure (Outdoor work)	Present Study
2.	Household cook exposure	Biomass (656), electric stoves (80), LPG (101) and kerosene stoves (169)				Personal exposure (Indoor work)	Pokhrel et al. (2015)
3.	Women exposure (Nepal)	211.93±133.44				Personal exposure (Indoor cooking, Midhill region of Nepal)	Kc et al., (2020)
4.	Traffic police exposure (Kathmandu)	123.51±38.66 (Spring) 45.21±24.09 (Summer)		18.80±7.68 (spring) 16.46±7.52 (summer)		Personal exposure (Outdoor work)	Shakya et al. (2019)
5.	Textile industry workers (Ethiopia)				1,300	Personal exposure (Indoor)	Tefera et al., (2020)
6.	Occupational exposure (Brazil)	~5 to 25		~1 to 5		Personal exposure (Seven different types of occupations)	Pattinson et al. (2018)
7.	E-waste dismantling workers (Thailand)	2,774±4,713	3,215±4,858			$PM_{2.5}$ from e-waste dismantling workers to be	Bungadaeng et al. (2019)
8.	Brick industry, Egypt		844.0±536.0			Breathing Zone (Outdoor work)	Zawilla et al. (2014)
9.	Fixed traditional kiln Marquez brick kiln Durango, Mexico	75.59 37.93					Ortínez-Alvarez et al. (2017)

2015; K. M. Shakya et al., 2016), but higher in better managed places (Pattinson et al., 2018). Personal exposure to PM_{10} in brick kilns is higher than many other occupations like that in e-waste handling or cement industries, while textile industries had a higher dust exposure. We also agree that for better comparisons, focus must be applied to personal exposure studies for various occupations in the developing world as air pollution is a larger problem here. Given the limitation of even finding such studies based on working environment, this comparison is highly useful.

Comparison of Personal Exposure in Different Kiln Types

Our earlier studies on fuel-based emission factors (g/kg of fuel) showed that FCBTKs produced higher levels of $PM_{2.5}$ and BC compared to IDZKs (Nepal et al., 2019). While another study comparing clamp kilns and IDZKs did not show any difference in particular (Stockwell et al., 2016). Hence, we took the opportunity to quantify the changes in personal exposure to particulate pollutants due to changes in kiln types. The average personal exposure of workers to BC, $PM_{2.5}$ and PM_{10} based on kiln types has been shown in Figure 5.4. Personal exposure to BC was observed to be lower in NDZKs (16±5 μg/m³), followed by IDZKs, and highest in FCBTKs (28±20 μg/m³). While

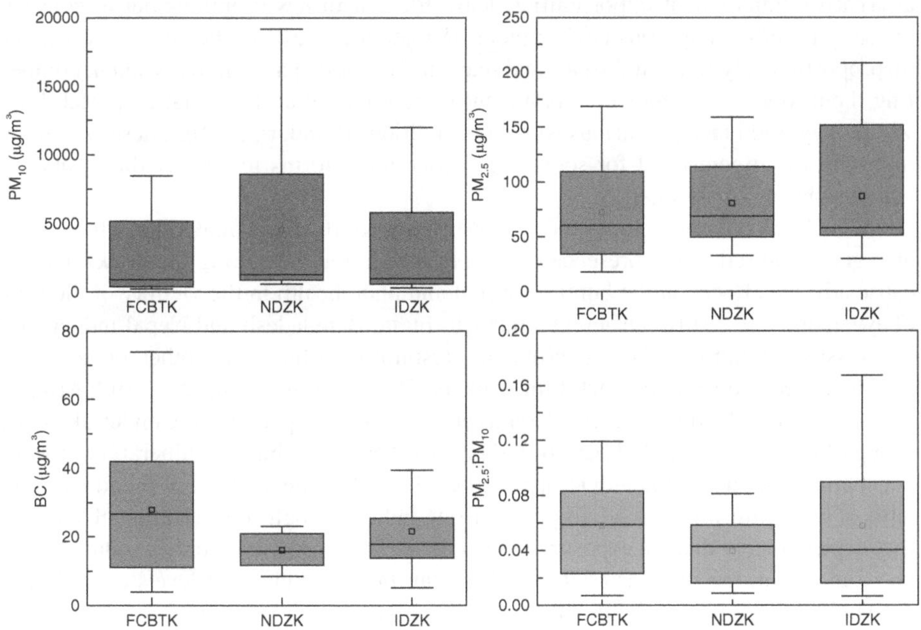

FIGURE 5.4 Average personal exposure of pollutants in three major categories of kilns. NDZK refers to Natural Draught Zig-Zag Kilns, IDZK refers to Induced Draught Zig-Zag Kilns and FCBTK refers to Fixed Chimney Bull's Trench kilns.

personal exposure to $PM_{2.5}$ and PM_{10} was observed to be lower in FCBTKs compared to NDZKs and IDZKs. The BC levels were comparable to traffic police exposures observed in the Kathmandu valley (Shakya et al., 2019), while higher to that observed in different occupations in Brazil (Pattinson et al., 2018). Such variations might also depend upon the overall ambient air quality of the region and the challenges observed in individual occupations. At this stage, the exact reasons for changes in exposure pattern are not predictable, warranting the need for a greater number of such studies to be conducted in the region.

SOCIAL ISSUES, POSSIBLE HEALTH IMPLICATIONS AND REMEDIAL MEASURES

Brick kiln workers often face severe social issues that compel them to endure hazardous working conditions with high levels of air pollution exposure. Many of these workers are often trapped in cycles of debt (discussed in a study encompassing workers from brick kilns, agricultural and tanning sectors in Pakistan, Wahid et al., 2024), where they are forced to work to repay loans with exorbitant interest rates, often inherited across generations. Added to that, often lower educational qualifications, coupled with lack of alternative employment opportunities, leave them with fewer options but to continue working in these dangerous environments (Sanjel et al., 2016). The agrarian crisis has disproportionately impacted small and marginal farmers, particularly women, compelling them to seek alternative employment in non-agricultural sectors, i.e., brick kilns where they often face precarious working conditions (Sundari, 2020). These conditions highlight the urgent need for social and economic reforms to protect the rights and health of brick kiln workers.

Clay brick baking includes the use of different kinds of raw materials, which might also get denatured in the presence of high heat required for baking the bricks and can cause adverse effects on the human, animal and plant health in the vicinity of the kilns (Thygerson et al., 2016). Different studies in India, Bangladesh and Nepal indicate the progression of musculoskeletal problems, respiratory illness and other occupational health-related problems in brick kiln workers (Biswas et al., 2018; Das, 2014; Mikrani et al., 2017; Sain & Meena, 2019; Thomas et al., 2014). A separate study undertaken by a group of doctors in Nepal observed that kiln workers had a higher vulnerability toward occupational safety and health (Rupakheti et al., 2018). The analysis of blood, hair and nails of brick kiln workers in Lahore, Pakistan, indicated high concentration of lead (Pb) that depicts recent dietary exposure (Saeed et al., 2024). Another study conducted only on women working in kilns of Rawat, Pakistan indicated elevated levels of Cadmium (Cd) in blood again, highlighting the public health issues surrounding this activity. Only a handful number of studies involves actual physical measurements of some indicative parameters associated with environmental health. A study of soil quality in agricultural fields near the brick kilns in Bhaktapur, Nepal, indicates that the quality of soil is directly proportional to the distance from kilns (Bisht & Neupane, 2015). Another study in Sheikhupura and Pattoki cities of Pakistan observed the soil near brick kilns to be acidic along with the accumulation of heavy metals (Hamid et al., 2023). A group of researchers in Pakistan observed that pollutants emitted from the brick kilns can cause significant DNA (deoxyribonucleic acid) damage among the workers (Khisroon et al., 2018). Few other studies also quantified exposure to PM in brick kilns but with a smaller sample size (Sanjel et al., 2016, 2018; Thygerson et al., 2016, 2019). Visual observations while conducting the present study also report the risk of certain occupational health hazards that might occur due to the complete absence of or extremely low use of personal protective equipment (protective clothing, good quality masks, goggles, helmets, shoes, etc.). Although the above-mentioned studies indicate certain occupational risks, health issues or vulnerabilities of brick kiln workers, a comprehensive evaluation of occupational health effects of brick kiln workers is still inadequate. Considering the high levels of exposure to particulate matter (observed in the present study) and different fumes, the use of appropriate personal protective equipment (PPE) could help alleviate the health hazards to a certain extent. A recent study in Nepal reports that only 28.9% workers used a respiratory PPE while at work (Gaviola et al., 2024), indicating the low usage rate and awareness levels. Moreover, a cost-benefit analysis (factoring social and private aspects) done by Croitoru and Sarraf (2012) suggests that implementation of cleaner technologies would be beneficial. Similarly, another study indicates the use of PV system installation at Hoffman kiln that can reduce 66.3 tons of CO_2 annually. While the payback time was three years and the benefits to cost ratio was 12.2. Another study by Sharma et al. (n.d.) conducted in Nepal indicates a good contribution of the brick sector to the national GDP, thus indicating the profitable nature of this sector. Hence, expectation is that creating healthy living conditions for the workers by investing some more money than normal might also provide a gain for the kiln owners. Creating proper awareness about existing problems and possible mitigation pathways among the workers and owners could lead a way to healthy and sustainable brick kiln operation. Moreover, best practices from nearby regions should be adopted for better management of brick kiln workers' health.

CONCLUSIONS

As far as the authors are aware, this is the inaugural study that estimates brick kiln workers' personal exposure to BC, $PM_{2.5}$ and PM_{10} during real-time working conditions. The brick kiln workers' average personal exposure to $PM_{2.5}$, PM_{10} and BC was observed to be approximately 80±48 μg/m³, 4,094±5,860 μg/m³ and 23±15 μg/m³, respectively. Our study indicates that the brick kiln workers are at a high risk of exposure to particulate matter in different forms but majorly to coarse particles. The evidence generated from this study could be used to highlight the importance of and promote the use of PPE for people working in the brick kilns. Further investigations on air pollution exposure and comprehensive health risk assessment integrated with socioeconomic facet in the regional context are necessary to have a better understanding of this situation.

ACKNOWLEDGMENTS

This study was partially supported by core funds of ICIMOD contributed by the governments of Afghanistan, Australia, Austria, Bangladesh, Bhutan, China, India, Myanmar, Nepal, Norway, Pakistan, Sweden and Switzerland. The views and interpretations in this publication are those of the authors and are not necessarily attributable to ICIMOD. We also acknowledge the team on field that included Sagar Adhikari, Sujan Shrestha, Mahendra Chitrakar, Shyam Maharjan, Roshan Kafle, Amit Chaudhary, Ashish Rai, Prakash Sharma, Sangeet Nepal and Dilasha Joshi. Special thanks to CBI, FNBI, GoN and DFID.

DECLARATIONS

The views and interpretations in this publication are those of the authors and are not necessarily attributable to their organizations.

REFERENCES

Adhikari, S., Mahapatra, P. S., Pokheral, C. P., & Puppala, S. P. (2020). Cookstove smoke impact on ambient air quality and probable consequences for human health in rural locations of Southern Nepal. *International Journal of Environmental Research and Public Health, 17*(2), 550. https://doi.org/10.3390/ijerph17020550

Ali, M. A., Ali, M. V., Abbas, F., & Chamberlain, J. M. (2017). Hidden hazardous child labor as a complex human rights phenomenon: A case study of child labor in Pakistan's brick-making industry. *Cogent Social Sciences*, *3*(1), 1369486. https://doi.org/10.1080/23311886.2017.1369486

Ball, C. (2024). *The Harmful Effects of Living in Brick Kiln Communities in the South Asia Region – Ballard Brief.* https://ballardbrief.byu.edu/issue-briefs/the-harmful-effects-of-living-in-brick-kiln-communities-in-the-south-asia-region

Baumgartner, J., Carter, E., Schauer, J. J., Ezzati, M., Daskalopoulou, S. S., Valois, M. F., Shan, M., & Yang, X. (2018). Household air pollution and measures of blood pressure, arterial stiffness and central haemodynamics. *Heart*, *104*(8), 1515–1521. https://doi.org/10.1136/heartjnl-2017-312595

Bisht, G., & Neupane, S. (2015). Impact of brick kilns' emission on soil quality of agriculture fields in the vicinity of selected Bhaktapur Area of Nepal. *Applied and Environmental Soil Science*, *2015*(1), 1–8. https://doi.org/10.1155/2015/409401

Biswas, A. A. A., Saha, M. K., Hasan, I., Faisal, M., & Prokash, J. (2018). Occupational musculoskeletal and respiratory illness among brick kiln industry workers in Bangladesh. *Caspian Journal of Health Research*, *3*(3), 80–85. https://doi.org/10.29252/cjhr.3.3.80

Borghi, F., Spinazzè, A., Fanti, G., Campagnolo, D., Rovelli, S., Keller, M., Cattaneo, A., & Cavallo, D. M. (2020). Commuters' personal exposure assessment and evaluation of inhaled dose to different atmospheric pollutants. *International Journal of Environmental Research and Public Health, 17*(10), 3357. https://doi.org/10.3390/ijerph17103357

Bungadaeng, S., Prueksasit, T., & Siriwong, W. (2019). Inhalation exposure to respirable particulate matter among workers in relation to their e-waste open burning activities in Buriram Province, Thailand. *Sustainable Environment Research, 1*(1), 1–12. https://doi.org/10.1186/s42834-019-0030-7

Croitoru, L., & Sarraf, M. (2012). Benefits and costs of the informal sector: The case of brick kilns in Bangladesh. *Journal of Environmental Protection, 03*(06), 476–484. https://doi.org/10.4236/jep.2012.36058

Daly, A., Hillis, A., Shrestha, S. M., & Shrestha, B. K. (2020). Bricks in the wall: A review of the issues that affect children of in-country seasonal migrant workers in the brick kilns of Nepal. *Geography Compass, 14*(12), 1–13. https://doi.org/10.1111/gec3.12547

Das, B. (2014). Assessment of occupational health problems and physiological stress among the brick field workers of West Bengal, India. *International Journal of Occupational Medicine and Environmental Health, 27*(3), 413–425. https://doi.org/10.2478/s13382-014-0262-z

Gaviola, C., Nicolaou, L., Sharma, A. K., Chandyo, R., Parker, D., Shrestha, L., Das, S. K., Ramachandran, G., Thygerson, S. M., Beres, L., & Checkley, W. (2024). Knowledge, attitudes and practices regarding respirable silica exposure and personal protective equipment use among brick kiln workers in Nepal. *Occupational and Environmental Medicine, 81*(6), 287–295. https://doi.org/10.1136/oemed-2024-109516

Guttikunda, S. K., Begum, B. A., & Wadud, Z. (2013). Particulate pollution from brick kiln clusters in the Greater Dhaka region, Bangladesh. *Air Quality, Atmosphere and Health, 6*(2), 357–365. https://doi.org/10.1007/s11869-012-0187-2

Hamid, A., Riaz, A., Noor, F., & Mazhar, I. (2023). Assessment and mapping of total suspended particulate and soil quality around brick kilns and occupational health issues among brick kilns workers in Pakistan. *Environmental Science and Pollution Research, 30*(2), 3335–3350. https://doi.org/10.1007/s11356-022-22428-8

Health Effects Institute. (2020). *State of Global Air 2020. Special Report.*

Herath, I., Karunaratne, S., & Dharmarathna, D. (2021). Global warming potential of English brick manufacturing in Sri Lanka: A cradle to gate analysis. *2021 Moratuwa Engineering Research Conference (MERCon)*, Moratuwa, pp. 251–256. https://doi.org/10.1109/MERCon52712.2021.9525760

Jayasinghe, C. (2007). Comparative performance of burnt clay bricks and compressed stabilized earth bricks and blocks. *The Engineer*, 2, 33–40.

Kamal, A., Malik, R. N., Martellini, T., & Cincinelli, A. (2014). Cancer risk evaluation of brick kiln workers exposed to dust bound PAHs in Punjab province (Pakistan). *Science of The Total Environment, 493*, 562–570. https://doi.org/10.1016/j.scitotenv.2014.05.140

Kc, B., Mahapatra, P. S., Thakker, D., Henry, A. P., Billington, C. K., Sayers, I., Puppala, S. P., & Hall, I. P. (2020). Proinflammatory effects in ex vivo human lung tissue of respirable smoke extracts from indoor cooking in Nepal. *Annals of the American Thoracic Society, 17*, 688–698. https://doi.org/10.1513/AnnalsATS.201911-827OC

Khan, M. W., Ali, Y., De Felice, F., Salman, A., & Petrillo, A. (2019). Impact of brick kilns industry on environment and human health in Pakistan. *Science of The Total Environment, 678*, 383–389. https://doi.org/10.1016/j.scitotenv.2019.04.369

Khisroon, M., Khan, A., Imran, M., Zaidi, F., Ahmadullah, F., & Fatima, S. H. (2018). Biomonitoring of DNA damage in individuals exposed to brick kiln pollution from Peshawar, Khyber Pakhtunkhwa, Pakistan. *Archives of Environmental & Occupational Health, 73*(2), 115–120. https://doi.org/10.1080/19338244.2017.1304881

Kingsley, S. L., Eliot, M. N., Glazer, K., Awad, Y. A., Schwartz, J. D., Savitz, D. A., Kelsey, K. T., Marsit, C. J., & Wellenius, G. A. (2017). Maternal ambient air pollution, preterm birth and markers of fetal growth in Rhode Island: Results of a hospital-based linkage study. *Journal of Epidemiology and Community Health*, *71*(12), 208963. https://doi.org/10.1136/jech-2017-208963

Maithel, S., Kumar, S., Lalchandani, D., & Kanungo, S. (2014). *Factsheets About Brick Kilns in South and Sout-East Asia: Fixed Chimney Bull's Trench Kiln*. UNEP, Nairobi, Kenya.

Mikrani, T., Manandhar, N., & Joshi, S. K. (2017). Musculoskeletal symptoms among the Brick Kiln workers of Kathmandu valley. *International Journal of Occupational Safety and Health*, *7*(1), 7–11. https://doi.org/10.3126/ijosh.v7i1.22760

Nepal, S., Mahapatra, P., Adhikari, S., Shrestha, S., Sharma, P., Shrestha, K., Pradhan, B., & Puppala, S. (2019). A comparative study of stack emissions from straight-line and zigzag brick kilns in Nepal. *Atmosphere*, *10*(3), 107. https://doi.org/10.3390/atmos10030107

Ortínez-Alvarez, A., Peralta, O., Alvarez-Ospina, H., Martínez-Arroyo, A., Castro, T., Páramo, V. H., Ruiz-Suárez, L. G., Garza, J., Saavedra, I., de la Luz Espinosa, M., De Vizcaya-Ruiz, A., Gavilan, A., Basaldud, R., & Munguía-Guillén, J. L. (2017). Concentration profile of elemental and organic carbon and personal exposure to other pollutants from brick kilns in Durango, Mexico. *Air Quality, Atmosphere and Health*, *11*(3), 285–300. https://doi.org/10.1007/s11869-017-0539-z

Pattinson, W., Targino, A. C., Gibson, M. D., Krecl, P., Cipoli, Y., & Sá, V. (2018). Quantifying variation in occupational air pollution exposure within a small metropolitan region of Brazil. *Atmospheric Environment*, *182*, 138–154. https://doi.org/10.1016/j.atmosenv.2018.03.011

Pokhrel, A. K., Bates, M. N., Acharya, J., Valentiner-Branth, P., Chandyo, R. K., Shrestha, P. S., Raut, A. K., & Smith, K. R. (2015). PM 2.5 in household kitchens of Bhaktapur, Nepal, using four different cooking fuels. *Atmospheric Environment*, *113*, 159–168. https://doi.org/10.1016/j.atmosenv.2015.04.060

Pokhrel, A. K., Smith, K. R., Khalakdina, A., Deuja, A., & Bates, M. N. (2005). Case – control study of indoor cooking smoke exposure and cataract in Nepal and India. *International Journal of Epidemiology*, *34*, 1–7. https://doi.org/10.1093/ije/dyi015

Rajagopalan, S., & Brook, R. D. (2012). Air pollution and type 2 diabetes mechanistic insights. *Diabetes*, *61*, 3037–3045. https://doi.org/10.2337/db12-0190

Rupakheti, D., Pradhan, P. M. S., & Basel, P. (2018). Occupational safety and health vulnerability among brick factory workers in Dhading district, Nepal. *Annals of Global Health*, *84*(3), 481–487. https://doi.org/10.29024/aogh.2313

Saeed, T., Abbasi, N. A., Zahid, M. T., Fatima, N., Ullah, K., & Khokhar, M. F. (2024). Toxicological profile and potential health concerns through metals and trace elements exposure in brick kiln workers from Lahore, Pakistan. *Environmental Geochemistry and Health*, *46*(5), 150. https://doi.org/10.1007/s10653-024-01937-0

Sain, M. K., & Meena, M. (2019). Identifying musculoskeletal issues and associated risk factors among clay brick kiln workers. *Industrial Health*, *57*(3), 381–391. https://doi.org/10.2486/indhealth.2018-0096

Sanjel, S., Khanal, S. N., Thygerson, S. M., Carter, W., Johnston, J. D., & Joshi, S. K. (2018). Exposure to respirable silica among clay brick workers in Kathmandu valley, Nepal. *Archives of Environmental & Occupational Health*, *73*(6), 347–350. https://doi.org/10.1080/19338244.2017.1420031

Sanjel, S., Khanal, S. N., Thygerson, S. M., Khanal, K., Pun, Z., Tamang, S., & Joshi, S. K. (2016). Airborne particulate matter and health condition in brick kiln workers in Kathmandu Valley, Nepal. *Kathmandu University Medical Journal*, *14*(54), 159–166.

Shakya, K. M., Rupakheti, M., Aryal, K., & Peltier, R. E. (2016). Respiratory effects of high levels of particulate exposure in a cohort of traffic police in Kathmandu, Nepal. *Journal of Occupational and Environmental Medicine*, *58*(6), e218–e225. https://doi.org/10.1097/JOM.0000000000000753

Shakya, K., Peltier, R., Zhang, Y., & Pandey, B. (2019). Roadside exposure and inflammation biomarkers among a cohort of traffic police in Kathmandu, Nepal. *International Journal of Environmental Research and Public Health*, *16*(3), 377. https://doi.org/10.3390/ijerph16030377

Sharma, S., Gurung, K., Mishra, A., Bajracharya, S., Mathema, L., Hussain, A., & Pradhan, B. B. (n.d.). Industry under the open sky Industry under the open sky An exploration of the political. Report number: Working Paper 2019/7.

Smith, K. R., Samet, J. M., Romieu, I., & Bruce, N. (2000). Indoor air pollution in developing countries and acute lower respiratory infections in children. *Thorax*, *55*, 518–532.

Stockwell, C. E., Christian, T. J., Goetz, J. D., Jayarathne, T., Bhave, P. V., Praveen, P. S., Adhikari, S., Maharjan, R., DeCarlo, P. F., Stone, E. A., Saikawa, E., Blake, D. R., Simpson, I. J., Yokelson, R. J., & Panday, A. K. (2016). Nepal Ambient Monitoring and Source Testing Experiment (NAMaSTE): Emissions of trace gases and light-absorbing carbon from wood and dung cooking fires, garbage and crop residue burning, brick kilns, and other sources. *Atmospheric Chemistry and Physics*, *16*(17), 11043–11081. https://doi.org/10.5194/acp-16-11043-2016

Sughis, M., Nawrot, T. S., Ihsan-ul-Haque, S., Amjad, A., & Nemery, B. (2012). Blood pressure and particulate air pollution in schoolchildren of Lahore, Pakistan. *BMC Public Health*, *12*(1), 378. https://doi.org/10.1186/1471-2458-12-378

Sundari, S. (2020). Structural changes and quality of women's labour in India. *Indian Journal of Labour Economics*, *63*(3), 689–717. https://doi.org/10.1007/s41027-020-00245-2

Tefera, Y., Schlünssen, V., Kumie, A., Deressa, W., Moen, B. E., & Bråtveit, M. (2020). Personal inhalable dust and endotoxin exposure among workers in an integrated textile factory. *Archives of Environmental and Occupational Health*, *75*(7), 1–7. https://doi.org/10.1080/19338244.2020.1743958

Thomas, B. E., Charles, N., Watson, B., Chandrasekaran, V., Senthil Kumar, R., Dhanalakshmi, A., Wares, F., & Swaminathan, S. (2014). Prevalence of chest symptoms amongst brick kiln migrant workers and care seeking behaviour: A study from South India. *Journal of Public Health*, *37*(4), fdu104. https://doi.org/10.1093/pubmed/fdu104

Thygerson, S. M., Beard, J. D., House, M. J., Smith, R. L., Burbidge, H. C., Andrus, K. N., Weber, F. X., Chartier, R., & Johnston, J. D. (2019). Air-quality assessment of on-site brick-kiln worker housing in Bhaktapur, Nepal: Chemical speciation of indoor and outdoor $PM_{2.5}$ pollution. *International Journal of Environmental Research and Public Health*, *16*(21), 4114. https://doi.org/10.3390/ijerph16214114

Thygerson, S. M., Sanjel, S., & Johnson, S. (2016). Occupational and environmental health hazards in the brick manufacturing industry in Kathmandu Valley, Nepal. *Occupational Medicine & Health Affairs*, *4*(05), 1000248. https://doi.org/10.4172/2329-6879.1000248

Tibrewal, K., Maithel, S., & Venkataraman, C. (2019). A state-level methodology for estimating present-day emissions of short-lived climate pollutants from fired brick production in India. In Venkataraman, C., Mishra, T., Ghosh, S., & Karmakar, S. (eds) *Climate Change Signals and Response* (pp. 211–229). Springer, Singapore. https://doi.org/10.1007/978-981-13-0280-0_13

Toren, K., Bergdahl, I. A., Nilsson, T., & Jarvholm, B. (2007). Occupational exposure to particulate air pollution and mortality due to ischaemic heart disease and cerebrovascular disease. *Occupational and Environmental Medicine*, *64*(8), 515–519. https://doi.org/10.1136/oem.2006.029488

Valdes, H., Vilches, J., Felmer, G., Hurtado, M., & Figueroa, J. (2020). Artisan brick kilns: State-of-the-art and future trends. *Sustainability*, *12*(18), 7724. https://doi.org/10.3390/su12187724

Wahid, A., Mantell, N. H., & Mumtaz, M. Z. (2025). The dynamics of debt bondage in Pakistan: Causes and consequences. *Fudan Journal of the Humanities and Social Sciences*, 18, 585–609. https://doi.org/10.1007/s40647-024-00427-y

Weyant, C., Athalye, V., Ragavan, S., Rajarathnam, U., Lalchandani, D., Maithel, S., Baum, E., & Bond, T. C. (2014). Emissions from South Asian brick production. *Environmental Science & Technology*, *48*(11), 6477–6483. https://doi.org/10.1021/es500186g

Weyant, C., Kumar, S., Maithel, S., Thompson, R., Floess, E., Ravi, A., Adhikari, S., Gautam, S., Shrestha, S., Baum, E., & Bond, T. (2016). Emissions and energy performance in South Asian brick kilns. *Environmental Science & Technology*, *48*(11), 6477–6483.

Yang, N., Fu, R., Chao, Y., Liu, H., & Ma, X. (2020). Quantitative assessment of environmental exposure of delivery men in Wuhan. *Archives of Environmental & Occupational Health*, *75*(8), 1–19. https://doi.org/10.1080/19338244.2020.1743959

Zawilla, N., Taha, F., & Ibrahim, Y. (2014). Liver functions in silica-exposed workers in Egypt: Possible role of matrix remodeling and immunological factors. *International Journal of Occupational and Environmental Health*, *20*(2), 146–156. https://doi.org/10.1179/2049396714Y.0000000061

Zhang, X., Chen, X., & Zhang, X. (2018). The impact of exposure to air pollution on cognitive performance. *Proceedings of the National Academy of Sciences*, *115*(37), 9193–9197. https://doi.org/10.1073/pnas.1809474115

Health Implications of Microplastic Pollution in Air and Water

6

Ishrat Mumtaz, Khursheed Ahmad Dar, Tauseef Ahmad Bhat, Syed Andleeba Jan, Tahira Mumtaz, Bisma Jan, Tajalee Gulshan, Ruqqaya Javid, and Insha Ali

INTRODUCTION

The annual production of plastics, which are organic polymers with high molecular weight produced by linking organic monomers through polymerization, has reached to 31.1 billion tons. However, about only 60% of these plastics undergo recycling and reuse (Dong et al., 2020). The accumulation of plastic waste in both aquatic and terrestrial environments has become a major environmental concern (Zhao et al., 2022). Microplastics refer to plastic particles that are smaller than 5 mm behave as ubiquitous pollutants deriving from primary and secondary origins (Loganathan & Kizhakedathil, 2023). Microplastics may be generated through primary and secondary sources. First in primary sources, the microplastics are intentionally made while secondary sources include the degradation of macro-size plastics to micro-size plastics. Primary sources

DOI: 10.1201/9781003613183-6

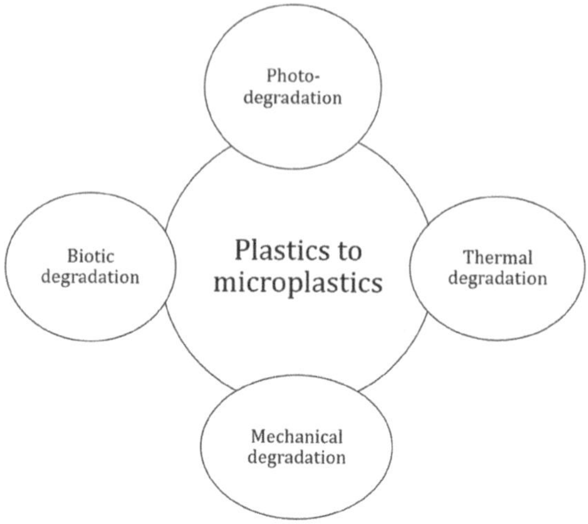

FIGURE 6.1 Different processes through which plastics get degraded into microplastics.

include microbeads in cosmetics, industrial abrasive substances, and cleaning agents, while secondary sources include the disintegration of plastic waste like discarded bottles, bags, and fishing nets into microplastics through different processes (Figure 6.1). Over decades, the relentless rise in plastic production and improper waste management has allowed these minute particles to infiltrate every corner of the planet, including the atmosphere, rivers, oceans, and even polar ice caps (Stefánsson et al., 2021). Microplastics enter the air through processes such as wind abrasion, industrial emissions, and tire wear, becoming airborne contaminants that are increasingly detected in urban, rural, and remote regions (Zhao et al., 2023). In water, they accumulate through urban runoff, wastewater discharge, marine debris, and agricultural leaching (Bigalke et al., 2022; Wei et al., 2023; Sun et al., 2024). From there, they permeate drinking water supplies and aquatic ecosystems. Microplastics present in air and water create complex pathways via which humans and wildlife can be exposed, making it urgent to address this pollution.

The persistent nature of microplastics is alarming, as they can act as vectors for pollutants such as heavy metals and persistent organic pollutants (Khan et al., 2024). These pollutants combine with microplastics to create a "cocktail effect", worsening their impact on living organisms. (Gao et al., 2022). Latest research reveals that contaminated seafood and water act as pathways for microplastics to get included in the food chain (Eze et al., 2024) and infiltrate the human respiratory, circulatory, reproductive, and digestive systems exposing individuals to potential health risks (Enyoh et al., 2023; Saha & Saha, 2024). Exposure of humans to microplastics is connected to a variety of health issues, including respiratory inflammation, oxidative stress, gastrointestinal distress, and potential genotoxicity. Oxidative stress and genotoxicity may contribute to chronic conditions like cardiovascular disease or cancer (Mas-Bargues et al., 2022). Beyond their physical and chemical toxicity, microplastics may also serve as vectors for pathogens by which different microbial contaminants are introduced into human and animal systems. These impacts are disproportionately felt by vulnerable populations, such as those near polluted water bodies or industrial zones, highlighting microplastic pollution as an environmental justice concern.

Despite growing awareness of this problem, there are significant gaps that have remained in our understanding of the long-term health impacts of microplastic exposure. More research is required to elucidate the mechanisms by which microplastics penetrate biological barriers, accumulate within tissues, and affect cellular functions over time. Addressing microplastic pollution requires scientific advancements, robust policy interventions, and active public engagement to reduce their presence and mitigate their impacts. This chapter delves into the pathways of microplastic pollution in air and water and the mechanisms through which they impact human health. By fostering a deeper understanding of these issues, this chapter aims to highlight the critical need for holistic strategies to reduce microplastic pollution and safeguard both public health and ecological integrity.

PATHWAYS OF MICROPLASTIC EXPOSURE IN HUMANS

Microplastics enter the human body via multiple pathways, posing significant health risks. Dietary intake is a primary exposure route, with seafood, salt, and drinking water being major contributors. Aquatic organisms, like shellfish and fish, frequently consumed by humans, accumulate microplastics in their gastrointestinal tracts and tissues. According to Cho et al. (2019), humans consume 123–11,000 microplastic particles from seafood. Contamination levels ranging from 6.74 to 9.12 microplastic particles per individual fish have been reported in samples from the coast of Mumbai (Gurjar et al., 2022). Kosuth et al. (2018) estimated that humans ingest approximately 4,400–5,800 microplastic particles annually through tap water. Sea salt has been found to contain up to 1,653 microplastic particles per kilogram (Di Fiore et al., 2023). Inhalation also plays a critical role, especially in urban and industrial areas, where airborne microplastics originate from synthetic textiles, road dust, and industrial emissions. Dermal exposure to substances like cosmetics and synthetic fabrics is another pathway. After entering the body, microplastics can cross the intestinal barrier, enter the bloodstream, and accumulate in organs such as the liver and kidneys. Prolonged exposure to microplastics has been associated with inflammation, oxidative stress, and long-term health risks, highlighting the widespread presence of microplastic pollution and its effects on human well-being. Microplastics enter the human body mainly via three primary routes: ingestion, inhalation, and skin contact (Figure 6.2). These exposure routes are influenced by the ubiquity of microplastics in air, water, and consumer products, making them nearly unavoidable in current environments (Singh et al., 2024).

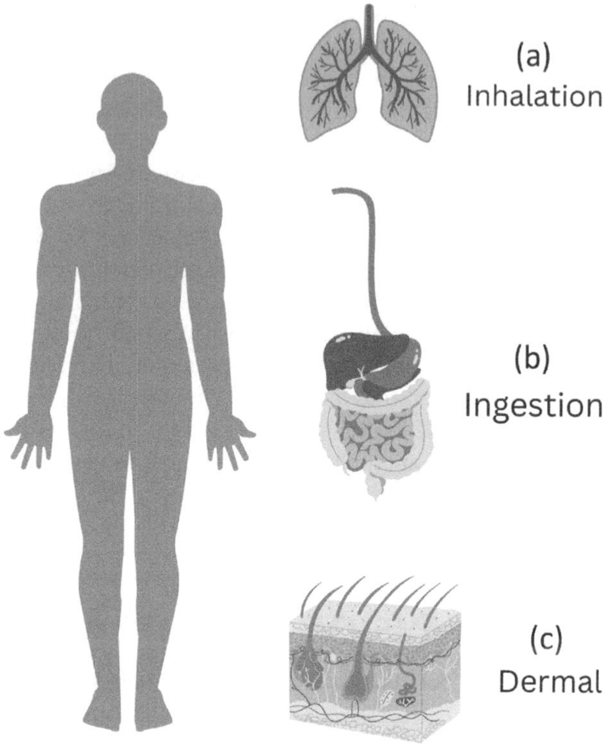

FIGURE 6.2 Different routes of microplastic exposure to humans.

Inhalation of Microplastics

Airborne microplastics are major route of exposure resulting from the suspension of microplastics in the atmosphere (López et al., 2023). Atmospheric microplastics is an emerging concern due to their ability to remain in air and travel over considerable distances (Tatsii et al., 2023; Xiao et al., 2023). Microplastics released into the atmosphere are transported by wind, rain, and snow, allowing them to travel long distances (Haque & Fan, 2023; Sathyamohan et al., 2023). This results in their widespread distribution and increases the likelihood of human exposure. The inhalation pathway underscores the omnipresence of microplastics in both outdoor and indoor environments where prolonged exposure is unavoidable (O'Brien et al., 2023).

Microplastics enter the air through several mechanisms, each contributing uniquely to their presence in the atmosphere. Industrial emissions are a major source, with factories and manufacturing plants releasing microplastic particles during the production, processing, and recycling of plastics (Noorimotlagh et al., 2024). Activities such as grinding, cutting, moulding, and shredding plastics lead to the creation of fine plastic particles, especially in facilities with insufficient control. Raw plastic pellets and dust, generated during transportation and handling, also contribute to atmospheric microplastic pollution. Thermal and mechanical processing in industrial settings further exacerbates this issue by breaking down polymer chains under heat and pressure, producing microplastic fragments. Synthetic textiles are another significant source due to their widespread use in modern life (Kaur & Dandautiya, 2024). Washing and drying synthetic fabrics like acrylic, nylon, and polyester releases microplastic fibres. These fibres are released with wastewater leading to water contamination, and drying allows the dispersal of these fibres directly into the air. Everyday wear and tear also causes fibres to detach from clothing and circulate in both outdoor and indoor environments. Enclosed spaces like homes and offices often have higher concentrations of airborne microplastics due to limited ventilation and accumulation over time.

Environmental degradation of plastic materials is a major contributor to the release of microplastics into the atmosphere. Plastics exposed to ultraviolet radiation undergo photodegradation, breaking down into microplastic particles as their polymer structure weakens (Key, 2023; Di Giulio et al., 2024). Temperature changes and mechanical forces, such as wind and rain, accelerate this process, making plastics brittle and prone to fragmentation. Mechanical weathering from wind abrasion, rainfall, and sand erosion, along with biotic degradation through enzymatic breakdown by microorganisms like bacteria and fungi and fragmentation by organisms such as worms and insects, contributes to the generation of fine microplastic particles (Jarosz et al., 2024). Open landfills and dumping sites are hotspots for such degradation, combining UV exposure, temperature shifts, and microbial activity. Tire wear and road dust are prominent contributors to airborne microplastics, especially in urban areas with high vehicular traffic (Giechaskiel et al., 2024). Tires made from synthetic rubber and polymers release tiny particles through friction with road surfaces. These particles mix with road dust, easily resuspended into the air by wind and vehicle movement. High-traffic zones and industrialized regions often exhibit elevated levels of airborne microplastics from these sources.

Atmospheric transport plays a critical role in dispersing microplastics far beyond their original sources. Strong winds transport microplastic particles from urban centres, industrial zones, and roadways over long distances. Vertical air turbulence can lift microplastics high into the atmosphere, enabling their movement through atmospheric currents such as jet streams and seasonal wind patterns. Jet streams, with their high-speed winds, transport microplastics across vast distances, while seasonal wind patterns influence their deposition and redistribution between land and sea. This results in microplastic deposition in remote areas including mountain ranges and polar ice caps, demonstrating the microplastic pollution as "Pollution with no boundaries" and "Global reach".

Ingestion of Microplastics

Ingestion is a major route through which humans are exposed to microplastics (Al Mamun et al., 2023). This occurs mainly via contaminated food and water. Various food items have been reported to contain microplastics with seafood being a significant contributor due to its role in bioaccumulation (Unuofin & Igwaran, 2023). Aquatic organisms like fish, shellfish, and crustaceans consume microplastics present in polluted water, which then enter the human food chain. Marine organisms frequently ingest microplastics, mistaking them for food, leading to their accumulation in digestive systems and eventual transfer to humans through consumption. A detailed survey found microplastics in all commercially available table salts, with concentrations ranging from 20 to 125 microplastic particles per kilogram (Li et al., 2022). Investigations were carried out regarding the presence of microplastic particles in honey, revealing a significant level of contamination (Lin et al., 2022). Drinking water is another critical source of ingestion (Oßmann, 2021). Studies reported that microplastics occur in both tap and bottled water (Weisser et al., 2021). The contamination in tap water often originates from degraded plastics in the environment or particles entering the water system through pipes and treatment plants. Bottled water, on the other hand, may also contain microplastics introduced during the manufacturing and bottling processes or from the degradation of the plastic bottles themselves. Research indicates that bottled water tends to have higher levels of microplastic contamination compared to tap water, likely due to plastic packaging (Gambino et al., 2020; Tosun et al., 2020). Given the heavy dependence on bottled water for drinking, this contamination poses serious public health concerns and highlights the need to encourage consumers to opt for tap water to reduce both microplastic exposure and plastic waste (Gambino et al., 2022).

Other sources of ingestion include agricultural produce irrigated with contaminated water or grown in microplastic contaminated soils (Tang, 2023). Agricultural practices have significant participation in the introduction of microplastic particles into food products. Microplastics may be added by the use of plastic mulches, organic and inorganic fertilizers, surfacer runoff, and erosion (Wu et al., 2024). Processed and packaged foods can be contaminated with microplastics shed from their plastic packaging (Sharma, 2024). Due to their minute size and low weight, microplastics are commonly found in the human diet. Their presence in staple food items and drinking water represents a widespread and unavoidable route of exposure for global populations.

Dermal Exposure to Microplastics

Dermal exposure to microplastics occurs when the skin comes into direct contact with contaminated water, soil, or cosmetic products containing microplastic particles (Emenike et al., 2023). While dermal exposure contributes less to overall microplastic exposure compared to ingestion and inhalation, it may still pose risks, particularly with prolonged or repeated contact. Personal care and cosmetic items like toothpaste, exfoliants, and facial scrubs frequently include microbeads made from polyethylene or other polymers (Singh & Mishra, 2023). During use these products leave microplastic residues on the skin and some particles can penetrate the epidermal barrier, especially when the skin is damaged (Akbari et al., 2024). Microplastics can penetrate the epidermal barrier via hair follicles or micro-abrasions, raising concerns about localized inflammation or systemic absorption. Showering or bathing in water contaminated with microplastics also exposes the skin to these particles.

Environmental contact can further contribute to dermal exposure (Boccia et al., 2024). Individuals working in industries dealing with plastic production, waste management, or recycling may come into regular contact with microplastic particles through contaminated surfaces or airborne dust settling on the skin. Similarly, recreational activities in polluted water bodies such as swimming expose individuals to microplastics suspended in the water or deposited along shorelines. Although the skin serves as a strong barrier, prolonged or repeated exposure to microplastics especially nano-sized particles may increase the risk of penetration (Schmidt et al., 2023). These particles may interact with the lipid layers of the skin, potentially entering the body via hair follicles or sweat glands. Additional studies are required to assess the potential risks of occupational and recreational exposure to microplastics through the skin, particularly their long-term health effects. Dermal exposure, while less direct than ingestion or inhalation, underscores the pervasive nature of microplastics and highlights the need for further investigation into their interactions with human skin.

MECHANISMS OF HEALTH IMPACTS

Microplastics can trigger a variety of biological and biochemical responses, ranging from direct physical irritation to cellular inflammation and oxidative stress, leading to a cascade of harmful physiological effects. These effects range from respiratory and gastrointestinal disturbances to cellular damage, underscoring the multidimensional nature of microplastic exposure.

Respiratory Effects

The respiratory system functions as a key point of interaction with the environment, making it susceptible to airborne pollutants, including microplastics. Recent studies have highlighted potential adverse effects of inhaling these microplastic particles, which can penetrate deep into the respiratory tract and trigger a range of adverse health outcomes (Riaz et al., 2024). Fine microplastic particles have small size and low mass due to which they exhibit specific aerodynamic properties. Because of their high surface area relative to volume, these particles can stay in air for extended periods, increasing the likelihood of deeper inhalation into the lungs. Their minute size facilitates them to bypass the natural filtering mechanisms in the upper respiratory tract, such as nasal hairs and mucus, and travel to the alveolar area of lungs. The aerodynamic properties of these particles enable them to penetrate alveolar spaces, where gas exchange occurs, potentially leading to inflammation, impaired lung function, and increased susceptibility to respiratory illnesses like asthma, chronic obstructive pulmonary disease (COPD), and lung cancer (Vattanasit et al., 2023). Biological responses to inhaled microplastics often begin with inflammation.

Upon deposition in the lungs, microplastics interact with epithelial cells and alveolar macrophages, triggering the production of cytokines-pro-inflammatory signalling molecules (Tomonaga et al., 2024). This immune response is the body's attempt to isolate and neutralize foreign particles, but prolonged inflammation can damage lung tissue, impairing respiratory function. Oxidative stress is another critical mechanism associated with inhalation of these particles (Kadac-Czapska et al., 2024). Microplastics containing reactive surface groups or adsorbed pollutants can interact with lung cells to produce reactive oxygen species (ROS). Excessive ROS production exhausts the defence system of antioxidants resulting in cellular damage, tissue damage, further inflammation, and the promotion of chronic respiratory conditions. Conditions such as asthma and COPD can worsen due to prolonged exposure to inhaled microplastics (Saha & Saha, 2024). Microplastics can also act as adjuvants, increasing immune sensitivity and aggravating pre-existing conditions. Chronic exposure has been linked to reduced lung capacity and fibrosis, which permanently impair lung elasticity and restrict breathing (Zhang et al., 2024).

Gastrointestinal and Systemic Effects

Ingestion of microplastics occurs through contaminated food and water, introducing these particles into the gastrointestinal tract. Upon entry, microplastics can physically interact with the mucosal lining and trigger the localized inflammation (Bruno et al., 2024). Persistent exposure may lead to chronic gastrointestinal irritation, including symptoms such as abdominal pain, diarrhoea, or constipation (Nwanaforo et al., 2024). A particularly concerning aspect of microplastic ingestion is their capacity to disrupt gut microbiota (Wang et al., 2023). When microplastics reach the gut, they can alter the gut flora composition by favouring the growth of opportunistic or pathogenic bacteria. This disruption may occur due to adsorbed pollutants or toxins that are often carried by microplastics, interfering with the microbial activity and nutrient availability (Tamargo et al., 2022). Microplastic surface also carries biofilms, preserving clusters of microbes (Tu et al., 2020). Biofilms protect the foreign microbes from immune system, making them more persistent and virulent. Physical presence of microplastics in gut can also influence the gut lining, causing inflammation and variations in microbial diversity (Yan et al., 2021). This leads to impaired digestion, weakened immunity, and systemic inflammation. The translocation of these microplastic particles adds another segment to their effects on human body. Studies show that smaller microplastics can cross intestinal barrier, passing into bloodstream and potentially reaching systemic circulation (Yee et al., 2021). Upon entering the bloodstream, these particles may be circulated to multiple tissues and organs, where they accumulate and cause localized effects (Prata, 2023). Tissue deposition has been observed in critical organs, like brain, heart, liver, and kidneys, where microplastic accumulation can disrupt normal function. Over time, this may trigger oxidative stress, inflammatory responses, and functional impairments, contributing to chronic health conditions (Enyoh et al., 2023; La Porta et al., 2023).

Cellular and Molecular Effects

Microplastics exhibit cytotoxicity by directly damaging cell membranes or interfering with cellular processes (He et al., 2023). They can induce apoptosis (programmed cell death) or necrosis, depending on the particle size, shape, and surface chemistry. Membranes are easily penetrated by smaller particles, having high ratio of surface area and volume generating oxidative stress and organelle damage. Physical damage to cell membranes is mostly caused by the irregular shaped microplastics such as fragments (Choi et al., 2021). The nature of reactive functional groups and adsorbed toxins regulates the pathway of cellular disruption. Another critical concern related to microplastics is genotoxicity, specifically those containing chemical additives like phthalates or bisphenol A (BPA) (Campanale et al., 2020). These chemical additives can interfere with DNA replication and repair mechanisms. Persistent organic pollutants and heavy metals aggravate the genotoxicity by inducing oxidative stress, which damages DNA and other genetic materials (Hajam et al., 2022). Such damages may increase the risk of mutations, potentially leading to carcinogenesis. Adsorbed pollutants carried by microplastics may also interfere with intracellular signalling pathways, amplifying their toxicological effects (Kumar et al., 2024). Microplastics and additives used during plastic production may get released inside humans and disrupt endocrine signalling, subsequently affecting hormonal balance (Ullah et al., 2023). They function as endocrine disrupting chemicals (EDCs) by interfering with hormone receptors and disrupting their synthesis, release, and translocation. Important hypothalamic axes, including the HPA, HPT, and HPG, are affected. Disruption causes imbalances in thyroid hormones, stress hormones, testosterone, and oestrogen levels (Kehinde et al., 2024). Microplastics are implicated in cellular damage by accumulating within mitochondria, disrupting their electron transport chain, and damaging mitochondrial membranes. Disruptions to the mitochondrial membrane potential impair the electron transport chain, resulting in excessive generation of ROS. They cause oxidative stress, which damages essential biomolecules like DNA, proteins, and lipids. This damage ultimately depletes the cell's antioxidant defences, triggering various signalling pathways that culminate in cell death (Das, 2023).

Microplastics as Vectors for Pathogens

Microplastics often act as carriers for microbial pathogens and biofilms (Wang et al., 2021). Their hydrophobic surfaces provide a conducive environment for microbial adhesion, allowing pathogenic bacteria, viruses, and fungi to colonize them. This "hitchhiking" phenomenon poses a significant threat to aquatic ecosystems, where microplastics serve as floating reservoirs of disease-causing organisms (Sooriyakumar et al., 2022). When humans are exposed to microplastics, these pathogens can bypass natural barriers and infect tissues. Ingestion of contaminated water or food may introduce these microorganisms directly into the gastrointestinal tract, leading to infections and systemic inflammation. Furthermore, biofilms on microplastics can enhance the persistence of pathogens by shielding them from environmental stresses such as UV radiation or desiccation (Olaimat et al., 2024). This enhances their survival and geographic distribution, promoting the spread of antibiotic-resistant bacteria. The role of microplastics as vectors for pathogens adds a significant dimension to their health impacts, linking them not only to physical and chemical toxicity but also to biological risks.

SUSCEPTIBLE POPULATIONS

Certain populations are more vulnerable to the health risks posed by microplastics due to a combination of environmental, socioeconomic, and physiological factors. These groups are at greater risk of exposure to microplastics and consequently suffer more from their associated health effects. Identifying these vulnerable populations is essential for formulating targeted approaches to reduce exposure and health risks of microplastic pollution.

Communities in Industrial and Polluted Areas

Individuals living in or near industrial zones, landfills, and areas with significant plastic waste accumulation are at high risk of exposure to microplastics (Horton & Barnes, 2020). Mokhtarzadeh et al. (2022) performed a study to find out about microplastic load in industrial and urban soils of Iran. The concentration of microplastics ranged from 17 to 347 particles/100 g soil, highlighting significant risks of exposure in these areas. Another study from Shaoxing area of China revealed high concentrations of microplastics in freshwater and sediment samples with concentrations ranging from 2.10 to 71.0 particles/L water and 16.70 to 1,323.3 particles/kg sediment (Deng et al., 2020). This area represents a textile industrial area contributing to microplastic pollution in nearby water and sediment environments. Factories and waste degradation processes release microplastic particles into the water, air, and soil, generating a constant risk for residents. Coastal and riverine communities are also vulnerable as plastics in water bodies degrade and enter the food chain, contaminating local seafood and drinking water. Areas with elevated industrial activity, tire wear, waste generation, and dense populations act as hotspots for microplastic pollution. Tire wear of vehicles and industrial emissions release microplastics into air, making urban environments particularly hazardous. People living in cities are more likely to breathe in contaminated air and consume microplastics.

Occupationally Exposed Groups

The level of exposure to microplastics is substantially higher in the workplace environments as compared to general public, highlighting the occupational exposure as a critical pathway (Dris et al., 2017; Stefaniak et al., 2019). The National Institute for Occupational Safety and Health (NIOSH) recognized multiple pathways via which humans can get exposed at their workplaces (Glassford et al., 2020). Workers working in plastic manufacturing industries, recycling facilities, textiles industries, and waste management organizations are exposed to higher concentrations of microplastics (Shahsavaripour et al., 2023). For instance, textile workers experience frequent exposure to synthetic fibres released during garment production and washing, while waste management personnel handles plastic debris that releases microplastics during decomposition. Individuals taking part in nylon flocking, which involves the cutting of polymer fibres, directly expose them to plastics (Boccia et al., 2024). Employees engaged in plastic processing and plastic smoothening are exposed to microplastics through working hours (Murashov et al., 2021). The lack of adequate protective measures in these industries increases the risk of chronic microplastic exposure via inhalation and dermal contact, potentially leading to significant health complications.

Pregnant Women and Children

Pregnant women and children are sensitive to the harmful effects of microplastics due to their developing bodies and immune systems (Segovia-Mendoza et al., 2020; Panneerselvam et al., 2024). Microplastics' ability to cross the placental barrier raises concerns for fetal health, potentially leading to developmental delays or long-term health implications. Children with increased hand-to-mouth activities face greater risk of ingesting microplastics. These early-life exposures can have profound effects on their growth, immune function, and overall health.

CHALLENGES IN UNDERSTANDING HEALTH IMPACTS

The growing concern over the health impacts of microplastic pollution is met with significant scientific challenges. These challenges stem from a combination of knowledge gaps and methodological limitations that impede the ability to thoroughly analyse and minimize the risks presented by microplastics. Addressing these issues is essential for developing a robust understanding of how microplastics interact with biological systems and affect human health. A major barrier lies in the lack of long-term data on microplastic exposure. Most studies are short-term, capturing only immediate effects without accounting for the cumulative impacts of prolonged exposure. This limitation makes it difficult to predict chronic outcomes, such as potential links between microplastic accumulation and long-term diseases like cancer or neurological disorders. Additionally, the effects of microplastics are often confounded by their association with other contaminants, including heavy metals, persistent organic pollutants (POPs), and microbial contaminants. Isolating the specific impacts of microplastics from these co-occurring toxicants remains a persistent challenge, leading to uncertainties in attributing health outcomes directly to microplastic exposure. Another layer of complexity arises from the inherent variability of microplastics themselves. These particles differ widely in shape, size, and chemical composition, which govern their behaviour, environmental persistence, and toxicity. For instance, smaller size microplastics can penetrate the biological barriers and reach the brain, blood, and even cross the placenta (Zhu et al., 2023; Bai et al., 2024). Irregular shaped microplastics are able to mechanically damage the tissues and accumulate in them. The chemical composition including the pollutants, heavy metals, poly hydrocarbons, and other additives governs the impact they have on exposed individuals (Ricardo et al., 2021).

Methodological challenges also hinder progress in comprehending the health risks associated with microplastics. One major challenge is the accurate identification of polymers, coupled with inconsistencies in how microplastic morphologies are classified. Furthermore, data availability is considerably limited, hindering comprehensive understanding. Key processes such as the fate, transport, and weathering of microplastics and their surface contaminants remain poorly understood. Detecting and quantifying microplastics in biological samples such as blood, tissues, and excreta is a technically demanding task. Current analytical techniques are time-intensive and often lack standardized protocols. The lack of universal guidelines for sampling, analysis, and reporting leads to inconsistencies across studies, making it difficult to compare findings or draw definitive conclusions. Furthermore, laboratory experiments frequently use concentrations of microplastics that may not accurately represent real-world exposure, raising questions about the relevance of these studies to actual human health risks. Inconsistent classification, limited data about the sources, and complications in detecting microplastics in biological samples pose substantial obstacles in accurate risk assessment and framing policies. Determining thresholds becomes strenuous, and the recognition of high-risk materials is hindered. Uncertainty is created for policymakers due to the absence of standardized protocols, and observations from test setups undermine their applicability to real-world situations.

To advance our vision and understanding, it is crucial to address these intertwined challenges through collaborative, multidisciplinary research efforts. Long-term studies that integrate environmental, toxicological, and epidemiological perspectives are essential for elucidating the full spectrum of microplastic health impacts. At the same time, developing standardized methodologies and improving detection techniques including the use of emerging technologies such as high-resolution mass spectrometry, remote sensing, and microplastic-specific sensors will enhance the reliability and comparability of findings. With the progressed research, sound data on primary sources of microplastics, pathways of movement in the environment, and their travelling across the food chains will be compiled. This database will support in the development of strong regulations related to plastic production and usage, unfolding of comprehensive monitoring programmes, and targeted solutions. These innovations will ultimately provide the scientific foundation needed to shape regulations and measures focused on minimizing microplastic pollution hazards.

CONCLUSION

Microplastic pollution in air and water depicts a critical human health and ecological concern with profound implications. This chapter explores three major pathways of human exposure—inhalation, ingestion, and dermal contact—and the unique risks associated with each. Inhaled microplastics can reach deep into the lungs, causing respiratory inflammation, oxidative stress, and other chronic conditions such as asthma and fibrosis. Ingested microplastics disrupt gut microbiota, translocate to vital organs, and trigger systemic effects, including immune dysfunction and organ damage. Microplastics act as amplifiers for their harmful impact by acting as carriers for toxic chemicals and microbial pathogens. At the cellular level, microplastics generate ROS, resulting in oxidative stress, DNA damage, and protein and lipid oxidation with eventual cell death. Vulnerable populations, including urban residents, industrial workers, pregnant women, and children, face heightened risks due to increased exposure or physiological sensitivity, emphasizing the inequitable nature of this threat. Despite the growing information, knowledge gaps remain related to long-term health effects, differential impacts of particle size and chemical composition, and the combined effects of microplastics with co-occurring pollutants. Addressing these concerns mandates a multidisciplinary plan that incorporates standardized methodologies, advanced detection technologies, and integrated research across health, environmental, and policy domains. Mitigation strategies must prioritize reducing microplastic release at the source, enhancing waste management practices, and developing public health interventions to minimize exposure. Industries should develop and adopt advanced filtration technologies that can capture the microplastics. Biodegradable alternatives should be promoted, and regulations must be enforced to limit the release of microplastics from personal care products. The utilization of microorganisms for the degradation of plastic polymers should be encouraged. The use of plastic mulches in agriculture may be replaced with biodegradable mulches. Campaigns regarding the public awareness should run tirelessly, encouraging the general public to reduce their use of single-use plastics and support more sustainable choices. By confronting the risks posed by microplastics with a collaborative and proactive approach, we can work towards safeguarding both human health and the integrity of ecosystems in the face of this emerging crisis.

REFERENCES

Akbari, A., Taghavi, K., & Jaafari, J. (2024). Sources of microplastics in the environment and human exposure routes: A review. *Caspian Journal of Health Research*, *9*(3), 163–176.

Al Mamun, A., Prasetya, T. A. E., Dewi, I. R., & Ahmad, M. (2023). Microplastics in human food chains: Food becoming a threat to health safety. *Science of The Total Environment*, *858*, 159834.

Bai, C. L., Wang, D., Luan, Y. L., Huang, S. N., Liu, L. Y., & Guo, Y. (2024). A review on micro-and nanoplastics in humans: Implication for their translocation of barriers and potential health effects. *Chemosphere, 361*, 142424.

Bigalke, M., Fieber, M., Foetisch, A., Reynes, J., & Tollan, P. (2022). Microplastics in agricultural drainage water: A link between terrestrial and aquatic microplastic pollution. *Science of the Total Environment, 806*, 150709.

Boccia, P., Mondellini, S., Mauro, S., Zanellato, M., Parolini, M., & Sturchio, E. (2024). Potential effects of environmental and occupational exposure to microplastics: An overview of air contamination. *Toxics, 12*(5), 320.

Bruno, A., Dovizio, M., Milillo, C., Aruffo, E., Pesce, M., Gatta, M., & Ballerini, P. (2024). Orally ingested micro-and nano-plastics: A hidden driver of inflammatory bowel disease and colorectal cancer. *Cancers, 16*(17), 3079.

Campanale, C., Massarelli, C., Savino, I., Locaputo, V., & Uricchio, V. F. (2020). A detailed review study on potential effects of microplastics and additives of concern on human health. *International Journal of Environmental Research and Public Health, 17*(4), 1212.

Cho, Y., Shim, W. J., Jang, M., Han, G. M., & Hong, S. H. (2019). Abundance and characteristics of microplastics in market bivalves from South Korea. *Environmental Pollution, 245*, 1107–1116.

Choi, D., Hwang, J., Bang, J., Han, S., Kim, T., Oh, Y., & Hong, J. (2021). In vitro toxicity from a physical perspective of polyethylene microplastics based on statistical curvature change analysis. *Science of the Total Environment, 752*, 142242.

Das, A. (2023). The emerging role of microplastics in systemic toxicity: Involvement of reactive oxygen species (ROS). *Science of the Total Environment, 895*, 165076.

Deng, H., Wei, R., Luo, W., Hu, L., Li, B., & Shi, H. (2020). Microplastic pollution in water and sediment in a textile industrial area. *Environmental Pollution, 258*, 113658.

Di Fiore, C., Sammartino, M. P., Giannattasio, C., Avino, P., & Visco, G. (2023). Microplastic contamination in commercial salt: An issue for their sampling and quantification. *Food Chemistry, 404*, 134682.

Di Giulio, T., De Benedetto, G. E., Ditaranto, N., Malitesta, C., & Mazzotta, E. (2024). Insights into plastic degradation processes in marine environment by X-ray photoelectron spectroscopy study. *International Journal of Molecular Sciences, 25*(10), 5060.

Dong, Y., Gao, M., Song, Z., & Qiu, W. (2020). As (III) adsorption onto different-sized polystyrene microplastic particles and its mechanism. *Chemosphere, 239*, 124792.

Dris, R., Gasperi, J., Mirande, C., Mandin, C., Guerrouache, M., Langlois, V., & Tassin, B. (2017). A first overview of textile fibers, including microplastics, in indoor and outdoor environments. *Environmental Pollution, 221*, 453–458.

Emenike, E. C., Okorie, C. J., Ojeyemi, T., Egbemhenghe, A., Iwuozor, K. O., Saliu, O. D., & Adeniyi, A. G. (2023). From oceans to dinner plates: The impact of microplastics on human health. *Heliyon, 9*(10), e20440.

Enyoh, C. E., Devi, A., Kadono, H., Wang, Q., & Rabin, M. H. (2023). The plastic within: Microplastics invading human organs and bodily fluids systems. *Environments, 10*(11), 194.

Eze, C. G., Nwankwo, C. E., Dey, S., Sundaramurthy, S., & Okeke, E. S. (2024). Food chain microplastics contamination and impact on human health: A review. *Environmental Chemistry Letters, 22*(4), 1889–1927.

Gambino, I., Bagordo, F., Coluccia, B., Grassi, T., Filippis, G. D., Piscitelli, P., & Leo, F. D. (2020). PET-bottled water consumption in view of a circular economy: The case study of Salento (South Italy). *Sustainability, 12*(19), 7988.

Gambino, I., Bagordo, F., Grassi, T., Panico, A., & De Donno, A. (2022). Occurrence of microplastics in tap and bottled water: Current knowledge. *International Journal of Environmental Research and Public Health, 19*(9), 5283.

Gao, N., Yang, L., Lu, X., Duan, Z., Zhu, L., & Feng, J. (2022). A review of interactions of microplastics and typical pollutants from toxicokinetics and toxicodynamics perspective. *Journal of Hazardous Materials, 432*, 128736.

Giechaskiel, B., Grigoratos, T., Mathissen, M., Quik, J., Tromp, P., Gustafsson, M., & Dilara, P. (2024). Contribution of road vehicle tyre wear to microplastics and ambient air pollution. *Sustainability, 16*(2), 522.

Glassford, E., Dunn, K., Dunn, K., Hammond, D., & Tyrawski, J. (2020). *3D Printing with Filaments: Health and Safety Questions to Ask.* National Institute for Occupational Safety and Health (NIOSH), Cincinnati, OH.

Gurjar, U. R., Xavier, K. M., Shukla, S. P., Jaiswar, A. K., Deshmukhe, G., & Nayak, B. B. (2022). Microplastic pollution in coastal ecosystem off Mumbai coast, India. *Chemosphere, 288*, 132484.

Hajam, Y. A., Rani, R., Ganie, S. Y., Sheikh, T. A., Javaid, D., Qadri, S. S., & Reshi, M. S. (2022). Oxidative stress in human pathology and aging: Molecular mechanisms and perspectives. *Cells, 11*(3), 552.

Haque, F., & Fan, C. (2023). Fate of microplastics under the influence of climate change. *iScience, 26*(9), 107649.

He, T., Qu, Y., Yang, X., Liu, L., Xiong, F., Wang, D., & Sun, R. (2023). Research progress on the cellular toxicity caused by microplastics and nanoplastics. *Journal of Applied Toxicology, 43*(11), 1576–1593.

Horton, A. A., & Barnes, D. K. (2020). Microplastic pollution in a rapidly changing world: Implications for remote and vulnerable marine ecosystems. *Science of the Total Environment, 738*, 140349.

Jarosz, K., Borek-Dorosz, A., Drozdek, M., Rokicińska, A., Kiełbasa, A., Janus, R., & Michalik, M. (2024). Abiotic weathering of plastic: Experimental contributions towards understanding the formation of microplastics and other plastic related particulate pollutants. *Science of the Total Environment, 917*, 170533.

Kadac-Czapska, K., Ośko, J., Knez, E., & Grembecka, M. (2024). Microplastics and oxidative Stress—current problems and prospects. *Antioxidants, 13*(5), 579.

Kaur, G., & Dandautiya, R. (2024). *Microplastics in Synthetic Textile Industries: Origin, Occurrence, Routes and Their Fates on Environment and Human.* In *E3S Web of Conferences* (Vol. *509*, p. 01010). EDP Sciences, Les Ulis, France.

Kehinde, S. A., Fatokun, T. P., Olajide, A. T., Praveena, S. M., Sokan-Adeaga, A. A., Adekunle, A. P., & Papadakis, M. (2024). Impact of polyethylene microplastics exposure on kallikrein-3 levels, steroidal-thyroidal hormones, and antioxidant status in murine model: Protective potentials of naringin. *Scientific Reports, 14*(1), 23664.

Key, S. (2023). *Investigation of Factors Affecting the Breakdown of Plastics in the Environment* (Doctoral dissertation, University of Leicester).

Khan, S., Mumtaj, Z. A., Khan, A. R., Alkahtani, M. Q., Aleya, E., Louzon, M., & Aleya, L. (2024). Reviewing the role of microplastics as carriers for microorganisms in absorbing toxic trace elements. *Environmental Science and Pollution Research, 31*(34), 46806–46819.

Kosuth, M., Mason, S. A., & Wattenberg, E. V. (2018). Anthropogenic contamination of tap water, beer, and sea salt. *PLoS One, 13*(4), e0194970.

Kumar, N., Lamba, M., Pachar, A. K., Yadav, S., & Acharya, A. (2024). Microplastics–A growing concern as carcinogens in cancer etiology: Emphasis on biochemical and molecular mechanisms. *Cell Biochemistry and Biophysics, 5*(2), 1–13.

La Porta, E., Exacoustos, O., Lugani, F., Angeletti, A., Chiarenza, D. S., Bigatti, C., & Ghiggeri, G. M. (2023). Microplastics and kidneys: An update on the evidence for deposition of plastic microparticles in human organs, tissues and fluids and renal toxicity concern. *International Journal of Molecular Sciences, 24*(18), 14391.

Li, H., Wu, Q., Ng, J., Yu, D., Chan, S. H., & Li, A. (2022). Identification and quantification of common microplastics in table salts by a multi-technique-based analytical method. *Analytical and Bioanalytical Chemistry*, *414*(22), 6647–6656.

Lin, Q., Zhao, S., Pang, L., Sun, C., Chen, L., & Li, F. (2022). Potential risk of microplastics in processed foods: Preliminary risk assessment concerning polymer types, abundance, and human exposure of microplastics. *Ecotoxicology and Environmental Safety*, *247*, 114260.

Loganathan, Y., & Kizhakedathil, M. P. J. (2023). A review on microplastics-an indelible ubiquitous pollutant. *Biointerface Research in Applied Chemistry*, *13*(2), 126.

López, A. P. A., Trilleras, J., Arana, V. A., Garcia-Alzate, L. S., & Grande-Tovar, C. D. (2023). Atmospheric microplastics: Exposure, toxicity, and detrimental health effects. *RSC Advances*, *13*(11), 7468–7489.

Mas-Bargues, C., Alique, M., Barrús-Ortiz, M. T., Borrás, C., & Rodrigues-Díez, R. (2022). Special issue "Oxidative stress in aging and associated chronic diseases". *Antioxidants*, *11*(4), 701.

Mokhtarzadeh, Z., Keshavarzi, B., Moore, F., Busquets, R., Rezaei, M., Padoan, E., & Ajmone-Marsan, F. (2022). Microplastics in industrial and urban areas in South-West Iran. *International Journal of Environmental Science and Technology*, *19*(10), 10199–10210.

Murashov, V., Geraci, C. L., Schulte, P. A., & Howard, J. (2021). Nano-and microplastics in the workplace. *Journal of Occupational and Environmental Hygiene*, *18*(10–11), 489–494.

Noorimotlagh, Z., Hopke, P. K., & Mirzaee, S. A. (2024). A systematic review of airborne microplastics emissions as emerging contaminants in outdoor and indoor air environments. *Emerging Contaminants*, *10*(4), 100372.

Nwanaforo, E., Obasi, C. N., Frazzoli, C., Bede-Ojimadu, O., & Orisakwe, O. E. (2024). Exposure to environmental pollutants and risk of diarrhea: A systematic review. *Environmental Health Insights*, *18*, 11786302241304539.

O'Brien, S., Rauert, C., Ribeiro, F., Okoffo, E. D., Burrows, S. D., O'Brien, J. W., & Thomas, K. V. (2023). There's something in the air: A review of sources, prevalence and behaviour of microplastics in the atmosphere. *Science of the Total Environment*, *874*, 162193.

Olaimat, A. N., Ababneh, A. M., Al-Holy, M., Al-Nabulsi, A., Osaili, T., Abughoush, M., & Holley, R. A. (2024). A review of bacterial biofilm components and formation, detection methods, and their prevention and control on food contact surfaces. *Microbiology Research*, *15*(4), 1973–1992.

Oßmann, B. E. (2021). Microplastics in drinking water? Present state of knowledge and open questions. *Current Opinion in Food Science*, *41*, 44–51.

Panneerselvam, D., Murugesan, A., Raveendran, S. K., Kumar, J. S., & Venkataraman, P. (2024). Examining the hidden dangers: Understanding how microplastics affect pregnancy. *European Journal of Obstetrics & Gynecology and Reproductive Biology*, *304*(4), 53–62.

Prata, J. C. (2023). Microplastics and human health: Integrating pharmacokinetics. *Critical Reviews in Environmental Science and Technology*, *53*(16), 1489–1511.

Riaz, H. H., Lodhi, A. H., Munir, A., Zhao, M., Farooq, U., Qadri, M., & Islam, M. S. (2024). Breathing in danger: Mapping microplastic migration in the human respiratory system. *Physics of Fluids*, *36*(4), 043338.

Ricardo, I. A., Alberto, E. A., Júnior, A. H. S., Macuvele, D. L. P., Padoin, N., Soares, C., & Trovo, A. G. (2021). A critical review on microplastics, interaction with organic and inorganic pollutants, impacts and effectiveness of advanced oxidation processes applied for their removal from aqueous matrices. *Chemical Engineering Journal*, *424*, 130282.

Saha, S. C., & Saha, G. (2024). Effect of microplastics deposition on human lung airways: A review with computational benefits and challenges. *Heliyon*, *10*(2), e24355.

Sathyamohan, G., Sewwandi, M., Ambade, B., & Vithanage, M. (2023). Sources and circulation of microplastics in the aerosphere–atmospheric transport of microplastics. In Meththika, V., & Majeti, N. V. P. (eds), *Microplastics in the Ecosphere: Air, Water, Soil, and Food* (pp. 125–146). Wiley, New York.

Schmidt, A., da Silva Brito, W. A., Singer, D., Mühl, M., Berner, J., Saadati, F., & Bekeschus, S. (2023). Short-and long-term polystyrene nano-and microplastic exposure promotes oxidative stress and divergently affects skin cell architecture and Wnt/beta-catenin signaling. *Particle and Fibre Toxicology*, *20*(1), 3.

Segovia-Mendoza, M., Nava-Castro, K. E., Palacios-Arreola, M. I., Garay-Canales, C., & Morales-Montor, J. (2020). How microplastic components influence the immune system and impact on children health: Focus on cancer. *Birth Defects Research*, *112*(17), 1341–1361.

Shahsavaripour, M., Abbasi, S., Mirzaee, M., & Amiri, H. (2023). Human occupational exposure to microplastics: A cross-sectional study in a plastic products manufacturing plant. *Science of The Total Environment*, *882*, 163576.

Sharma, P. (2024). Microplastic contamination in food processing: Role of packaging materials. *Food Science and Engineering*, *5*(2), 271–287.

Singh, A., & Mishra, B. K. (2023). Microplastic in personal care products: An overlooked environmental concern. *Journal of Cleaner Production*, *472*, 139082.

Singh, A., Joseph, A., Madaiah, T. B., & Kunnel, S. G. (2024). Micro-and nanoplastics—an invisible threat to human health. In Inamuddin, A., Tariq, A.,& Fernandes, V. C. (eds), *Toxic Effects of Micro-and Nanoplastics: Environment, Food and Human Health* (pp. 61–71). Wiley, New York.

Sooriyakumar, P., Bolan, N., Kumar, M., Singh, L., Yu, Y., Li, Y., & Siddique, K. H. (2022). Biofilm formation and its implications on the properties and fate of microplastics in aquatic environments: A review. *Journal of Hazardous Materials Advances*, *6*, 100077.

Stefaniak, A. B., Bowers, L. N., Knepp, A. K., Luxton, T. P., Peloquin, D. M., Baumann, E. J., & Virji, M. A. (2019). Particle and vapor emissions from vat polymerization desktop-scale 3-dimensional printers. *Journal of Occupational and Environmental Hygiene*, *16*(8), 519–531.

Stefánsson, H., Peternell, M., Konrad-Schmolke, M., Hannesdóttir, H., Ásbjörnsson, E. J., & Sturkell, E. (2021). Microplastics in glaciers: First results from the Vatnajökull ice cap. *Sustainability*, *13*(8), 4183.

Sun, H., Hu, J., Wu, Y., Gong, H., Zhu, N., & Yuan, H. (2024). Leachate from municipal solid waste landfills: A neglected source of microplastics in the environment. *Journal of Hazardous Materials*, *465*, 133144.

Tamargo, A., Molinero, N., Reinosa, J. J., Alcolea-Rodriguez, V., Portela, R., Bañares, M. A., & Moreno-Arribas, M. V. (2022). PET microplastics affect human gut microbiota communities during simulated gastrointestinal digestion, first evidence of plausible polymer biodegradation during human digestion. *Scientific Reports*, *12*(1), 528.

Tang, K. H. D. (2023). Microplastics in agricultural soils in China: Sources, impacts and solutions. *Environmental Pollution*, *322*, 121235.

Tatsii, D., Bucci, S., Bhowmick, T., Guettler, J., Bakels, L., Bagheri, G., & Stohl, A. (2023). Shape matters: Long-range transport of microplastic fibers in the atmosphere. *Environmental Science & Technology*, *58*(1), 671–682.

Tomonaga, T., Higashi, H., Izumi, H., Nishida, C., Kawai, N., Sato, K., & Morimoto, Y. (2024). Investigation of pulmonary inflammatory responses following intratracheal instillation of and inhalation exposure to polypropylene microplastics. *Particle and Fibre Toxicology*, *21*(1), 29.

Tosun, J., Scherer, U., Schaub, S., & Horn, H. (2020). Making Europe go from bottles to the tap: Political and societal attempts to induce behavioral change. *Wiley Interdisciplinary Reviews: Water*, *7*(3), e1435.

Tu, C., Zhou, Q., Zhang, C., Liu, Y., & Luo, Y. (2020). Biofilms of microplastics. In Defu, H., & Yongming, L. (eds), *Microplastics in Terrestrial Environments: Emerging Contaminants and Major Challenges* (pp. 299–317). Springer, Cham.

Ullah, S., Ahmad, S., Guo, X., Ullah, S., Ullah, S., Nabi, G., & Wanghe, K. (2023). A review of the endocrine disrupting effects of micro and nano plastic and their associated chemicals in mammals. *Frontiers in Endocrinology*, *13*, 1084236.

Unuofin, J. O., & Igwaran, A. (2023). Microplastics in seafood: Implications for food security, safety, and human health. *Journal of Sea Research*, *194*, 102410.

Vattanasit, U., Kongpran, J., & Ikeda, A. (2023). Airborne microplastics: A narrative review of potential effects on the human respiratory system. *Science of The Total Environment*, *904*, 166745.

Wang, F., Zhang, Q., Cui, J., Bao, B., Deng, X., Liu, L., & Guo, M. Y. (2023). Polystyrene microplastics induce endoplasmic reticulum stress, apoptosis and inflammation by disrupting the gut microbiota in carp intestines. *Environmental Pollution*, *323*, 121233.

Wang, J., Guo, X., & Xue, J. (2021). Biofilm-developed microplastics as vectors of pollutants in aquatic environments. *Environmental Science & Technology*, *55*(19), 12780–12790.

Wei, H., Wang, J., Pan, S., Liu, J., Ding, H., Smith, K., & Gao, S. (2023). Are wastewater treatment plants as the source of microplastics in surface water and soil?. *Journal of Hazardous Materials*, *459*, 132154.

Weisser, J., Beer, I., Hufnagl, B., Hofmann, T., Lohninger, H., Ivleva, N. P., & Glas, K. (2021). From the well to the bottle: Identifying sources of microplastics in mineral water. *Water*, *13*(6), 841.

Wu, J. Y., Gao, J. M., Pei, Y. Z., Luo, K. Y., Yang, W. H., Wu, J. C., & Luo, Y. (2024). Microplastics in agricultural soils: A comprehensive perspective on occurrence, environmental behaviors and effects. *Chemical Engineering Journal*, *489*, 151328.

Xiao, S., Cui, Y., Brahney, J., Mahowald, N. M., & Li, Q. (2023). Long-distance atmospheric transport of microplastic fibres influenced by their shapes. *Nature Geoscience*, *16*(10), 863–870.

Yan, Z., Liu, Y., Zhang, T., Zhang, F., Ren, H., & Zhang, Y. (2021). Analysis of microplastics in human feces reveals a correlation between fecal microplastics and inflammatory bowel disease status. *Environmental Science & Technology*, *56*(1), 414–421.

Yee, M. S. L., Hii, L. W., Looi, C. K., Lim, W. M., Wong, S. F., Kok, Y. Y., & Leong, C. O. (2021). Impact of microplastics and nanoplastics on human health. *Nanomaterials*, *11*(2), 496.

Zhang, J., Du, J., Liu, D., Zhuo, J., Chu, L., Li, Y., & Dong, H. (2024). Polystyrene microplastics induce pulmonary fibrosis by promoting alveolar epithelial cell ferroptosis through cGAS/STING signaling. *Ecotoxicology and Environmental Safety*, *277*, 116357.

Zhao, M., Huang, L., Arulmani, S. R. B., Yan, J., Wu, L., Wu, T., & Xiao, T. (2022). Adsorption of different pollutants by using microplastic with different influencing factors and mechanisms in wastewater: A review. *Nanomaterials*, *12*(13), 2256.

Zhao, X., Zhou, Y., Liang, C., Song, J., Yu, S., Liao, G., & Wu, C. (2023). Airborne microplastics: Occurrence, sources, fate, risks and mitigation. *Science of The Total Environment*, *858*, 159943.

Zhu, L., Xie, C., Chen, L., Dai, X., Zhou, Y., Pan, H., & Tian, K. (2023). Transport of microplastics in the body and interaction with biological barriers, and controlling of microplastics pollution. *Ecotoxicology and Environmental Safety*, *255*, 114818.

Evaluating Compound Dry and Hot Events in Bhopal

Past Trends and Insights from Future Climate Projections

7

Chaitanya Raj, Vaishnavi Sahu,
Somil Swarnkar, and Vikas Poonia

INTRODUCTION

In recent decades, climate change has emerged as a critical global concern, fundamentally altering weather patterns and increasing the frequency and intensity of extreme weather events. Among these phenomena, Compound Dry and Hot Events (CDHEs)—characterized by simultaneous or sequential occurrences of extreme dryness and high temperatures—have gained attention due to their severe cumulative impacts on ecosystems and human activities (Zhang et al., 2024). Unlike isolated extreme events, CDHEs

DOI: 10.1201/9781003613183-7

amplify socioeconomic and environmental challenges by combining the stresses of low rainfall and high temperatures, significantly straining water resources and agricultural systems (Barriopedro et al., 2011; Zscheischler and Fischer, 2020; Swarnkar et al., 2021; He et al., 2024). The concurrent occurrence of droughts and heatwaves intensifies their individual effects, elevating risks such as food insecurity, reduced agricultural productivity, and heightened health hazards (Mukherjee and Mishra, 2021; Guntu and Agarwal, 2024; Maurya et al., 2024).

Heatwaves are defined by prolonged periods of abnormally high temperatures, while droughts involve extended low precipitation (Dalezios and Eslamian, 2017; Domeisen et al., 2023; Ghosal et al., 2024). Their interaction exacerbates impacts: insufficient rainfall depletes soil moisture, reducing evapotranspiration—a natural cooling mechanism—thus elevating temperatures (Vautard et al., 2007; Herold et al., 2016). Conversely, elevated temperatures accelerate soil and vegetation drying, worsening drought conditions (Feng et al., 2020).

Globally, research on CDHEs has primarily focused on regions such as North America, Europe, and Asia, where these events are increasingly frequent, intense, and prolonged due to shifting climate patterns. CDHEs threaten agriculture, water resources, public health, and socioeconomic stability. For instance, Europe and western Russia faced extreme summers in 2003, 2010, 2015, and 2018, with unprecedented environmental and economic consequences (Fink et al., 2004; Ionita et al., 2017; Liu et al., 2020). Similarly, China's southwestern and northern regions experienced severe CDHEs during the summers of 2006, 2009–2010, and 2014 (Wu et al., 2019; Ye et al., 2019). Other examples include Australia's record-breaking temperatures in 2013 (Cowan et al., 2014) and the USA's prolonged heat and dryness from 2012 to 2014, which severely impacted agriculture and water resources (Diaz & Wahl, 2015; Griffin & Anchukaitis, 2014; Williams et al., 2015). Advanced monitoring and forecasting systems are now being developed in regions like the USA and Australia to improve the prediction and mitigation of CDHEs (Tavakol et al., 2020; Wang et al., 2024; Ridder et al., 2022).

India's unique climatic conditions, characterized by monsoon variability and increasing temperatures, make it highly vulnerable to CDHEs. Despite this vulnerability, research on CDHEs in India remains limited, even though frequent droughts and heatwaves pose severe threats to the country's food and water security.

A stark example of the devastating effects of CDHEs is the 2015 heatwave in Andhra Pradesh and Telangana, where extreme temperatures coupled with prolonged dry conditions led to significant socioeconomic losses and increased mortality. A comprehensive analysis by Prabhakar et al. (2023) utilized the Standardized Compound Event Indicator (SCEI) and the Standardized Dry and Hot Index (SDHI) to evaluate CDHEs across India (Prabhakar et al., 2023). Similarly, Kumar et al. (2024) investigated the compound precipitation-temperature extremes across India. The study utilizes various statistical methods to analyze the occurrence of CDHE, revealing significant spatial patterns and dependencies in extreme weather events across different months.

The central Indian region, particularly Madhya Pradesh, is especially susceptible to CDHEs. While extensive studies have examined hydro-climatic extremes in the region, they primarily focus on individual events such as droughts (Sharma & Sen, 2021;

Tripathi et al., 2016) or heatwaves (Singh et al., 2022; Das et al., 2020), overlooking their combined impacts. Bhopal, the state capital, faces heightened vulnerability due to its dependence on rain-fed agriculture and its exposure to rising summer temperatures, which deplete water resources and reduce crop yields (Kumar et al., 2024; Surjibhai et al., 2024). These compounded challenges exacerbate food insecurity, economic instability, and migration, highlighting the urgent need for research on CDHEs in this region.

This chapter aims to comprehensively evaluate the increasing risk of CDHEs in the Bhopal district by integrating historical trends and future climate projections. Using monthly data from TerraClimate, we examine historical changes in the frequency and severity of CDHEs from 1958 to 2022. Additionally, we analyze scenarios of global mean temperature rises by +2°C and +4°C relative to pre-industrial levels to assess future threats. By bridging historical analysis with future projections, the study aims to achieve the following objectives:

1. Identify patterns and anomalies in the occurrence and severity of CDHEs in Bhopal from 1958 to 2022.
2. Evaluate the potential impacts of climate change on CDHEs under +2°C and +4°C warming scenarios.
3. Highlight cascading impacts on agriculture, water resources, public health, and regional livelihoods.

This integrated approach provides valuable insights for policymakers to develop adaptive measures tailored to central India's unique climatic and socioeconomic context.

STUDY AREA

Bhopal, the capital of Madhya Pradesh, is located at approximately 23.16°N latitude and 77.36°E longitude, covering an area of 285.9 km^2 at an average elevation of 500 m. The city's landscape is characterized by rolling hills, lakes, and urban forests, contributing to its ecological significance. Bhopal experiences a humid subtropical climate with mild, dry winters, hot summers, and a monsoon season that brings 80%–90% of its average annual rainfall of 1,200 mm, largely influenced by the southwest monsoon.

The city's hydrology is shaped by small rivers and tributaries, with the Betwa and Kolar Rivers playing a crucial role in water security and agriculture. Bhopal's dependence on monsoonal rainfall makes it highly sensitive to climate variability, particularly in the availability of surface water resources. Key water bodies like Upper Lake (Bhojtal) and Lower Lake serve as vital sources for drinking water and biodiversity but are vulnerable to prolonged dry spells and rising temperatures. Erratic rainfall patterns and increasing heatwaves intensify evaporation rates and threaten water availability. As urbanization expands, the city faces challenges in balancing development with climate resilience, making it an important area for studying rainfall variability, temperature trends, and the impacts of CDHE (Figure 7.1).

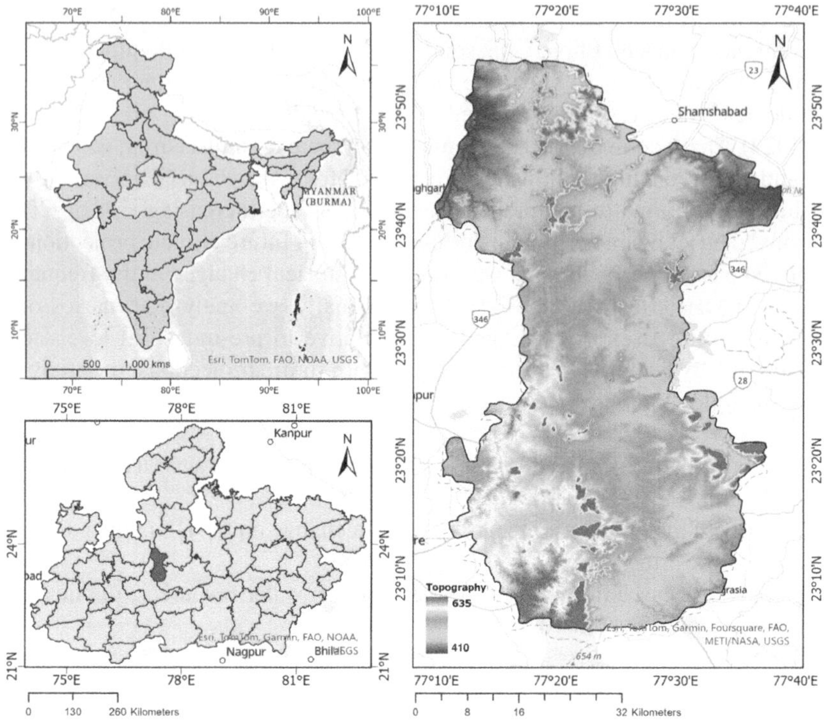

FIGURE 7.1 The geographical study area of Bhopal encompasses the district of Bhopal, located in the central Indian state of Madhya Pradesh.

INPUT DATASET

Monthly precipitation and temperature data were used to define CDHEs, following methodologies from Hao et al. (2019a). A 65-year dataset (1958–2022) for the Bhopal district was sourced from TerraClimate (Abatzoglou et al., 2018), which is a high-resolution (~4 km) gridded climate dataset designed for ecological and hydrological modeling. TerraClimate enables detailed climate analysis, making it ideal for assessing climate extremes. Future climate projections were incorporated to evaluate shifts in precipitation and temperature under 2°C and 4°C warming scenarios, ensuring consistency with historical data. TerraClimate data, accessed via Google Earth Engine, was selected for its accuracy and reliability in India. Studies such as Dubey et al. (2021) confirm its superiority over CHIRPS, TRMM, and ERA5 for capturing precipitation trends. This study leverages TerraClimate's strengths to analyze the impacts of CDHEs on agriculture and water resources in Bhopal, addressing critical climate resilience challenges.

METHODOLOGY

Standardized Indexes Computation

The Standardized Precipitation Index (SPI) was used to assess drought conditions by analyzing cumulative precipitation over various time scales, including 3-month intervals. This multi-scalar approach distinguishes short-term droughts affecting soil moisture from long-term droughts impacting groundwater and reservoirs (Wu et al., 2018; Aghelpour et al., 2021). The Standardized Temperature Index (STI) was applied to evaluate extreme heat, adapting the SPI framework for temperature data. Both indices standardize precipitation and temperature anomalies by fitting probability distributions to historical data and transforming them to a normal scale, allowing consistent comparisons over time and regions. The combined use of SPI and STI provides a robust methodology for analyzing CDHEs. This dual-index approach helps track extreme events' spatial and temporal evolution, supporting the development of adaptation strategies to mitigate climate change impacts on water resources and agriculture.

Assessment of the Severity and Frequency of CDHE

This study utilized two distinct indicators to evaluate the severity and frequency of CDHEs. The first indicator, known as the SCEI, is a multivariate, standardized measure introduced by Hao et al. (2019b) to quantify the severity of compound events. The SCEI is derived from the bivariate distribution function of precipitation (P) and temperature (T), integrating the SPI and the STI. Mathematically, SCEI is expressed as:

$$\text{SCEI} = \varnothing^{-1}(F(P(X \leq x, Y > y))) \tag{7.1}$$

Here, $\varnothing^{-1}$ denotes the inverse of the standard normal cumulative distribution function applied to the bivariate distribution of SPI (X) and STI (Y). The SCEI quantifies the severity of CDHEs, enabling direct comparisons with severity levels derived from individual indices such as SPI and STI (Muthuvel & Mahesha, 2021; Wu et al., 2020; Hao et al., 2019a, 2019b). SCEI offers a comprehensive framework for evaluating the combined impact of dry and hot conditions, providing a reliable measure of the intensity of CDHEs (Guntu & Agarwal, 2024). Lower SCEI values correspond to more severe CDHEs, as they reflect the simultaneous occurrence of reduced precipitation (low SPI) and elevated temperatures (high STI) (Hao et al., 2019a). This makes SCEI a critical tool for understanding and addressing the increasing risks associated with compound climate extremes.

The second indicator employed in this study is CDHE Frequency, which quantifies the total number of CDHEs occurring within a specific time period. This indicator is treated as a binary variable, representing the presence or absence of a CDHE on any given day. Mathematically, it is defined as:

$$F = \sum_{i=1}^{N} O_i \tag{7.2}$$

where N is the number of months, $O_i=1$ represents a CDHE occurrence, and $O_i=0$ denotes non-occurrence. We define $O_i=1$ for months meeting both low precipitation (P) and high temperature (T) thresholds (based on SPI and STI)

Here, N signifies the number of months, $O_i=1$ represents a CDHE occurrence, and $O_i=0$ denotes non-occurrence. We define $O_i=1$ for months meeting both low precipitation (P) and high temperature (T) thresholds (based on SPI and STI), as follows:

$$CDHE = \begin{cases} 1, & \text{if SPI} \leq \text{p and STI} > \text{t} \\ 0, & \text{otherwise} \end{cases} \tag{7.3}$$

The thresholds for defining CDHEs are determined based on percentile values. Precipitation levels identify dry periods at or below the P30 (moderate) or P10 (severe) thresholds, while hot periods correspond to temperature values above the T70 (moderate) or T90 (severe) thresholds. As the thresholds become more stringent (e.g., shifting from P30/T70 to P10/T90), the frequency of CDHE occurrences decreases due to the increasing rarity of extreme conditions. In this study, the P30/T70 thresholds were applied exclusively to classify events as 'High CDHEs,' without incorporating more extreme thresholds. It is important to note that the P30/T70 threshold encompasses the range of values defined by the more stringent P10/T90 threshold; therefore, the decision was made to proceed with the P30/T70 threshold to ensure comprehensive analysis. The level of severity—ranging from high to extreme—is determined by these percentile-based thresholds, enabling a nuanced categorization of CDHEs based on their intensity (Guntu and Agarwal, 2024). This framework provides valuable insights into the varying impacts of compound extremes under different climatic conditions (Figure 7.2).

Future Projections under 2°C and 4°C Warming

To analyze future climate projections under 2°C and 4°C global warming scenarios, precipitation and temperature data for the period 1985–2015 were utilized. The 2°C scenario represents a moderate warming trajectory, while the 4°C scenario reflects a high-end warming pathway associated with limited mitigation efforts. Both scenarios were applied to calculate the SPI and STI, using methodologies consistent with those applied to historical data. These indices were employed to evaluate changes in the frequency, intensity, and spatial distribution of CDHEs. Comparing the outcomes across the two warming scenarios allowed for the identification of potential climatic risks and

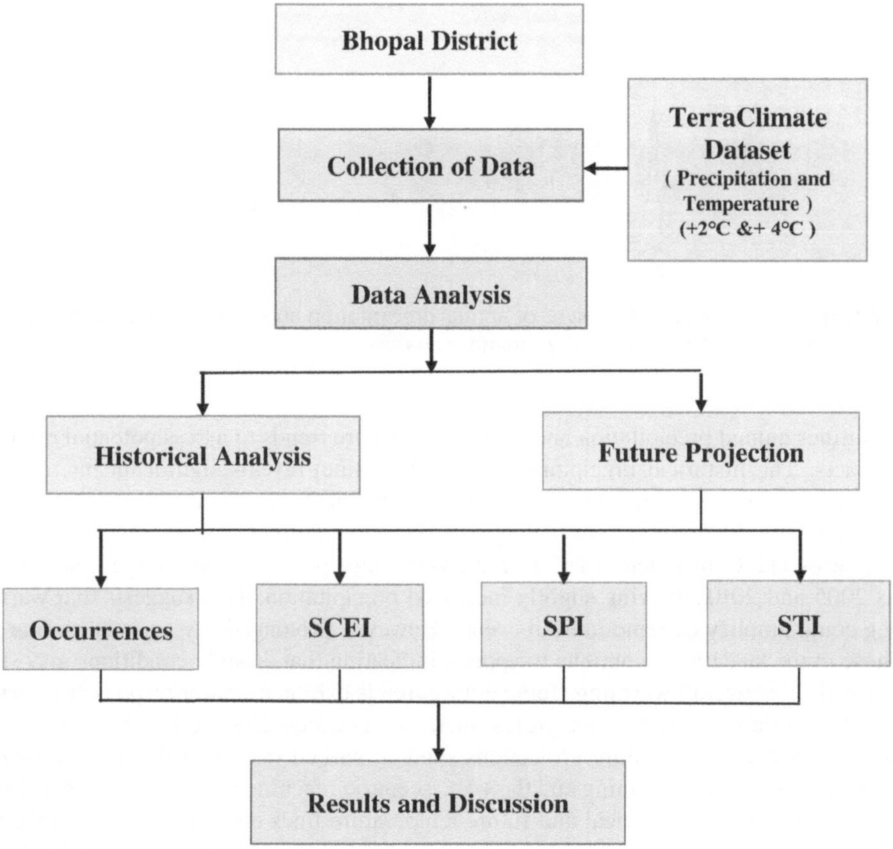

FIGURE 7.2 Workflow for analyzing compound dry-hot events (CDHE).

vulnerabilities. This analysis provides valuable insights into the projected evolution of CDHEs under varying levels of global warming, enhancing our understanding of the implications for ecosystems, water resources, and human systems.

RESULTS AND DISCUSSION

Historical and +2°C/+4°C Future Scenario Hydrometeorological Assessment

Figure 7.3a and b presents a comparative hydrometeorological analysis of the Bhopal district under historical conditions (1958–2022) and projected future climate scenarios (+2°C and +4°C warming, based on 1985–2015 pseudo-year data). The analysis

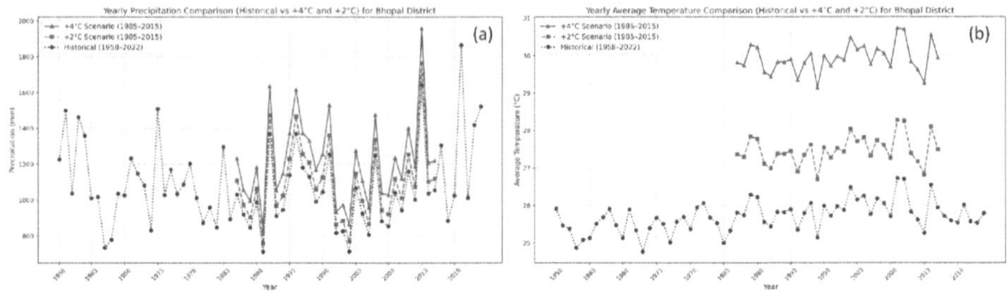

FIGURE 7.3 Comparative analysis of annual precipitation and average temperature trends under historical, +2°C, and +4°C warming scenarios.

examines annual precipitation and mean temperature trends to assess potential climate impacts. The historical precipitation trend (blue line) reveals significant interannual variability, ranging from 800 to 1800 mm. Notable fluctuations, such as in 1965, 1973, and the early 2000s, highlight the region's irregular rainfall patterns. Under warming scenarios (+2°C in green, +4°C in red), variability persists, with wetter years, such as 2005 and 2010, showing slightly increased precipitation. This suggests that warming could amplify extreme rainfall events. However, prolonged dry periods, similar to those in the late 1990s, continue to appear, indicating that drought conditions may still occur despite overall warming. Temperature trends exhibit a clearer response to warming. Historically, annual temperatures fluctuated between 25°C and 27.5°C, gradually increasing over time. Future projections show a marked rise, with the +2°C scenario displaying moderate warming and the +4°C scenario reaching nearly 31°C. The distinct separation between historical and future temperature lines underscores the significant impact of warming, increasing the risk of heat-related stress. While precipitation variability remains similar across all scenarios, intensified rainfall extremes may challenge water management and flood control. More pronounced temperature increases highlight the urgent need for adaptation strategies to mitigate heat stress and its cascading effects on agriculture, water resources, and public health in Bhopal.

SPI, STI, CDHE Frequency, and SCEI of the Historical Period in the Bhopal District

Figure 7.4a–d provides a detailed analysis of key hydrometeorological indicators for the Bhopal district, based on historical data (1958–2022). These panels illustrate (a) the SPI, (b) the STI, (c) the frequency of CDHEs, and (d) the SCEI. These metrics offer insights into the trends, intensity, and temporal variability of dry, hot, and compound extreme events. The SPI measures precipitation anomalies, with positive values indicating wetter-than-average conditions and negative values signaling dry periods. The data show highly seasonal variability, with extreme wet years (SPI > 3) in the early 1960s, 1980, and 2015, and prolonged drought periods (SPI < −1) in the 1970s and early 2000s.

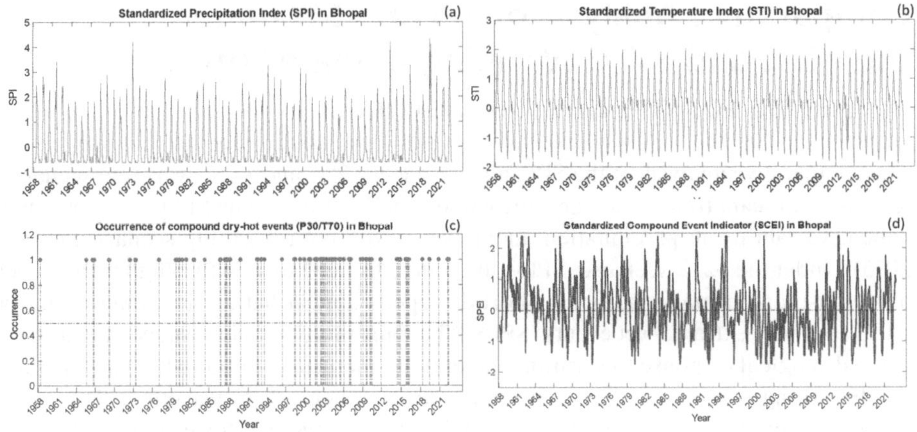

FIGURE 7.4 Time series of standardized climate indices and compound event occurrences in the Bhopal district from 1958 to 2022. (a) The Standardized Temperature Index (STI) indicates temperature anomalies over time, (b) the Standardized Precipitation Index (SPI) reflects precipitation deviations, (c) the occurrence of compound dry-hot events (CDHEs) displays the frequency of concurrent dry and hot days, and (d) the Standardized Compound Event Indicator (SCEI) assesses the combined severity of drought and heat events.

These variations highlight the cyclic nature of wet and dry spells, influencing water availability and hydrological stability. The STI captures temperature anomalies, where positive values represent hotter-than-normal conditions and negative values indicate cooler periods. The STI shows an increasing trend in extreme heat events, particularly in the early 2000s and late 2010s (STI>2). Earlier decades (1950s–1970s) exhibited lower temperature variability, reinforcing concerns over rising temperatures and more frequent heat extremes in recent decades. The frequency of CDHEs, defined using the P30/T70 threshold (precipitation below the 30th percentile and temperature above the 70th percentile), is depicted as red markers. A clear increase in CDHEs is observed from the 1990s onward, persisting through the 2000s and 2010s. These events were relatively rare before the 1970s, suggesting that their growing occurrence is likely linked to anthropogenic climate change. The SCEI integrates SPI and STI to assess compound drought-heat stress severity. Negative SCEI peaks (SCEI<–1) indicate periods of extreme climatic stress, where low precipitation and high temperatures coincide. Such events have become more frequent since the 1990s, with pronounced negative peaks clustering in the 1990s and 2000s. These episodes intensify vulnerability by depleting soil moisture, increasing evapotranspiration, and exacerbating agricultural and water resource challenges. The findings highlight a rising trend in the frequency and severity of compound dry-hot events in Bhopal, posing significant environmental and socioeconomic risks. The increasing prevalence of these extremes underscores their cascading impacts, necessitating integrated adaptation strategies to safeguard vulnerable sectors. Climate resilience measures must address the interconnected effects of drought and heat stress to mitigate future risks effectively.

Comparative Analysis of SPI and STI under the +2°C and +4°C Warming Climate Future Scenarios for the Bhopal District

The comparative analysis of the SPI and STI under +2°C and +4°C warming scenarios reveals significant trends and variability in Bhopal's climate. The SPI plots show inter-annual variability in precipitation anomalies across both warming scenarios (1985–2015). Under the +2°C scenario, SPI values range from −1 to 4, with a higher frequency of wetter-than-normal conditions. Short dry spells appear but are less intense. In the +4°C scenario, extreme wet events (SPI > 3) become more frequent, suggesting a stronger hydrological response to warming, potentially increasing flood risks. The STI plots exhibit strong seasonal cycles in both scenarios. Under +2°C warming, STI values fluctuate between −2 and 2, with occasional warm anomalies. However, the +4°C scenario shows a marked increase in the frequency and intensity of extreme heat events. Positive STI anomalies become more pronounced, indicating significantly heightened heat stress under higher warming levels. Compound dry and hot events, identified by simultaneous negative SPI and positive STI anomalies, occur in both scenarios but become more frequent and severe under +4°C warming. This trend underscores the growing risk of concurrent drought and heat stress, which can amplify vulnerabilities in agriculture, water resources, and public health. While precipitation variability remains consistent, the +4°C scenario intensifies extreme rainfall and heat events. The increasing frequency of compound extremes highlights the urgent need for targeted climate adaptation measures in Bhopal.

Figure 7.4 shows the time series of standardized climate indices and compound event occurrences in the Bhopal district from 1958 to 2022. (a) The STI indicates temperature anomalies over time, (b) the SPI reflects precipitation deviations, (c) the occurrence of CDHEs displays the frequency of concurrent dry and hot days, and (d) the SCEI assesses the combined severity of drought and heat events. These plots collectively illustrate trends in climate variability and the frequency of extreme events in the district (Figures 7.5 and 7.6).

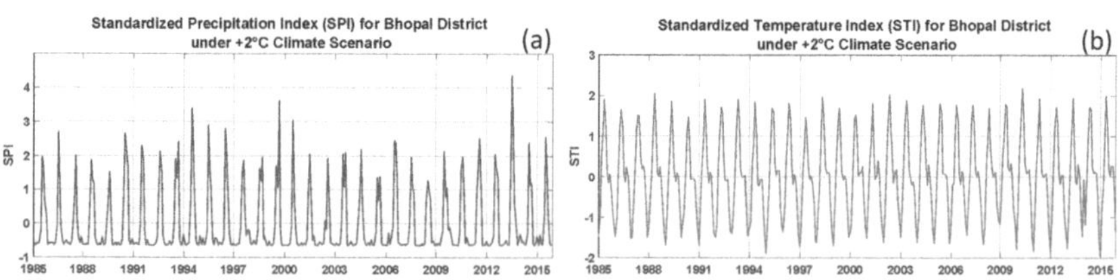

FIGURE 7.5 Time series plot of the Standardized Precipitation Index (SPI) and the Standardized Temperature Index (STI) from 1985 to 2015 for the Bhopal district under the +2°C scenario.

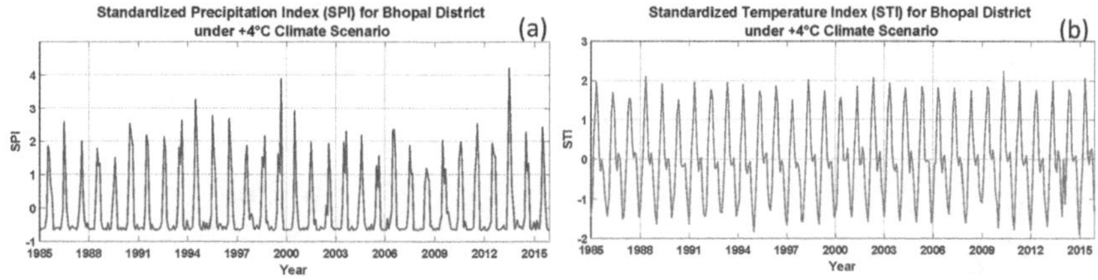

FIGURE 7.6 Time series plot of the Standardized Precipitation Index (SPI) and the Standardized Temperature Index (STI) from 1985 to 2015 for the Bhopal district under the +4°C scenario.

Comparative Analysis of CDHE Occurrences: Historical vs +2°C and +4°C Future Scenarios

The bar chart illustrates the average number of months per year affected by CDHEs under historical conditions, +2°C, and +4°C warming scenarios. The results highlight a significant increase in CDHE frequency with rising temperatures, underscoring future climatic risks for Bhopal. Historically, CDHEs occurred for an average of 1.6 months per year, indicating that such extreme events were relatively rare. Under a +2°C warming scenario, their frequency doubles to 3.29 months per year, suggesting that nearly one-third of the year could be affected by dry-hot conditions. This increase poses severe challenges for agriculture, water resources, and public health, reinforcing concerns about rising climate variability. Under the +4°C scenario, CDHEs average 2.68 months per year. While slightly lower than in the +2°C case, this remains significantly higher than historical levels. This apparent decrease may result from shifts in precipitation patterns, where occasional increased rainfall temporarily offsets dry conditions despite extreme warming. However, the overall trend confirms a heightened risk of compound climate extremes. The findings indicate that even a +2°C temperature rise—considering a critical threshold under the Paris Agreement—would substantially increase extreme events, with severe implications for Bhopal's rain-fed agriculture and heat-vulnerable population. The escalation of CDHEs emphasizes the urgent need for climate adaptation measures, improved water management strategies, and policies to mitigate the cascading impacts of these extreme conditions in the future (Figures 7.7 and 7.8).

Comparative Analysis of SCEI: Historical vs +2°C and +4°C Scenarios

The box plot presents the distribution of SCEI values under three climate scenarios: historical, +2°C warming, and +4°C warming. The SCEI integrates precipitation and temperature anomalies to quantify the severity of CDHEs, with lower values indicating more extreme conditions. This analysis highlights the increasing severity and frequency of CDHEs in Bhopal under future warming scenarios. During the historical

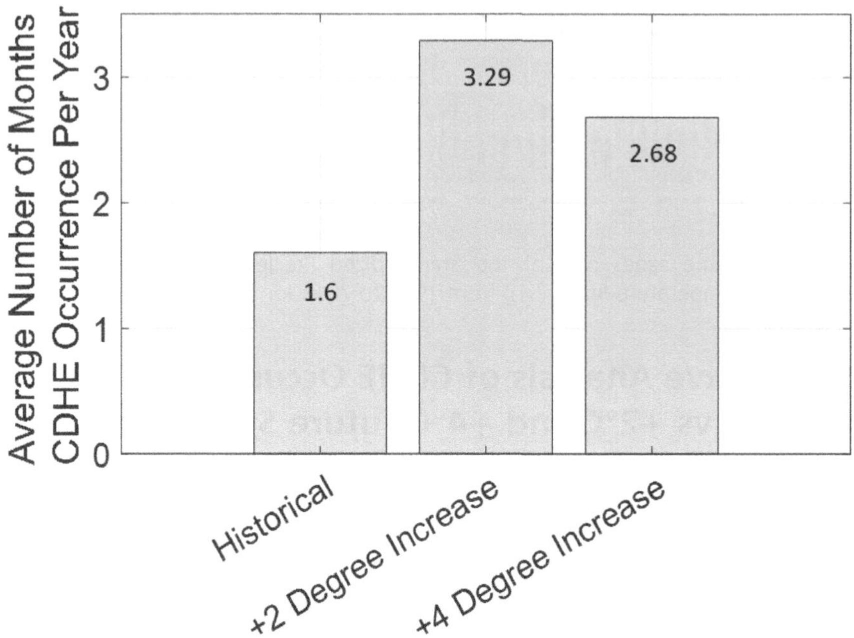

FIGURE 7.7 Bar chart comparing the average number of months per year with CDHE occurrences in Bhopal under historical conditions, +2°C, and +4°C climate scenarios.

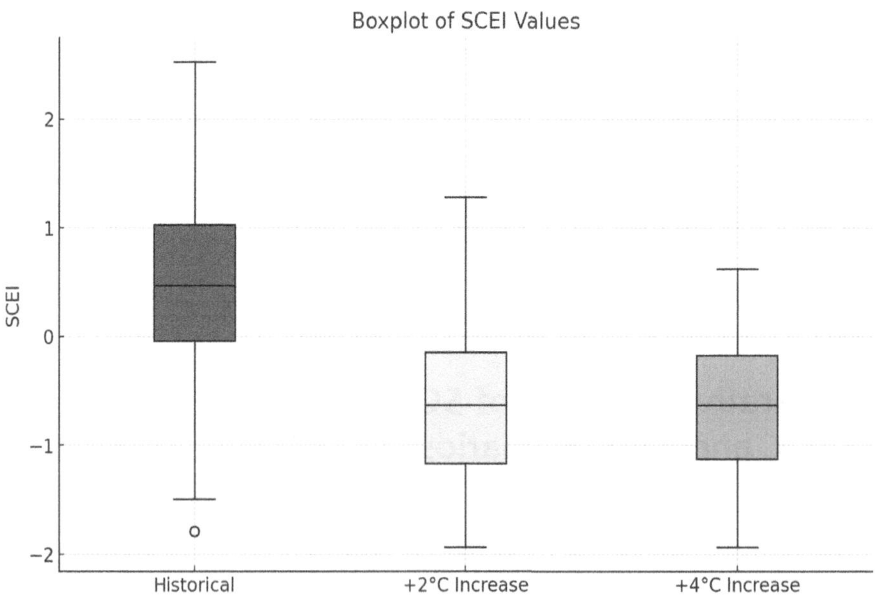

FIGURE 7.8 Boxplot of the Standardized Compound Event Index (SCEI) values under historical conditions, +2°C, and +4°C climate scenarios.

period, the SCEI distribution is broad, with a median value near zero, indicating a climate characterized by mild to moderate compound events interspersed with occasional extremes. Negative outliers (approaching –2) represent severe CDHEs, while positive outliers indicate wetter-than-average conditions. This variability reflects a balance between climate stress and favorable conditions, typical of pre-1990 climate patterns that were less influenced by anthropogenic warming. Under the +2°C warming scenario, the SCEI distribution shifts downward, with a median slightly below zero. This suggests that CDHEs become both more frequent and somewhat more severe. The interquartile range narrows, indicating fewer mild events and a greater concentration of moderate to severe extremes. Although extreme negative values persist, their frequency appears slightly reduced, suggesting a stabilization of the most severe CDHEs at this warming level. In the +4°C scenario, the SCEI median drops further into the negative range, signifying an even higher frequency and intensity of severe compound events. The distribution becomes narrower, with values clustering around more negative SCEI levels, indicating that extreme CDHEs dominate. The upper whisker shortens, showing that milder conditions become infrequent, while extreme positive values (indicative of favorable conditions) nearly disappear. While negative outliers remain, their magnitude appears to plateau, suggesting a possible upper limit to extreme CDHE severity even under heightened warming. Figure 7.9 visualizes extreme events' temporal patterns and intensity and SCEI values across historical and future climate scenarios. Panels (a), (b), and (c) illustrate the occurrence of extreme events, showing an increasing frequency and spatial-temporal spread under warming scenarios. This confirms that as global temperatures rise, the risk of extreme events escalates. Panels (d), (e), and (f) depict the SCEI, highlighting greater variability and intensity of extremes under the +4°C scenario, with certain months exhibiting stronger anomalies due to seasonal sensitivities and feedback mechanisms. These findings indicate that warming scenarios not only increase the frequency of extreme events but also amplify

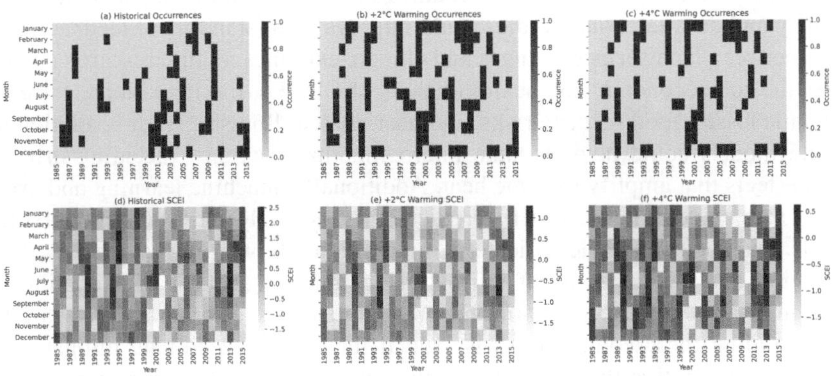

FIGURE 7.9 Heatmaps illustrating occurrences and Standardized Climate Event Indices (SCEI) under historical and warming scenarios: (a) Historical occurrences of extreme events, (b) occurrences under +2°C warming scenario, (c) occurrences under +4°C warming scenario, (d) SCEI for historical conditions, (e) SCEI under +2°C warming scenario, and (f) SCEI under +4°C warming scenario.

their intensity. The analysis of SCEI distributions reveals a clear trend of worsening CDHEs with rising global temperatures. The +2°C scenario already signals a substantial increase in compound event severity, posing serious risks to agriculture, water resources, and public health in Bhopal. The +4°C scenario exacerbates this trend, with severe compound events becoming a dominant feature of the local climate. The shrinking range of SCEI values suggests that Bhopal's climate is shifting toward a prolonged state of extreme stress, with fewer periods of relief. This trend underscores the urgent need for targeted climate adaptation strategies. Proactive measures such as improved water resource management, enhanced agricultural resilience, and heat mitigation strategies are crucial to coping with escalating CDHE impacts. Additionally, the findings highlight the necessity of limiting global temperature rise to below 2°C to mitigate severe socioeconomic and environmental consequences. Without intervention, the increasing frequency and intensity of compound extremes will significantly challenge climate resilience efforts in the Bhopal district.

FUTURE ANALYSIS

While this study projects future CDHEs for Bhopal, several research directions could enhance the understanding and applicability of the findings. Expanding the analysis to diverse regions—such as arid Rajasthan, coastal eastern India, or high-altitude Himalayan areas—could reveal how CDHEs vary with climate, topography, and land use. A comparative approach would improve insights into regional vulnerabilities and adaptation needs. Another key area is the role of land-atmosphere interactions in modulating CDHEs. Integrating land surface models and soil moisture dynamics could clarify how vegetation cover, urbanization, and irrigation impact extreme event intensity and duration. Such studies could improve understanding of feedback mechanisms between land surface changes and climate extremes. Future research could also assess the socioeconomic and health impacts of CDHEs, linking climate extremes to economic losses, food security risks, and heat-related illnesses. High-resolution urban climate models could help evaluate microclimatic variations, including urban heat island effects that amplify extreme heat. Additionally, machine learning and artificial intelligence could enhance CDHE predictability by analyzing satellite data, climate models, and socioeconomic indicators, improving early warning systems. A detailed attribution analysis using event attribution techniques and causal modeling could differentiate the influence of anthropogenic climate change from natural variability. Understanding these contributions is crucial for informing climate mitigation policies. By incorporating these extensions, the study can provide a more comprehensive and actionable perspective on CDHEs, supporting both scientific advancements and policy-driven climate adaptation strategies.

CONCLUSION

This study analyzed CDHEs in Bhopal under historical and future climate scenarios using key indices such as the SPI, STI, SCEI, and CDHE frequency. The findings indicate a substantial rise in both the frequency and severity of CDHEs as global temperatures increase. Analysis of the SCEI index reveals a significant rise in negative peaks (SCEI<−1) since the 1990s, reflecting growing vulnerability to simultaneous high temperatures and low precipitation. Future projections suggest that this trend will intensify, with CDHE duration increasing from 1.6 months per year historically to 3.29 months under +2°C warming and 2.68 months under +4°C. This underscores a future marked by prolonged extreme events. The SCEI box plots further highlight this trend, showing a downward shift in median values and a narrowing interquartile range under warming scenarios, suggesting more frequent and persistent compound extremes. Hydrometeorological assessments confirm worsening drought severity and temperature extremes, exacerbating risks to water resources, agriculture, and public health in Bhopal. These findings emphasize the urgent need for climate adaptation strategies aligned with Sustainable Development Goals (SDGs), particularly SDG 6 (Clean Water and Sanitation), SDG 11 (Sustainable Cities and Communities), and SDG 13 (Climate Action). Key actions include improving water conservation, promoting climate-resilient agriculture with drought-resistant crops, and developing early warning systems. Implementing these strategies will enhance resilience and safeguard vulnerable communities against escalating climate risks in Bhopal.

REFERENCES

Abatzoglou, J. T., Dobrowski, S. Z., Parks, S. A., & Hegewisch, K. C. (2018). TerraClimate, a high-resolution global dataset of monthly climate and climatic water balance from 1958–2015. *Scientific Data*, 5(1), 1–12.

Aghelpour, P., Kisi, O., & Varshavian, V. (2021). Multivariate drought forecasting in short-and long-term horizons using MSPI and data-driven approaches. *Journal of Hydrologic Engineering*, 26(4), 04021006.

Barriopedro, D., Fischer, E. M., Luterbacher, J., Trigo, R. M., & García-Herrera, R. (2011). The hot summer of 2010: Redrawing the temperature record map of Europe. *Science*, 332(6026), 220–224.

Cowan, T., Purich, A., Perkins, S., Pezza, A., Boschat, G., & Sadler, K. (2014). More frequent, longer, and hotter heat waves for Australia in the twenty-first century. *Journal of Climate*, 27(15), 5851–5871.

Dalezios, N. R., & Eslamian, S. (2017). Drought assessment and management for heat waves monitoring. In Eslamian, S., & Eslamian, F. A. (eds), *Handbook of Drought and Water Scarcity* (pp. 235–260). CRC Press, Boca Raton, FL.

Das, P.K., Podder, U., Das, R., Kamalakannan, C., Rao, G.S., Bandyopadhyay, S., & Raj, U. (2020). Quantification of heat wave occurrences over the Indian region using long-term (1979–2017) daily gridded (0.5×0.5) temperature data—A combined heat wave index approach. *Theoretical and Applied Climatology*, 142, 497–511.

Diaz, H. F., & Wahl, E. R. (2015). Recent California water year precipitation deficits: A 440-year perspective. *Journal of Climate*, 28(12), 4637–4652.

Domeisen, D. I., Eltahir, E. A., Fischer, E. M., Knutti, R., Perkins-Kirkpatrick, S. E., Schär, C., Seneviratne, S. I., Weisheimer, A., & Wernli, H. (2023). Prediction and projection of heat-waves. *Nature Reviews Earth & Environment*, 4(1), 36–50.

Dubey, S., Gupta, H., Goyal, M. K., & Joshi, N. (2021). Evaluation of precipitation datasets available on Google earth engine over India. *International Journal of Climatology*, 41(10), 4844–4863.

Feng, S., Wu, X., Hao, Z., Hao, Y., Zhang, X., & Hao, F. (2020). A database for characteristics and variations of global compound dry and hot events. *Weather and Climate Extremes*, 30, 100299.

Fink, A. H., Brücher, T., Krüger, A., Leckebusch, G. C., Pinto, J. G., & Ulbrich, U. (2004). The 2003 European summer heatwaves and drought-synoptic diagnosis and impacts. *Weather*, 59(8), 209–216.

Ghosal, M., Swarnkar, S., Poonia, V., Kundu, S., & Pandey, S. (2024). Hydrological dichotomy: Streamflow drought in dammed vs. undammed regions of the largest Indian Peninsular basin. *Journal of Arid Environments*, 225, 105273.

Griffin, D., & Anchukaitis, K. J. (2014). How unusual is the 2012–2014 California drought? *Geophysical Research Letters*, 41(24), 9017–9023.

Guntu, R. K., & Agarwal, A. (2024). Compound dry and hot extremes: A review and future research pathways for India. *Journal of Hydrology*, 635, 131199.

Hao, Z., Hao, F., Xia, Y., Singh, V. P., & Zhang, X. (2019a). A monitoring and prediction system for compound dry and hot events. *Environmental Research Letters*, 14(11), 114034.

Hao, Z., Hao, F., Singh, V. P., & Zhang, X. (2019b). Statistical prediction of the severity of compound dry-hot events based on El Niño-Southern Oscillation. *Journal of Hydrology*, 572, 243–250.

He, K., Chen, X., Zhou, J., Zhao, D., & Yu, X. (2024). Compound successive dry-hot and wet extremes in China with global warming and urbanization. *Journal of Hydrology*, 636, 131332.

Herold, N., Kala, J., & Alexander, L. V. (2016). The influence of soil moisture deficits on Australian heatwaves. *Environmental Research Letters*, 11(6), 064003.

Ionita, M., Tallaksen, L. M., Kingston, D. G., Stagge, J. H., Laaha, G., Van Lanen, H. A., Scholz, P., Chelcea, S. M., & Haslinger, K. (2017). The European 2015 drought from a climatological perspective. *Hydrology and Earth System Sciences*, 21(3), 1397–1419.

Kumar, A., Ray, T., & Mohanasundari, T. (2024). Exploring climatic dynamics in Madhya Pradesh, India utilizing long-term gridded data (1951–2021): An integrated statistical and GIS modules. In Yadav, A. K., Yadav, K., & Singh, V. P. (eds), *Integrated Management of Water Resources in India: A Computational Approach: Optimizing for Sustainability and Planning* (pp. 3–21). Springer Nature Switzerland, Cham.

Liu, X., He, B., Guo, L., Huang, L., & Chen, D. (2020). Similarities and differences in the mechanisms causing the European summer heatwaves in 2003, 2010, and 2018. *Earth's Future*, 8(4), e2019EF001386.

Maurya, A. K., Swarnkar, S., & Prakash, S. (2024). Hydrological impacts of altered monsoon rain spells in the Indian Ganga basin: A century-long perspective. *Environmental Research: Climate*, 3(1), 015010.

Mukherjee, S., & Mishra, A. K. (2021). Increase in compound drought and heatwaves in a warming world. *Geophysical Research Letters*, 48(1), e2020GL090617.

Muthuvel, D., & Mahesha, A. (2021). Spatiotemporal analysis of compound agrometeorological drought and hot events in India using a standardized index. *Journal of Hydrologic Engineering*, 26(7), 04021022.

Prabhakar, A., Mitra, S., & Varghese, F. C. (2023). Multi-index characterization of compound dry hot events in India. *International Journal of Climatology*, 43(13), 6243–6267.

Ridder, N. N., Pitman, A. J., & Ukkola, A. M. (2022). High impact compound events in Australia. *Weather and Climate Extremes*, 36, 100457.

Sharma, A., & Sen, S. (2021). Impact of drought on economy: A district level analysis of Madhya Pradesh, India. *Journal of Environmental Planning and Management*, 64(6), 1021–1043.

Singh, V. P., Mathew, J., & Verma, I. J. (2022). Inter-spatial heat vulnerability assessment of Summer-2018 over Madhya Pradesh using discomfort (wind and thermal) indices. *MAUSAM*, 73(1), 105–114.

Surjibhai, A. S., Nath, R., Singh, S., Swarnkar, S., & Patra, B. (2024). Unravelling surface water dynamics in semi-arid central Indian region for sustainable agricultural practices. *Environmental Monitoring and Assessment*, 196(9), 1–22.

Swarnkar, S., Prakash, S., Joshi, S. K., & Sinha, R. (2021). Spatio-temporal rainfall trends in the Ganga River basin over the last century: Understanding feedback and hydrological impacts. *Hydrological Sciences Journal*, 66(14), 2074–2088.

Tavakol, A., Rahmani, V., & Harrington Jr, J. (2020). Probability of compound climate extremes in a changing climate: A copula-based study of hot, dry, and windy events in the central United States. *Environmental Research Letters*, 15(10), 104058.

Tripathi, M. K., Srivastava, S. C., Singh, J., Yadava, H. S., & Tomar, S. S. (2016). Analysis of rainfall and drought occurrence in districts of Madhya Pradesh. *Environment & Ecology*, 34(1A), 372–376.

Vautard, R., Yiou, P., D'andrea, F., De Noblet, N., Viovy, N., Cassou, C., Polcher, J., Ciais, P., Kageyama, M., & Fan, Y. (2007). Summertime European heat and drought waves induced by wintertime Mediterranean rainfall deficit. *Geophysical Research Letters*, 34(7), 01.

Wang, J., DeFlorio, M. J., Gershunov, A., Guirguis, K., Delle Monache, L., & Ralph, F. M. (2024). Association of western US compound hydrometeorological extremes with Madden-Julian oscillation and ENSO interaction. *Communications Earth & Environment*, 5(1), 314.

Williams, A. P., Seager, R., Abatzoglou, J. T., Cook, B. I., Smerdon, J. E., & Cook, E. R. (2015). Contribution of anthropogenic warming to California drought during 2012–2014. *Geophysical Research Letters*, 42(16), 6819–6828.

Wu, J., Liu, Z., Yao, H., Chen, X., Chen, X., Zheng, Y., & He, Y. (2018). Impacts of reservoir operations on multi-scale correlations between hydrological drought and meteorological drought. *Journal of Hydrology*, 563, 726–736.

Wu, X., Hao, Z., Hao, F., Li, C., & Zhang, X. (2019). Spatial and temporal variations of compound droughts and hot extremes in China. *Atmosphere*, 10(2), 95.

Wu, X., Hao, Z., Zhang, X., Li, C., & Hao, F. (2020). Evaluation of severity changes of compound dry and hot events in China based on a multivariate multi-index approach. *Journal of Hydrology*, 583, 124580.

Ye, L., Shi, K., Xin, Z., Wang, C., & Zhang, C. (2019). Compound droughts and heat waves in China. *Sustainability*, 11(12), 3270.

Zhang, G., Zhang, S., Wang, H., Gan, T.Y., Fang, H., Su, X., Song, S., Feng, K., Jiang, T., Huang, J., & Xu, P. (2024). Biodiversity and wetting of climate alleviate vegetation vulnerability under compound drought-hot extremes. *Geophysical Research Letters*, 51(10), e2024GL108396.

Zheng, H., Chiew, F. H., Charles, S., & Podger, G. (2018). Future climate and runoff projections across South Asia from CMIP5 global climate models and hydrological modelling. *Journal of Hydrology: Regional Studies*, 18, 92–109.

Zscheischler, J., & Fischer, E. M. (2020). The record-breaking compound hot and dry 2018 growing season in Germany. *Weather and Climate Extremes*, 29, 100270.

A Multidimensional Tool to Reduce Greenhouse Gas Emissions at Urban Scale Against Climate Risks

8

Oznur Isinkaralar, Kaan Isinkaralar, and Hasan Umur Alsancak

GREENHOUSE GAS EMISSIONS AND IMPACTS

The rapid growth of urban areas has brought many environmental problems to the agenda. Environmental concerns were first formally presented in a conceptual context in the United Nations (UN) Brundtland Commission's 1987 report titled 'Our Common Future', emphasizing sustainability. The heat island effect resulting from urban activities has caused significant impacts on regional climates, where it is a primary concern. It is the temperature rise measured within cities compared with their rural surroundings due to anthropogenic emissions and the replacement of natural habitats with heat-absorbing artificial ones (Naserikia et al., 2022). The heat island effect can have massive environmental and socioeconomic impacts: increased energy consumption by urban citizens, increased greenhouse gas (GHG) emissions, and heat-related health

DOI: 10.1201/9781003613183-8

FIGURE 8.1 Benefits of reducing GHG emissions.

problems for citizens. Reducing GHG provides multidimensional environmental, economic, and social benefits. The main environmental benefits are climate change mitigation, improved air quality, and ecosystems protection. Among the economic benefits, reduced energy costs and enhanced business competitiveness are crucial. Improved public health and quality of life are the main social benefits in Figure 8.1.

These intensify with the increase in global temperatures under climate change. Hence, holistic urban planning approaches are necessary as part of mitigation measures against the urban heat island (UHI) effect (Dhakal, 2010; Srivanit & Jareemit, 2020). A significant approach to alleviating the UHI effect is intentionally incorporating greenspaces and vegetation in urban areas (Melaas et al., 2016; Rojas-Cortorreal et al., 2017). Indeed, studies have shown that strategically placing trees and other vegetative species can help lower surface and air temperatures, providing cooling benefits and enhancing thermal comfort for city citizens. Moreover, green roofs, reflective surfaces, and other environmentally aware design features can significantly alleviate the UHI phenomenon (Villefranque et al., 2022). In addition to isolated measures, a comprehensive, climate-responsive strategy in urban planning is essential for tackling the intricate relationships among urbanization, GHG emissions, and the UHI effect. This may include embedding climate risk assessments into land-use planning, emphasizing nature-based solutions, and fostering interdisciplinary engagement between urban planners, architects, and climate scientists. Herein, cities may reduce their GHG emissions and climatic impact by mitigating the UHI, which is the synergistic result of specific interventions and holistic urban planning methods, and create healthier and more resilient conditions for citizens.

In addition, urban morphology and wind patterns are drawing much attention nowadays in terms of their power to modulate the heat island effect of cities. Appropriate urban design, considering all those aspects enhancing natural ventilation, may reduce heat accumulation within the urban area and provide another approach to dealing with such a complex problem (Ku & Tsai, 2020). In other words, the UHI problem is so complex that it needs a comprehensive multi-discipline methodology to solve problems in urban design and planning. It is possible to manage the distance of the Earth from the Sun or the resistance of the landforms to heat. However, we can take precautions against climate change by controlling the proportions of the gases that make up the atmospheric layer that absorbs the heat coming into our world. These decisions are related to how we will approach the building blocks that make up urban land use. In the next section, land uses and systems where decisions are made in an urban plan. The main emphasis will be placed on measures that can be taken against the climate crisis.

LAND USES AND URBAN INFRASTRUCTURE AFFECTING URBAN GREENHOUSE GASES

As the global population grows and urbanization accelerates, understanding the relationship between land use and GHG emissions in urban areas has become increasingly important. Accounting for only 2% of the global surface area, urban areas contribute to about 75% of global carbon emissions and are a major driver of climate change (Feng et al., 2022).

The rapid expansion of urban centers, particularly in developing countries, has led to significant land use changes that can significantly impact urban GHG emissions. Urbanization often involves the conversion of carbon sink land, such as forests and wetlands, into carbon-emitting land uses like residential, commercial, and industrial areas in Figure 8.2 (Li et al., 2022). These land use changes not only increase direct emissions from energy consumption and transportation but also indirectly affect the carbon sequestration capacity of the urban ecosystem (Gao & O'Neill, 2020).

FIGURE 8.2 The role of urban uses in GHG control.

It is crucial to develop a comprehensive understanding of the complex interactions between urban land use and carbon metabolism to mitigate the impact of urbanization on GHG emissions. This study attempted to present a guiding document by defining the following land uses responsible for the city's GHG emissions. The following sections consist of headings on managing the urban uses in order. We can interpret the basic strategies for urban infrastructure as follows.

Urban Green Spaces and Urban Forests

Urban green spaces, encompassing urban forests, parks, roofs, and other green spaces, are crucial in climate change mitigation. These green elements offer many benefits, creating a more sustainable and resilient urban environment. According to carbon sequestration and storage, urban forests and green spaces act as carbon sinks, absorbing atmospheric CO_2 through photosynthesis. Trees and vegetation store this carbon in their biomass, effectively removing it from the carbon cycle and mitigating the greenhouse effect. The amount of carbon sequestered depends on factors like tree species, age, and management practices. The urban green infrastructure (UGI) can substantially minimize building energy consumption. Trees provide shade and help reduce the demand for air conditioning in hot weather. Green roofs and walls offer insulation, minimizing heat loss in winter and heat gain in summer. This reduced energy demand translates to lower GHG emissions from power plants. The urban environment often exhibits significantly higher temperatures than adjacent rural areas, owing to the prevalence of heat-absorbing surfaces like concrete and asphalt. This phenomenon, called the UHI effect, exacerbates heat waves and elevates energy consumption. Incorporating urban greenery and infrastructure, such as trees and parks, can help mitigate this effect by providing shade, facilitating evapotranspiration, and reducing the overall surface temperature within the city. Furthermore, urban forests and green spaces can filter air pollutants, removing particulate matter, nitrogen oxides, and other harmful substances from the atmosphere. This improvement in air quality translates to public health benefits and reduces the reliance on energy-intensive air purification technologies (Figure 8.3).

Green infrastructure, such as permeable pavements, green roofs, and rain gardens, helps to manage stormwater runoff by absorbing and filtering rainwater. This reduces the burden on urban drainage systems, minimizes flooding risks, and prevents water pollution. Indirectly, this can also reduce GHG emissions associated with wastewater treatment and flood management. Urban green spaces provide habitat for various species, promoting biodiversity within cities. These ecosystems offer valuable services, such as pollination, pest control, and nutrient cycling, contributing to a healthier and more resilient urban environment. By strategically integrating urban green infrastructure

FIGURE 8.3 The contribution of the UGI in mitigating GHG emissions.

and urban forests into urban planning and development, cities can effectively mitigate climate change impacts, improve environmental quality, and enhance the overall well-being of their residents.

Transportation Sector

The transportation sector significantly contributes to global GHG emissions, accounting for nearly one-quarter of the total. As the world grapples with the pressing issue of climate change, it is crucial to address the transportation sector's environmental impact and explore effective planning strategies to mitigate its carbon footprint. Road transport is the primary driver of these emissions, responsible for almost 70% of worldwide transport-related emissions. GHG emissions from road transportation are primarily driven by the combustion of fossil fuels, such as gasoline and diesel, used to power vehicles. By trapping heat in the atmosphere, these emissions, including carbon dioxide (CO_2), methane (CH_4), and nitrous oxide (N_2O), contribute to global warming and climate change. A multifaceted approach is required to tackle this challenge, involving implementing comprehensive planning strategies that focus on reducing emissions, promoting sustainable transportation alternatives, and transitioning to cleaner energy sources. One strategy that has gained significant traction is adopting alternative fuel vehicles, particularly electric models, including plug-in hybrids and hydrogen-powered vehicles (Lal & You, 2023).

Adopting electric vehicles presents significant opportunities for mitigating the transportation sector's environmental impact. As governments and automotive manufacturers collaborate to enhance the accessibility and affordability of electric vehicles, consumers are demonstrating a growing awareness of the advantages of this technology and are embracing it as a practical option for personal mobility (Gautam et al., 2022; Khiari & Olaverri-Monreal, 2023). Government incentives are critical for the transition to electric vehicles. Enhancing the electric vehicle charging network, encompassing fast-charging stations and widespread accessibility in urban and rural areas, is essential for increasing adoption.

Alongside the widespread implementation of electric vehicles, additional planning initiatives can help mitigate GHG emissions from the transportation sector. Behavioral modifications, such as promoting eco-driving techniques to enhance fuel efficiency and encouraging public transit, carpooling, or active modes of transportation like walking and cycling, can also significantly contribute to reducing emissions from the transportation domain (Sussman et al., 2020). Furthermore, implementing low-emission or zero-emission zones within urban centers can be crucial for fostering healthier and more sustainable cities. The appeal of urban centers like Copenhagen's bike lanes, where we see bicycle paths by reducing fossil fuel consumption, is indisputable. For example, people primarily own bicycles in traditional cycling countries like the Netherlands and Denmark. Broadway in New York encourages public use of designated areas for pedestrians and bicycle paths. These designated areas have significant potential to reduce emissions, particularly of particulate matter and nitrogen oxides, ultimately improving air quality and public health outcomes.

Production Areas

The GHG emissions linked to urban expansion and industrial operations have emerged as a growing concern in recent years as the effects of climate change have become more pronounced (Shi, 2016; Isinkaralar & Isinkaralar, 2023). As centers of economic and social activity, urban areas are major contributors to global GHG emissions, responsible for over one-third of the total emissions (Huang et al., 2020). An important facet of this matter is the carbon emissions associated with agricultural crop cultivation, which represents a substantial source of GHGs (Cheng et al., 2015).

In recent years, the issue of climate change has been extensively studied and discussed, with numerous factors contributing to the observed variations in global temperature, precipitation patterns, and extreme weather occurrences (Salinger, 2005; Nema et al., 2012). Among these factors, the role of urban areas, mainly industrial and commercial zones, has emerged as a prominent area of academic investigation. Existing research indicates that urban regions are responsible for a substantial fraction of global CO_2 emissions, with industrial and commercial operations primarily contributing to this phenomenon (Heidari & Pearce, 2016). The activities within urban areas, such as energy utilization, transportation networks, and alterations in land use, collectively contribute to the substantial carbon footprint of these metropolitan regions.

The influence of industrial and commercial zones on climate change is multifaceted. The growth of these areas, frequently involving the conversion of natural landscapes to artificial surfaces, can induce substantial alterations in the local microclimate, such as the UHI phenomenon (Chen & You, 2020). The concentration of energy-intensive industries and the high demand for transportation within urban areas can lead to increased GHG emissions, thereby intensifying the issue of climate change (Li et al., 2022). Researchers have highlighted the significance of adopting sustainable urban planning and design approaches to tackle the challenge of climate change in the context of cities (Woodcock et al., 2009; Säynäjoki et al., 2015; Hughes, 2017).

Residential Areas

Given the escalating challenges of climate change, the necessity for enacting effective strategies to mitigate GHG emissions has become increasingly paramount. One crucial approach to this challenge involves the thoughtful planning of residential areas, as the built environment significantly contributes to carbon footprints.

The construction and operations of buildings account for over one-third of global GHG emissions (Roh et al., 2018). The building industry possesses a significant opportunity to enact positive transformation by adopting sustainable construction methods and environmentally mindful residential area design. Researchers have investigated diverse strategies to enhance the energy efficiency of buildings and diminish their life-cycle environmental footprint (Mirabella et al., 2018). Research has shown that the residential density of an urban development can significantly influence its energy consumption and GHG emissions (Norman et al., 2006).

A range of strategies can be employed to address GHG emissions from residential settings:

Improve Building Energy Efficiency: This can involve better insulation, high-performance windows, energy-efficient appliances, and renewable energy sources like solar panels.

Optimize Site Planning and Building Design: Consider building orientation to maximize solar gain in winter and minimize it in summer. Promote natural ventilation and daylighting to reduce reliance on artificial lighting and HVAC systems.

Promote Sustainable Transportation: Encourage walking, cycling, and public transport by providing convenient access to these options. Incorporate electric vehicle charging infrastructure within residential areas.

Increase Residential Density: Higher-density developments tend to have lower per capita energy consumption due to shared walls and reduced land area per household.

Use Sustainable Building Materials: Choose materials with low embodied carbon, such as recycled or locally sourced materials. Minimize the use of concrete and steel, which have high carbon footprints.

Implement Green Infrastructure: Incorporate green roofs, urban forestry, and permeable pavements to reduce the UHI effect and absorb rainwater.

Encourage Sustainable Lifestyles: Educate residents about energy conservation practices and waste reduction strategies. Promote community gardens and composting programs.

Energy Management, Water Use, and Waste Management

The energy issue can be interpreted as a common need for all land uses described so far. From residential heating to transportation, production to lighting, we cannot imagine a city without energy. The energy issue is a fundamental topic of discussion that can harm nature, consume resources with its production, and cause gas emissions with its consumption. To reduce the GHG emissions associated with energy production and consumption, several strategies can be implemented as follows:

Switching to Renewable Energy Solutions: The transition from fossil fuels (coal, oil, and natural gas) to renewable energy sources such as solar, wind, hydro, geothermal, and biomass significantly reduces GHG emissions.

Improve Energy Efficiency: Implementing energy-efficient technologies in power generation, transmission, and end-use applications (e.g., appliances, lighting, industrial processes) reduces energy demand and associated emissions.

Develop and Deploy Carbon Capture and Storage Technologies: CCS technologies capture CO_2 emissions from power plants and industrial facilities and store them underground, preventing their release into the atmosphere. However, the scalability and long-term effectiveness of CCS are still under investigation.

Promote Energy Conservation: Encouraging behavioral changes to reduce energy consumption, such as using public transport, reducing heating and cooling needs, and adopting energy-saving practices, can reduce emissions.

Invest in Research and Development: Continued research and development of new energy technologies, including advanced energy storage systems, smart grids, and next-generation renewable energy technologies, is crucial for long-term emission reductions.

Implement Carbon Pricing Mechanisms: Putting a price on carbon emissions through carbon taxes or cap-and-trade systems can incentivize emission reductions by making high-emitting activities more expensive.

International Cooperation: Global cooperation and policy coordination are essential to address climate change effectively. International agreements and collaborative efforts can facilitate technology transfer, financial assistance, and shared emission reduction targets.

Water use, emissions, and planning are closely connected. How we manage water resources significantly impacts our carbon footprint and community sustainability. Wastewater irrigation can contribute to GHG emissions by releasing compounds like CH_4 and N_2O. Furthermore, poorly treated wastewater can lead to other environmental issues like heavy metal contamination (Singh, 2021). Optimizing pressure management and using efficient pumping systems in drinking water services is another key strategy to reduce emissions from water infrastructure (Zhang et al., 2021a).

Reducing GHG emissions is essential for climate change mitigation, and the water sector plays an important role. Increasing access to sanitation and wastewater treatment is critical for public health and emissions reductions, as wastewater systems are a notable source of CH_4, N_2O, and CO_2. Despite the importance of water and wastewater management for climate change, there are gaps in how these issues are addressed in global climate policy and financing (Dickin et al., 2020). Considering the water-climate nexus, careful planning is necessary to create more sustainable and resilient communities. All important steps are integrating GHG accounting into water infrastructure decision-making, investing in efficiency and renewable energy, and building robust adaptation strategies.

Waste management is also a pivotal component of environmental sustainability, as the disposal and treatment of waste can substantially contribute to GHG emissions. Sustainable waste management strategies that emphasize resource efficiency, recycling, and alternative treatment methods can play a vital role in mitigating the environmental impact of waste (Zorpas, 2020). Developing effective waste management strategies requires a holistic approach that considers the entire waste management lifecycle, from generation to disposal. Municipal authorities and policymakers need to engage with various stakeholders, including waste generators, processors, and relevant organizations, to ensure the implementation of sustainable waste management practices (Joseph, 2006).

Recent studies have highlighted the potential environmental benefits of implementing integrated solid waste management systems over traditional landfilling or incineration methods. These integrated systems can reduce GHG emissions from 33% to 154%, depending on the specific waste management approaches used (Zhang et al., 2021b). However, the success of these strategies depends on addressing key issues, such as the importance of targeted management for organic and recyclable waste, the environmental impact of waste collection and transportation, and the need for comprehensive assessments that consider economic and social factors in addition to environmental impacts (Zorpas, 2020). Developing effective waste management strategies that address these challenges can contribute to the transition toward a resource-efficient, green, and low-carbon economy (Ferronato & Torretta, 2019).

TOWARD SUSTAINABLE URBAN PLANNING AND DESIGN DECISIONS

Sustainable urban planning and design decisions encompass various sectors, all working toward creating more environmentally, socially, and economically resilient cities. According to energy, waste, and water management, transitioning to renewable energy sources (solar, wind, geothermal), improving energy efficiency in buildings and transportation, and implementing smart grids are key strategies. Decentralized energy systems and microgrids can enhance resilience and reduce transmission losses. Implementing water-sensitive urban design, including rainwater harvesting, graywater recycling, and permeable pavements, can conserve water resources and reduce stormwater runoff. Protecting and restoring natural water bodies enhances ecosystem health and provides recreational opportunities. Promoting waste reduction, reuse, and recycling, implementing composting programs, and adopting circular economy principles minimize waste generation and environmental impact. Waste-to-energy technologies can recover energy from waste, but careful consideration of environmental and health impacts is necessary. Prioritizing public transport, cycling, and walking through well-connected networks, promoting electric vehicles, and reducing reliance on private cars are crucial for sustainable mobility. Transit-oriented development and mixed-use zoning can minimize travel distances and encourage non-motorized transport. Integrating green spaces, urban forests, green roofs, and vertical gardens enhances biodiversity, improves air quality, mitigates the UHI effect, and provides esthetic and recreational benefits. Green infrastructure also provides an important role in stormwater management and carbon sequestration. By integrating these sustainable practices across different sectors, cities can build more livable, resilient, and environmentally responsible urban communities for present and future generations.

Cities today are influenced by many internal and external dynamics, where the relationships have become quite complex due to their networks. The spatial context of global dynamics and internal city interactions should be questioned based on analyses of their processes specific to each city. Understanding and interpreting these interconnected systems' horizontal and vertical relationships is impossible only with these limited dimensions. However, all our evidence regarding GHG emissions shows that some uses are responsible for the UHI effect due to poor management. Our study, which produces a comprehensive strategy set by approaching the components we know so far to produce decisions in a multidimensional manner, will guide decision-makers.

AS AN EPILOG

All these applications are now possible with new technologies. New technologies for climate action in smart cities are pretty advanced. Here are some key examples: The first could be AI-powered energy management. Smart grids: AI algorithms optimize energy distribution and can predict demand fluctuations. Building energy management can be provided. The high efficiency of new-generation solar panels can be used. Artificial intelligence can analyze large datasets to predict climate change mitigation like extreme weather events, heat waves, and sea level rise. It helps cities develop and evaluate climate adaptation plans and prioritize interventions to maximize resource allocation and resilience. All these advantages are vital to achieving climate goals and building a more sustainable future. However, the cost of public support issues is a matter of debate. Climate change is a pressing global issue that requires collective action to mitigate its effects. However, various barriers, such as the high cost of climate-friendly solutions and the lack of public support, have hindered the implementation of effective climate policies and initiatives. Despite significant theoretical background on the obstacles to energy saving and conservation, it seems that there is a need to overcome the apparent gap by ensuring policy frameworks and partnerships well beyond individual businesses (Meadowcroft, 2009). One of the primary challenges is the perceived high cost of transitioning to renewable energy sources and implementing energy-efficient technologies. The substantial financial burdens associated with adopting and implementing sustainable technologies can deter individuals, businesses, and governments from taking action. Another significant barrier is the lack of public support and engagement in climate change mitigation efforts.

Long-term sustainable development is significant for ecological security (Isinkaralar & Isinkaralar, 2025). It recognizes that climate change is a collective issue and cannot be solved without public agreement to make policy action easier and/or to make the necessary collective behavioral changes (Mendoza-Poudereux et al., 2021). The media's role in promoting or prohibiting these changes is of great importance. Furthermore, it should be recognized that scientific communication tends to be based on facts, whereas public communication combines facts and values. This represents a challenge for the encounter between science and society. For this reason, institutions have tried in recent decades to intensify public involvement initiatives. From consultations to deliberative processes, public participation initiatives have opened spaces for discussion and comparison and, in some cases, even the possibility of making decisions involving social partners. A multifaceted approach is necessary to overcome these barriers. Policymakers should introduce policies and incentives that enhance the financial feasibility of sustainable solutions, such as tax credits, subsidies, and investment programs.

REFERENCES

Chen, Ruoning, and You, Xue-Yi. Reduction of urban heat island and associated greenhouse gas emissions. *Mitigation and Adaptation Strategies for Global Change* 25,4: 689–711, 2020.

Cheng, Kai, Yan, M., Nayak, D., Pan, G. X., Smith, P., Zheng, J. F. and Zheng, J. W. Carbon footprint of crop production in China: An analysis of National Statistics data. *The Journal of Agricultural Science* 153,3: 422–431, 2015.

Dhakal, Shobhakar. GHG emissions from urbanization and opportunities for urban carbon mitigation. *Current Opinion in Environmental Sustainability* 2,4: 277–283, 2010.

Dickin, Sarah, Bayoumi, Moustafa, Garriga, Ricard, Andersson, Kim, and Jiménez, Alejandro. Sustainable sanitation and gaps in global climate policy and financing. *NPJ Clean Water* 3,1: 24, 2020.

Feng, Xinhui, et al. Carbon metabolism in urban production–living–ecological space based on ecological network analysis. *Land* 11,9: 1445, 2022.

Ferronato, Navarro, and Torretta, Vincenzo. Waste mismanagement in developing countries: A review of global issues. *International Journal of Environmental Research and Public Health* 16,6: 1060, 2019.

Gao, Jing, and O'Neill, Brian C. Mapping global urban land for the 21st century with data-driven simulations and shared socioeconomic pathways. *Nature Communications* 11,1: 2302, 2020.

Gautam, Abhinav K., et al. Hybrid sources powered electric vehicle configuration and integrated optimal power management strategy. *IEEE Access* 10: 121684–121711, 2022.

Heidari, Negin, and Pearce, Joshua M. A review of greenhouse gas emission liabilities as the value of renewable energy for mitigating lawsuits for climate change related damages. *Renewable and Sustainable Energy Reviews* 55: 899–908, 2016.

Huang, Bin, et al. Ecological–economic assessment of renewable energy deployment in sustainable built environment. *Renewable Energy* 161: 1328–1340, 2020.

Huang, Huanchun, et al. Spatial evolution of the effects of urban heat island on residents' health. *Tehnički Vjesnik* 27,5: 1427–1435, 2020.

Hughes, Sara. *Reducing Urban Greenhouse Gas Emissions: Effective Steering Strategies for City Governments*. Institute on Municipal Finance and Governance, Toronto, 2017.

Isinkaralar, Oznur, Isinkaralar, Kaan, Sevik, Hakan, and Küçük, Ömer. Thermal comfort modeling, aspects of land use in urban planning and spatial exposition under future climate parameters. *International Journal of Environmental Science and Technology* 22,1: 1–14. 2025.

Isinkaralar, Oznur, and Isinkaralar, Kaan. Sustainable urban growth boundaries for ecological protection via 2046 models of İzmir. *Spatial Information Research* 33,1: 1–13, 2025.

Isinkaralar, Oznur, and Isinkaralar, Kaan. Projection of bioclimatic patterns via CMIP6 in the Southeast Region of Türkiye: A guidance for adaptation strategies for climate policy. *Environmental Monitoring and Assessment* 195,12: 1448, 2023.

Joseph, Kurian. Stakeholder participation for sustainable waste management. *Habitat International* 30,4: 863–871, 2006.

Khiari, Jihed, and Olaverri-Monreal, Cristina. Uncertainty-aware vehicle energy efficiency prediction using an ensemble of neural networks. *IEEE Intelligent Transportation Systems Magazine* 15,5: 109–119, 2023.

Ku, Chia-An, and Tsai, Hung-Kai. Evaluating the influence of urban morphology on urban wind environment based on computational fluid dynamics simulation. *ISPRS International Journal of Geo-Information* 9,6: 399, 2020.

Lal, Apoorv, and You, Fengqi. Will reshoring manufacturing of advanced electric vehicle battery support renewable energy transition and climate targets?. *Science Advances* 9,24: eadg6740, 2023.

Li, Lingwei, et al. A low-carbon land use management framework based on urban carbon metabolism: A case of a typical coal resource-based city in China. *Sustainability* 14,21: 13854, 2022.

Meadowcroft, James. Climate Change Governance. World Bank Policy Research Working Paper, WPS 4941, 2009.

Melaas, Eli K., et al. Interactions between urban vegetation and surface urban heat islands: A case study in the Boston metropolitan region. *Environmental Research Letters* 11,5: 054020, 2016.

Mendoza-Poudereux, Isabel, Dzimińska, Malgorzata, Pellegrini, Giuseppe, and Rowland, Jussara. Climate change and public perception. Citizens' proposals for better communication and involvement. *Journal of Science Communication* 20,03: A09, 2021.

Mirabella, Nadia, et al. Strategies to improve the energy performance of buildings: A review of their life cycle impact. *Buildings* 8,8: 105, 2018.

Naserikia, Marzie, et al. Background climate modulates the impact of land cover on urban surface temperature. *Scientific Reports* 12,1: 15433, 2022.

Nema, Pragya, Nema, Sameer, and Roy, Priyanka. An overview of global climate changing in current scenario and mitigation action. *Renewable and Sustainable Energy Reviews* 16,4: 2329–2336, 2012.

Norman, Jonathan, MacLean, Heather L., and Kennedy, Christopher A. Comparing high and low residential density: Lifecycle analysis of energy use and greenhouse gas emissions. *Journal of Urban Planning and Development* 132,1: 10–21, 2006.

Roh, Seungjun, Tae, Sungho, and Kim, Rakhyun. Developing a Green Building Index (GBI) certification system to effectively reduce carbon emissions in South Korea's building industry. *Sustainability* 10,6: 1872, 2018.

Rojas-Cortorreal, Gilkauris, et al. Climate and urban morphology in the City of Barcelona: The role of vegetation. *Mediterranean Identities-Environment, Society, Culture* 1: 426, 2017.

Salinger, M. James. Climate variability and change: Past, present and future–an overview. *Climatic Change* 70,1: 9–29, 2005.

Säynäjoki, Eeva-Sofia, Heinonen, Jukka, and Junnila, Seppo. Role of urban planning in encouraging more sustainable lifestyles. *Journal of Urban Planning and Development* 141,1: 04014011, 2015.

Shi, Yubing. Climate change and international shipping: The regulatory framework for the reduction of greenhouse gas emissions. *Brill* 18: 498, 2016.

Singh, Ajay. A review of wastewater irrigation: Environmental implications. *Resources Conservation and Recycling* 168: 105454, 2021.

Srivanit, Manat, and Jareemit, Daranee. Modeling the influences of layouts of residential townhouses and tree-planting patterns on outdoor thermal comfort in Bangkok suburb. *Journal of Building Engineering* 30: 101262, 2020.

Sussman, Reuven, Tan, Li Qin, and Kormos, Christine E. Behavioral interventions for sustainable transportation: An overview of programs and guide for practitioners. *Transport and Energy Research* 2020: 315–371, 2020.

Villefranque, Najda, et al. The teapot in a city: A paradigm shift in urban climate modeling. *Science Advances* 8,27: eabp8934, 2022

Woodcock, James, et al. Public health benefits of strategies to reduce greenhouse-gas emissions: Urban land transport. *The Lancet* 374,9705: 1930–1943, 2009.

Zhang, Junting, Qin, Quande, Li, Guangming, and Tseng, Chao-Heng. Sustainable municipal waste management strategies through life cycle assessment method: A review. *Journal of Environmental Management* 287: 112238, 2021a.

Zhang, Qian, Smith, Kate, Zhao, Xu, Jin, Xinkai, Wang, Shuya, Shen, Junjie, and Ren, Zhiyong. Jason. Greenhouse gas emissions associated with urban water infrastructure: What we have learnt from China's practice. *Wiley Interdisciplinary Reviews: Water* 8,4: e1529, 2021b.

Zorpas, Antonis A. Strategy development in the framework of waste management. *Science of the Total Environment* 716: 137088, 2020.

Nanotechnology in Wastewater Treatment for Quality and Safety

9

Amol Kahandal, Aagam Bumb,
Aruna Sivaram, and Chandrakant K. Tagad

INTRODUCTION

Water quality and safety are critical for maintaining human health, environmental balance, and economic development. Clean water is not only vital for drinking but also indispensable for the welfare of agriculture, industry, and the ecosystem. However, the pressures of a growing population, rapid industrialization, and climate change have intensified challenges caused by the contamination of water (Dasgupta et al., 2017; Theron et al., 2008). Contaminants such as heavy metals, harmful pathogens, and persistent organic pollutants threaten safe water supplies, particularly in low-income and developing regions with limited access to modern water treatment systems. This issue links directly to Sustainable Development Goal (SDG) 6, which emphasizes clean water and sanitation for all (du Plessis, 2022; Qu et al., 2013). Approximately 2.2 billion people lack access to safe drinking water free from faecal and chemical contamination. Water pollution is a major environmental health risk, accounting for 1.4 million premature deaths annually (Ciccarelli et al., 2024). The cumulative and synergistic effects

 DOI: 10.1201/9781003613183-9

of prolonged chemical exposure, particularly from complex mixtures, remain poorly understood, posing significant yet underappreciated health concerns. Lead exposure causes 1.8 million premature deaths annually, highlighting the critical need for comprehensive risk assessment and mitigation efforts to protect global public health (Fuller et al., 2022; Naidu et al., 2021). This underscores the urgent need for improved water quality and safety management.

The reliability of traditional water treatment methods like chlorination and sedimentation has posed a major challenge in addressing the complexity of modern pollutants. Micropollutants, antibiotic-resistant bacteria (Naidu et al., 2021), and persistent organic chemicals often bypass these systems, leaving harmful residues (Madhura et al., 2019). The increasing variety and complexity of water pollutants stemming from industrial discharge, agricultural runoff, and urbanization further complicate water treatment. Contaminants like heavy metals, pharmaceuticals, and organic compounds are difficult to remove using conventional methods and leave residues that pose significant health risks through bioaccumulation. Besides the chemical contaminants, the rise of antibiotic-resistant bacteria and treatment-resistant viruses has also made disinfection increasingly problematic, raising concerns about waterborne illnesses. In addition to technological limitations, the scalability and accessibility of modern water treatment methods remain significant obstacles, especially in resource-constrained regions. Many rural areas rely on outdated infrastructure, which cannot meet current water safety standards. Moreover, energy-intensive processes like advanced oxidation and reverse osmosis are not economically viable for widespread implementation. Climate change exacerbates these issues by intensifying droughts, flooding, and water scarcity, further stressing already limited resources. This underscores the need for solutions that are not only effective but also affordable and sustainable (Wang et al., 2024).

The challenges associated with water treatment necessitate innovative treatment technologies that can effectively safeguard both human health and ecological systems (Liu et al., 2024; Wang et al., 2024).

With its innovative approaches, nanotechnologies are emerging as a key tool for enhancing water treatment processes and ensuring safe, clean water in the 21st century (du Plessis, 2022). Nanotechnology offers transformative solutions to these challenges by leveraging the unique properties of nanoparticles (NPs), such as high surface area, tunable reactivity, and enhanced selectivity. These attributes enable the efficient removal of heavy metals, organic pollutants, and microbial contaminants that traditional methods struggle to address. Metal oxide NPs like zinc oxide (ZnO) and titanium dioxide (TiO_2) exhibit photocatalytic activity that breaks down persistent pollutants into less harmful substances under light exposure (Mustapha et al., 2020). Similarly, materials such as graphene oxide (GO) and carbon nanotubes (CNTs) excel in adsorbing heavy metals and dyes, making them highly effective for decontamination. Furthermore, nanotechnology facilitates the precise design of membranes at the nanoscale, improving filtration efficiency while reducing fouling and enhancing permeability (Bhol et al., n.d.; Geldasa et al., 2023).

Water quality, being challenging at a large scale, needs innovative solutions to detect pollutants at low concentrations. Beyond contamination removal, nanotechnology is pivotal in water quality monitoring and desalination. Nanosensors provide real-time detection of trace contaminants and pathogens, enabling proactive water safety measures. Nanosensors are of different types depending upon their sensing mechanisms like electrochemical, optical, colorimetric, piezoelectric, etc. This leverages the potential of nanoparticles for their applications in water quality management (Vikesland, 2018). These sensors have advantages like rapid response time, portability, ergonomics, ease of use, and selectivity (Arellano Vidal & Govan, 2024; Zhou et al., 2024). The real-time monitoring integrated with machine learning assures quality management at the individual level, enabling a step towards a safer ecosystem (Figure 9.1).

Nanomaterials integrated into desalination systems, such as nanocomposite membranes, offer energy-efficient methods for converting saline water into freshwater, potentially addressing water scarcity in arid regions. Importantly, these nanotechnology-based solutions are scalable and adaptable, making them suitable for urbanized and resource-limited settings (Madhura et al., 2019; Saha et al., 2013; Theron et al., 2008).

By bridging technological and socioeconomic gaps, nanotechnology represents a transformative pathway towards sustainable and universal access to clean water. Its ability to address emerging pollutants, enhance filtration systems, and provide real-time

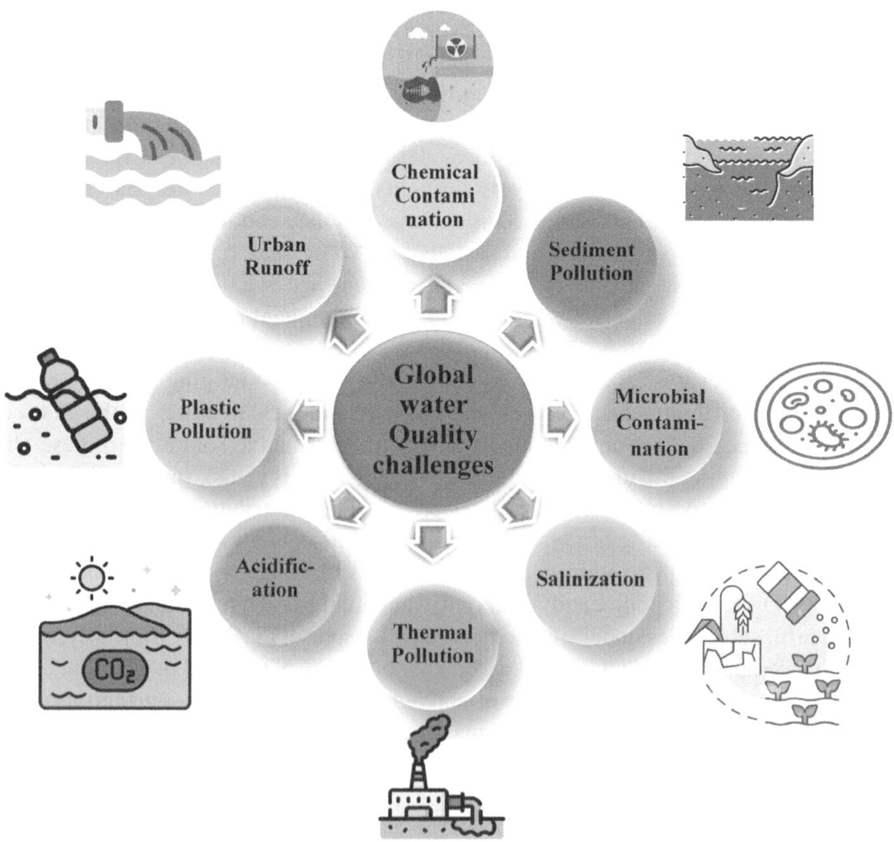

FIGURE 9.1 Global water quality challenges.

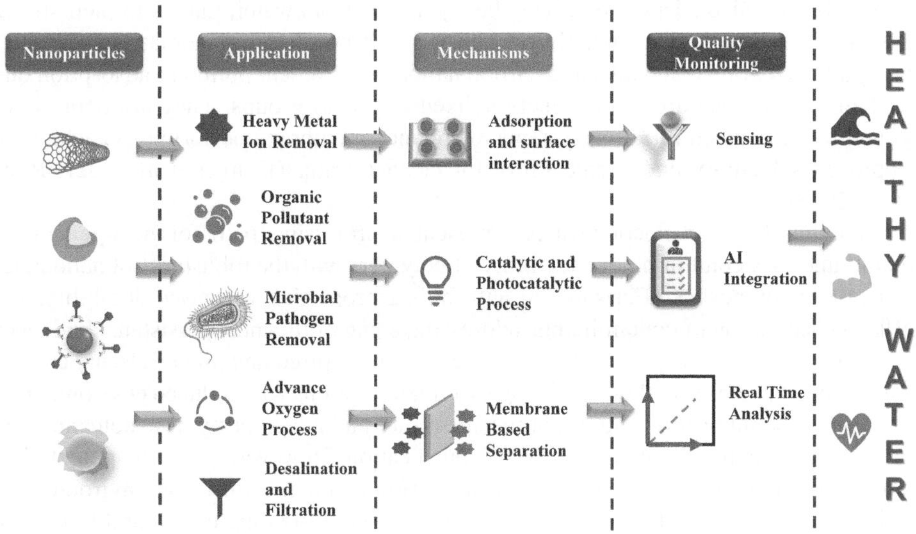

FIGURE 9.2 Overview of the chapter.

monitoring offers hope for overcoming the limitations of traditional methods. As we face the pressing global challenge of water scarcity and contamination, nanotechnology is a key player in ensuring a healthier and more sustainable future.

This chapter dwells on the role of nanotechnology in water quality and safety management by focusing on different types of nanomaterials used in water treatment, then their application in the removal of contaminants, organic pollutants, pathogens, and their mechanism of action, quality monitoring using sensors for heavy metal ion detection, real-time monitoring using machine learning, and then discussing its challenges, opportunities, and future scope (Figure 9.2).

NANOMATERIALS FOR WATER TREATMENT

Overview of Nanomaterials

Nanomaterials, with their structural dimensions ranging between 1 and 100 nm, bring transformative potential to water treatment. Their unique physicochemical properties, such as an exceptionally high surface area-to-volume ratio, tunable reactivity, and enhanced catalytic efficiency, enable innovative solutions for tackling complex water contamination challenges. Among the most effective materials, metal and metal oxide NPs like TiO_2 and Fe_2O_3 stand out for their photocatalytic prowess (Al-Mamun et al., 2019; Aragaw et al., 2021). When exposed to light, they generate reactive oxygen species (ROS) capable of breaking down harmful organic pollutants into safer compounds.

These NPs are also adept at removing heavy metals from water, thanks to their strong adsorption and reduction capabilities. Similarly, carbon-based nanomaterials, including graphene, graphene oxide, and carbon nanotubes, excel at pollutant adsorption due to their vast surface areas and functionalized chemical groups. They are being integrated into advanced filtration systems to enhance membrane performance, achieving improved selectivity and permeability for cleaner water (Danish et al., 2021; Raza et al., 2023).

Polymer-based nanocomposites represent a promising path for next-generation membranes. By combining the flexibility of polymers with the robustness of nanomaterials, these composites offer superior anti-fouling properties, increased durability, and efficient separation of contaminants, addressing some of the most persistent challenges in filtration. Eco-conscious alternatives, such as bio-inspired nanomaterials, are crafted using green synthesis methods, offering sustainable options with reduced environmental impact. These innovations highlight the versatility and adaptability of nanotechnology in developing holistic solutions for water purification. However, while the potential of nanomaterials is immense, important considerations remain. Ensuring environmental safety, achieving cost-effective scalability, and maintaining long-term stability require further research. Balancing these challenges with their benefits is essential for translating nanotechnology's promise into widespread, real-world applications that are both effective and sustainable (Kolya & Kang, 2023; Sahu et al., 2023).

Types of Nanomaterials Used in Water Treatment

Metal and Metal Oxide NPs

Metal and metal oxide NPs have revolutionized water treatment due to their unique properties, offering innovative solutions for various contaminants. For example, ZnO and TiO_2 are remarkable for their ability to harness ultraviolet or visible light to generate ROS. These ROS break down complex organic pollutants, such as dyes and pesticides, into harmless molecules, making them invaluable for sustainable wastewater treatment (Mustapha et al., 2020). Additionally, the inherent antibacterial properties of ZnO provide another layer of effectiveness, allowing it to target biological contaminants as well, doubling its utility (Tan et al., 2015). Iron oxide NPs, particularly magnetite (Fe_3O_4), stand out for their exceptional ability to adsorb heavy metals like lead, chromium, and arsenic. Their magnetic properties make them easy to recover and recycle, minimizing environmental impact while maintaining efficiency (Aragaw et al., 2021; Recillas et al., 2011). Silver NPs (AgNPs) disrupt microbial cell membranes and hinder critical enzymatic and DNA functions, effectively neutralizing a broad spectrum of pathogens, including bacteria, viruses, and fungi. This versatility makes NPs highly adaptable for applications ranging from industrial wastewater treatment to ensuring safe drinking water (Durán et al., 2007; Palani et al., 2023).

Additional contributions come from other metal oxides. Manganese oxide (MnO_2) is effective in removing toxic ions, such as arsenic and manganese, through redox reactions, while aluminium oxide (Al_2O_3) efficiently adsorbs fluoride and other anionic contaminants (Bhutto et al., 2023; Fouda et al., 2021; Saied et al., 2021). Cerium oxide (CeO_2), known for its high oxygen storage capacity, breaks down persistent organic pollutants and industrial effluents, even under harsh conditions (Barton et al., 2015; Ramirez Arenas et al., 2023). Copper oxide (CuO) rounds out this impressive roster, combining strong antimicrobial properties with catalytic capabilities to address organic and microbial contaminants effectively. However, these advancements come with challenges. The potential environmental risks posed by nanoparticle leaching, their toxicity to aquatic ecosystems, and the high costs associated with large-scale synthesis pose significant hurdles. Addressing these issues through green synthesis methods, enhanced recovery strategies, and rigorous environmental assessments is critical to unlocking their full potential in sustainable water treatment technologies. By tackling these challenges, we can ensure that these groundbreaking materials effectively serve environmental and societal needs (Eid et al., 2023; Sibhatu et al., 2022).

Carbon-Based Nanomaterials

GO is characterized by an abundance of oxygen-containing functional groups, such as hydroxyl, epoxy, and carboxyl groups, which are dispersed across its basal planes and edges. These functional groups enhance their interaction with pollutants through electrostatic attraction, hydrogen bonding, and π–π stacking interactions. The hydrophilic nature imparted by these groups, coupled with their role as active sites for binding, makes GO highly effective in adsorbing a variety of contaminants, including heavy metals, dyes, and hydrocarbons (Fu et al., 2020; Han et al., 2021). The large specific surface area of GO further contributes to its superior adsorption capacity, allowing it to capture significant quantities of pollutants while maintaining strong molecular interactions. For organic pollutants, π–π interactions between their aromatic structures and the conjugated π-electron system of GO play a key role in efficient adsorption. Inorganic contaminants, such as metal ions and oxygen functional groups, form strong coordination bonds, enhancing their overall performance in water purification (Fu et al., 2020; Liu et al., 2022a).

CNTs, with their unique cylindrical nanostructures and high aspect ratios, demonstrate remarkable potential for wastewater treatment. Single-walled CNTs and multi-walled CNTs are highly effective in adsorbing heavy metals, organic compounds, and biological contaminants. Functionalization of CNTs, achieved by introducing specific chemical groups, enhances their dispersion in aqueous systems and improves their interaction with various pollutants (Aslam et al., 2021). Their intrinsic antibacterial properties enable CNTs to disrupt microbial cell membranes, making them particularly suitable for disinfection. Moreover, both GO and CNTs have been integrated into electrochemical and photocatalytic devices, where their combined properties accelerate the degradation of pollutants into non-toxic byproducts, further boosting their versatility and efficiency (Aslam et al., 2021; Han et al., 2021; Yin et al., 2020).

Despite these advantages, challenges persist. The high production costs of these nanomaterials, environmental risks associated with nanoparticle leaching, and difficulties in large-scale deployment hinder their widespread application. Addressing these concerns through sustainable synthesis methods, comprehensive lifecycle assessments, and developing hybrid material systems can unlock their full potential. By overcoming these obstacles, carbon-based nanomaterials can be effectively scaled up to provide robust, eco-friendly solutions for wastewater treatment (Han et al., 2021).

Polymer-Based Nanocomposites

Polymer-based nanocomposites are gaining prominence as innovative materials for water treatment, offering a powerful combination of polymer flexibility and nanomaterial functionality to tackle a wide range of contaminants. The polymer matrix, which may include materials such as polyvinyl alcohol, polyethylene, or natural polymers like chitosan, serves as a durable scaffold that ensures mechanical strength, chemical resistance, and processability. NPs embedded within these matrices, such as TiO_2, AgNPs, and Fe_3O_4, introduce unique properties like photocatalysis, antimicrobial activity, and enhanced adsorption, making these composites highly versatile for water purification (Adeola & Nomngongo, 2022; Ray et al., 2020; Spoială et al., 2021).

In adsorption processes, polymer-based nanocomposites efficiently remove heavy metals and organic pollutants due to the high reactivity and surface area of embedded NPs. For instance, composites containing TiO_2 not only adsorb pollutants but also catalyse their breakdown under UV light, degrading toxic dyes, pesticides, and pharmaceutical residues into harmless byproducts. Similarly, chitosan-based nanocomposites with AgNPs are effective against biological contaminants, offering broad-spectrum antimicrobial activity to disinfect water and ensure safety (Khodakarami & Bagheri, 2021; Mecha et al., 2023).

Membrane technologies also benefit significantly from polymer nanocomposites, which enhance the performance of the filtration system. By integrating nanomaterials like silver or copper oxide into polymeric membranes, these composites improve anti-fouling and antimicrobial properties, extending their lifespan and effectiveness. Such membranes are utilized in ultrafiltration and reverse osmosis, effectively removing dissolved salts, pathogens, and micropollutants, ensuring high water quality. Magnetic NPs like Fe_3O_4 in polymer composites allow for the selective removal and recovery of heavy metals through magnetic separation, offering a sustainable approach to managing industrial effluents (Adeola & Nomngongo, 2022; Fang et al., 2021; Mecha et al., 2023).

Despite their advantages, challenges remain in scaling up the production of polymer-based nanocomposites and addressing potential environmental risks, such as nanoparticle leaching. Uniform dispersion of NPs within the polymer matrix is another critical area requiring optimization to ensure consistent performance. Advancing sustainable synthesis techniques, hybrid designs, and rigorous lifecycle assessments will be essential for overcoming these hurdles. These developments promise to make polymer-based nanocomposites a cornerstone of future water treatment technologies, balancing efficiency, scalability, and environmental safety (Adeola and Nomngongo, 2022).

Bio-inspired Nanomaterials

Bio-inspired nanomaterials are transforming water treatment technologies by combining sustainability with high efficiency in pollutant removal. Biological systems, mimicking structures, and processes such as plant surfaces, shells, and enzymatic activities inspire these materials. Their synthesis leverages green chemistry principles, often using plant extracts, microbes (bacteria, fungi, algae), or other biogenic sources. This approach minimizes the use of toxic chemicals and reduces energy demands, aligning with environmental sustainability goals (Kumar et al., 2020).

A notable application is the removal of heavy metals and organic pollutants through adsorption. Bio-derived silica and hydroxyapatite, modelled after natural structures like bones and shells, enhance adsorption due to their high surface area and functional groups. These nanomaterials bind effectively to toxic contaminants like arsenic, lead, and mercury, offering reusability and minimizing secondary waste. Bio-inspired membranes, modelled on natural protein channels like aquaporins, provide superior water permeability and selective rejection of salts, pathogens, and micropollutants. This enhances energy efficiency while ensuring high purification standards (Kumar et al., 2020; Liu et al., 2022a).

In catalytic applications, bio-inspired nanomaterials excel in advanced oxidation processes (AOPs). For instance, peroxidase-inspired nanostructures catalysed the degradation of persistent organic pollutants, such as dyes and pharmaceuticals, into non-toxic byproducts. Magnetic NPs, such as Fe_3O_4 synthesized via plant extract methods, facilitate the adsorption of heavy metals and their recovery through magnetic separation, thereby reducing secondary environmental impacts (Nguyen et al., 2021; Tripathy et al., 2024).

Despite their promise, challenges such as scalability, long-term stability, and cost-effective production remain. Addressing these issues through hybrid material designs and advancements in synthesis techniques can enhance their performance and practicality. As research progresses, bio-inspired nanomaterials hold great potential to revolutionize water treatment, delivering efficient and environmentally conscious solutions to pressing water quality challenges (Figure 9.3).

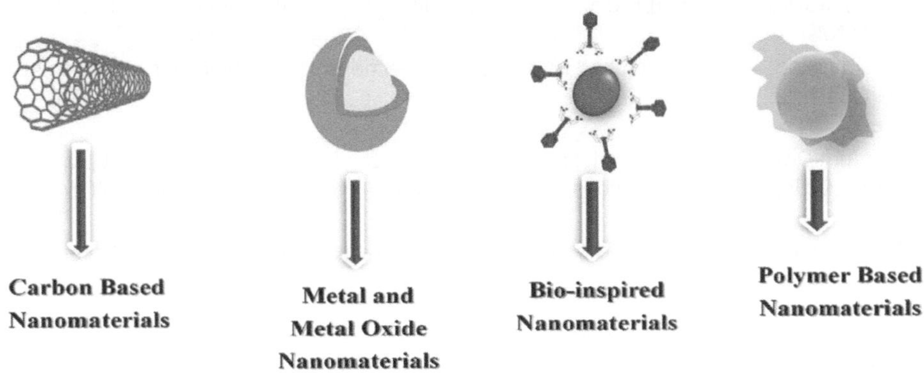

FIGURE 9.3 Types of nanomaterials used in water treatment.

APPLICATIONS OF NANOTECHNOLOGY IN WATER TREATMENT

Nanotechnology is critical for tackling global water concerns since it allows for the effective removal of impurities (Table 9.1).

TABLE 9.1 Application of nanotechnology in water treatment

NANOMATERIAL	*CHEMICAL SUBSTANCE SEPARATED*	*SEPARATION MECHANISM*	*APPLICATION AREA*	*REFERENCES*
Carbon Nanotubes (CNTs)	Heavy metals (Pb^{2+}, Cd^{2+}, Hg^{2+})	Adsorption and Filtration	Water purification	Khan et al. (2021)
Graphene Oxide (GO)	Organic dyes (MB, RhB)	π–π interactions and Adsorption	Wastewater treatment	Ucar et al. (2017)
Titanium Dioxide (TiO_2)	Organic pollutants (phenols)	Photocatalysis	Environmental Remediation	Barka et al. (2013)
Silver Nanoparticles (AgNPs)	Bacteria and pathogens	Antimicrobial interaction	Water disinfection	Bhardwaj et al. (2021)
Mesoporous Silica (MCM-41, SBA-15)	Pharmaceuticals (ibuprofen, antibiotics)	Adsorption and Entrapment	Drug removal from water	Chircov et al. (2020)
Magnetic Nanoparticles (Fe_3O_4)	Arsenic (As^{3+}, As^{5+})	Magnetic separation	Drinking water safety	Nikić et al. (2019)
Zinc Oxide (ZnO) NPs	Heavy metals and dyes	Adsorption and Photocatalysis	Water and wastewater treatment	Al-Arjan (2022)
Copper Oxide Nanoparticles (CuO NPs)	Pesticides and herbicides	Adsorption and Catalysis	Agricultural wastewater treatment	Alhalili (2023)
Cerium Oxide Nanoparticles (CeO_2)	Oxygen and NOx gases	Redox reactions	Air purification and catalysis	Kr Basu et al. (2024)
Quantum Dots (CdSe, CdTe, ZnS)	Heavy metal ions (Hg^{2+}, Pb^{2+})	Fluorescence detection	Sensing and water quality analysis	Gupta et al. (2024)
Metal-Organic Frameworks (MOFs)	Gases (CO_2, CH_4)	Adsorption and Selective trapping	Gas separation and purification	Yan et al. (2024a)
Alumina Nanoparticles (Al_2O_3 NPs)	Fluoride and phosphates	Adsorption	Drinking water safety	Sv et al. (2021)

Removal of Contaminants

Heavy Metals

Nanotechnology offers innovative and highly effective solutions for removing hazardous heavy metals such as lead (Pb), mercury (Hg), and arsenic (As) from water. Nanomaterials like GO, CNTs, and metal-organic frameworks (MOFs) are widely recognized for their exceptional adsorption capabilities owing to their large surface areas, versatile functional groups, and unique interaction mechanisms (Adel et al., 2022; Iqbal et al., 2023). These advanced materials bind metal ions through electrostatic attraction, ion exchange, or covalent bonding, ensuring remarkable removal efficiency (Adel et al., 2022). For instance, oxygen-containing functional groups of GO create strong binding sites for heavy metals, while CNTs provide a high surface area for pollutant interaction.

Magnetic NPs, such as Fe_3O_4, are particularly noteworthy due to their dual functionality; they efficiently adsorb arsenic ions and can be easily separated from water using external magnets, minimizing secondary waste (Nguyen et al., 2021). Functionalized like nanoscale zero-valent iron (nZVI) goes beyond adsorption by facilitating redox reactions, converting toxic metal ions into less harmful forms, such as reducing hexavalent chromium (Cr^{6+}) to its safer trivalent state (Cr^{3+}) (Liu et al., 2021; Zou et al., 2016). These nanomaterials demonstrate exceptional removal efficiencies and offer reusability and scalability, making them powerful tools for mitigating heavy metal contamination in water. Their versatility and sustainability position nanotechnology as a global cornerstone for advancing water purification technologies.

Organic Pollutants

Organic contaminants, such as pesticides, dyes, and pharmaceutics, pose a significant challenge to water safety because they are difficult to break down in the natural ecosystems. Advanced Oxidation Processes (AOPs) and adsorption are transformative technologies that effectively address this issue. AOPs rely heavily on photocatalytic NPs like TiO_2 and ZnO (Alavi et al., 2019; Athira et al., 2020; Mustapha et al., 2020). When exposed to light, these NPs generate ROS, such as OH, which actively break down organic pollutants into non-toxic byproducts. This method is particularly efficient against persistent organic contaminants, achieving complete mineralization while ensuring water remains pure and safe (Kahandal et al., 2023).

Carbon-based nanomaterials, such as CNTs and activated graphene, enhance photocatalysis through adsorption. By leveraging their π–π interactions with the aromatic structures of contaminants, these materials excel at capturing hydrophobic organic molecules (Aslam et al., 2021; Yin et al., 2020). As a result, they are highly effective at removing dyes, pesticides, and other contaminants that are difficult to eliminate using conventional techniques. Adsorption and photocatalysis work together in a powerful synergy. While photocatalysts chemically break down contaminants, carbon-based adsorbents capture residual molecules, ensuring complete removal. This dual-action approach makes nanomaterials indispensable in wastewater treatment systems, enabling them to address even the most complex mixtures of pollutants (Papanicolaou & Portan, 2015; Wang et al., 2021).

Microbial Pathogens

Nanomaterials with antibacterial characteristics, such as AgNPs, CuO, and ZnO, are critical in inactivating microbial pathogens such as bacteria, viruses, and fungi (Mustapha et al., 2020). To successfully eradicate infections, these NPs use multiple methods of action (Durán et al., 2007; Palani et al., 2023). They break microbial cell membranes, cause oxidative stress by producing ROS, and emit metal ions that disturb essential cellular functions like enzyme function and DNA replication. These effects, when combined, result in pathogen eradication that is both quick and permanent. AgNPs were highly valued for their broad-spectrum antibacterial efficacy and longevity. They are often employed in filters, coatings, and composite materials to provide long-term microbial resistance and inhibit biofilm growth. Their nanoscale size improves interactions with microbial cells, improving efficacy at lower concentrations while lowering environmental impact (Franco et al., 2022).

Advanced Oxidation Processes (AOPs)

Nanomaterial-catalysed AOPs are at the forefront of modern water filtration technology, offering a highly efficient solution for degrading persistent organic contaminants. These processes generate highly reactive OH groups, which are among the most powerful oxidizing agents, to break down complex pollutants such as dyes, pharmaceuticals, and endocrine disruptors into harmless byproducts like carbon dioxide and water.

Photocatalytic nanomaterials, such as TiO_2, Fe_2O_3, and Fe_3O_4, play pivotal roles in AOPs (Nguyen et al., 2021). TiO_2, in particular, is highly valued for its exceptional chemical stability, non-toxicity, and the ability to generate OH when activated by UV or visible light (Álvarez et al., 2010; Mustapha et al., 2020; Rottman et al., 2013). TiO_2 produces electron–hole pairs upon light exposure, generating ROS such as •OH and O_2^- •. These ROS exhibit extraordinary oxidative power, breaking down even the most resilient contaminants ensuring complete detoxification and mineralization of pollutants in water (Aragaw et al., 2021; Jena et al., 2022).

In contrast, iron-based nanocatalysts enable AOPs such as Fenton and Fenton-like reactions. In these processes, Fe^{2+} reacts with H_2O_2 to produce OH group. Nanoscale Fe materials improve reaction efficiency due to their large surface area and reactivity. Combining UV or visible light with these nanocatalysts considerably improves the degradation process (Aragaw et al., 2021; Recillas et al., 2011). Doping TiO_2 with elements like Nitrogen or Sulphur expands its activity into the visible light spectrum, making AOPs compatible with natural sunlight and more practical. These NPs not only effectively degrade pollutants but also reduce the development of harmful intermediates, resulting in safer water output. Their scalability, reusability, and compatibility with existing water treatment systems make them critical instruments for solving worldwide water contamination issues (Al-Mamun et al., 2019; Mustapha et al., 2020).

Desalination and Filtration Technologies

Graphene-based membranes and carbon nanotube filters revolutionize desalination and water filtration by offering exceptional water permeability and superior salt rejection. Nanocomposite membranes embedded with nanoparticles enhance resistance to fouling, extending operational lifespans and boosting efficiency. Additionally, aquaporin-inspired designs mimic natural water channels, significantly reducing energy requirements for desalination. These advanced technologies effectively remove salts, micropollutants, and other contaminants from brackish or seawater, providing a more sustainable and cost-effective approach to water treatment while addressing global freshwater scarcity challenges (Aslam et al., 2021; Bhol et al., n.d.; Han et al., 2021; Yin et al., 2020).

MECHANISMS OF ACTION IN WATER TREATMENT

Adsorption and Surface Interactions

One effective method by which nanomaterials purify water is through adsorption. Impurities are attracted to and bound by the surface of NPs via physical, chemical, or electrostatic forces. Materials with large surface areas and adaptable functional groups, such as GO and CNTs, can efficiently adsorb contaminants like heavy metals, dyes, and organic pollutants (Aslam et al., 2021; Han et al., 2021; Iqbal et al., 2023; Kyzas et al., 2014; Yin et al., 2020). Surface functionalization further enhances selectivity, allowing targeted removal of specific pollutants. For instance, magnetic nanoparticles like Fe_3O_4 effectively adsorb arsenic and can be easily retrieved with magnets, minimizing secondary contamination, while thiol-functionalized silica NPs form strong covalent bonds with Hg^{2+} (Aragaw et al., 2021; Recillas et al., 2011).

Catalytic and Photocatalytic Processes

Photocatalysis, a widely used application, leverages catalysts like TiO_2 NPs and other metal oxides, which utilize light energy to generate ROS, such as OH. These ROS can break down persistent organic pollutants into non-toxic byproducts. For example, under UV light, TiO_2 efficiently degrades dyes, pharmaceuticals, and pesticides in water (Álvarez et al., 2010; Rottman et al., 2013).

Similarly, bimetallic NPs, such as Au-Pd or Fe-Cu, exhibit enhanced catalytic activity for oxidation–reduction processes. These materials facilitate electron transfer, making them highly effective in breaking down pollutants. Additionally, zero-valent iron is particularly effective in environmental remediation; it can reduce toxic hexavalent chromium (Cr^{6+}) to the less hazardous trivalent state (Cr^{3+}), making it an excellent choice for in situ groundwater treatment (Li et al., 2024; Liu et al., 2021; Yan et al., 2024a).

Membrane-Based Separation

Clean water is now more accessible than ever due to revolutionary advancements in membrane-based separation techniques. Advanced nanomaterials, including GO, AgNPs, and CNTs, are being integrated into membranes to improve reverse osmosis and nanofiltration processes. These developments improve the remarkable precision of membranes in filtering water, guaranteeing that impurities are efficiently eliminated while preserving high flow rates (Aslam et al., 2021; Bhol et al., n.d.; Han et al., 2021; Iqbal et al., 2023; Yin et al., 2020). For instance, AgNPs have antibacterial qualities that prolong the membrane lifespan, prevent the accumulation of dangerous microorganisms, and improve filtration (Mecha et al., 2023; Palani et al., 2023). Bio-inspired membranes that imitate the effectiveness of natural water channels are based on aquaporin proteins found in nature. These designs offer sustainable desalination and wastewater treatment solutions with exceptional water permeability and energy savings. By connecting state-of-the-art research with real-world use, nanotechnology potentially opens the door to a time when everyone has access to clean water (Qu et al., 2013).

NANOTECHNOLOGY FOR WATER QUALITY MONITORING

Water quality to be contaminant-free is of utmost importance for human health, and to check its quality, various analytical tools like HPLC, Mass spectrometry, LC-MS, ICP-MS, etc., are used to detect contaminants. Although reliable, these techniques pose severe limitations like high cost, trained personnel, time consumption, the need for sample collection and preparation, etc. To address these limitations, nanosensors are paving the way for contaminant detection in water by being advantageous, such as rapid, selective, cost-effective, on-site monitoring, no sample preparation, and easy to use.

Nanosensors like electrochemical and optical sensors offer unique advantages in terms of selectivity towards the pollutant and sensitivity, offering the detection of trace amounts of contaminants.

Heavy Metal Ion Detection

Ghasemi et al. synthesized thiazolylazopyrimidine-functionalized TiO_2 nanosensor (TiO_2-TAP) for sensitive and selective detection of Cu^{2+} ions and getting a linear range with a limit of detection of 2.51 nM (Ghasemi & Mohammadi, 2020). Li et al. developed a sensor for colorimetric detection of Pb^{2+} ions using glucose-functionalized gold NPs, which was highly selective and sensitive with a limit of detection of 1.07 μM and a linear range between 10 and 80 μM (Li et al., 2022)

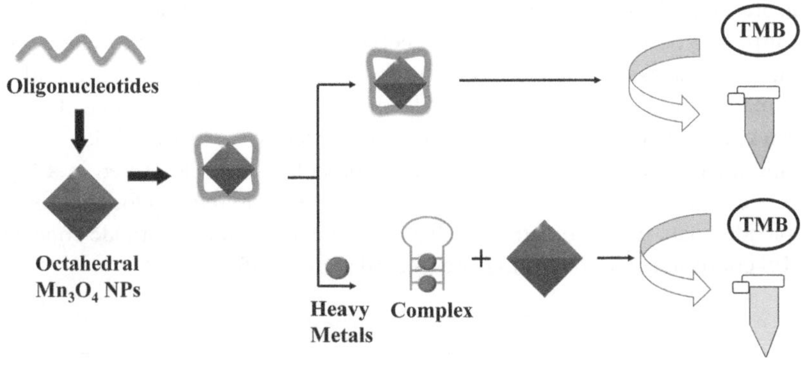

FIGURE 9.4 Heavy metals detection based on the oxidase-mimicking activity of octahedral Mn_3O_4 NPs (Wang et al., 2020).

Wang et al. developed a cost-effective and user-friendly colorimetric sensor for detecting Hg (II) and Cd (II) ions in water. The sensor utilized thymine-rich ssDNA (Hg) aptamers to regulate the oxidase-mimicking activity of Mn_3O_4 NPs (Mn_3O_4-NPs). Mn_3O_4-NPs oxidized 3,3′,5,5′-tetramethylbenzidine (TMB) in acidic conditions, producing a yellow colour. The aptamer initially adsorbed onto Mn_3O_4-NPs, suppressing their catalytic activity and altering the solution's colour to light green with reduced absorbance at 450 nm. Upon exposure to Hg (II) and Cd (II), the aptamer bound to these ions, restoring the Mn_3O_4-NPs' catalytic activity and the yellow colour, correlating with the metal concentrations. The method demonstrated detection limits of 3.8 μgL^{-1} for Hg (II) and 2.4 μgL^{-1} for Cd (II), with potential adaptability for other metals using specific aptamers (Wang et al., 2020) (Figure 9.4).

Real-time Monitoring Systems

The emergence of the Internet of Things (IoT) and wireless sensor networks, alongside advancements in data acquisition, has significantly enhanced water and air quality monitoring systems. Modern water quality monitoring approaches emphasize continuous sensing, integrating multiple sensors, automated controls, and wireless data acquisition mechanisms (Noorjannah Ibrahim et al., 2018).

Additionally, emerging technologies such as artificial intelligence, including machine learning, deep learning (Lalithadevi et al., 2019; Noorjannah Ibrahim et al., 2018), and fuzzy logic, when combined with IoT, are increasingly recognized as efficient tools for effective water quality monitoring (Elijah et al., 2018; Zulkifli et al., 2022). Huang et al. developed phosphorus-doped biochar–attapulgite/bismuth film electrodes (MBA-BiFE) with magnetic Fe_3O_4 NPs for detecting Cd (II), Pb (II), and Hg (II) at detection limits of 0.036, 0.003, and 0.011 nM, respectively. This study integrated a machine learning model using an artificial neural network for multi-metal analysis with data from a portable wireless smart sensor linked to the modified screen-printed electrode.

Sicard et al. developed a paper-based analytical device for detecting malathion in water, utilizing pesticides to inhibit the acetylcholinesterase enzyme. The device captured images of a test strip and control using an iPhone with a colour analysis algorithm. These images were geo-tagged to identify the testing location and uploaded to a web platform called Water Map Ca. The platform enabled users to submit location coordinates and measurement data via the HTTP protocol, which was stored in a MySQL 5 database. Users registered on the website to receive a random token for data submission. The uploaded data included details such as the date, username, pesticide concentration, and GPS coordinates of the testing site (Sicard et al., 2015).

CHALLENGES AND RISKS

Nanotechnology offers innovative solutions for water treatment; however, its implementation is accompanied by challenges and risks that must be addressed for safe and sustainable use. While NPs are highly effective, their release into the environment raises significant health and ecological concerns. Certain NPs, such as metal oxides and carbon-based materials, may exhibit toxicity to organisms and have the potential to bioaccumulate in food chains, posing long-term health risks to humans. This makes them an emerging class of contaminants, necessitating detailed studies on their environmental fate, transportation, and degradation. Economic feasibility also presents a significant hurdle. The synthesis, functionalization, and integration of NPs into large-scale water treatment systems remain cost-intensive, restricting their application in low-income regions or smaller systems.

Additionally, the lack of standardized guidelines and regulatory frameworks for the safe production, use, and disposal of nanomaterials limits their widespread adoption. Governments and international bodies must prioritize the development of safety protocols and regulatory standards to ensure the responsible implementation of nanotechnology in water treatment. From a technical perspective, challenges such as providing the stability, reproducibility, and efficiency of NPs under diverse real-world conditions must be addressed. Moreover, preventing secondary pollution by recovering or safely removing NPs from treated water is crucial. Despite significant advancements, gaps remain in understanding the long-term impacts of nanomaterials on ecosystems and human health. Addressing these uncertainties will require interdisciplinary research to develop eco-friendly, cost-effective, and sustainable nanotechnology-based water treatment solutions.

FUTURE PERSPECTIVES

Nanotechnology has the potential to revolutionize water treatment by enabling efficient removal of a broader spectrum of contaminants. Future innovations focused on developing cost-effective, environmentally friendly nanomaterials that can be integrated with advanced technologies such as artificial intelligence and machine learning. These integrations will enhance real-time monitoring, optimize water treatment processes, and allow adaptive responses to varying water quality conditions. As research advances, scaling up these technologies makes them more accessible and affordable, particularly in underserved or disaster-prone regions. Developing decentralized, portable water treatment systems powered by nanotechnology could also provide safe drinking water in areas lacking infrastructure. However, addressing the potential risks of nanomaterials to human health and the environment remains a key challenge. Ongoing research is essential to ensure the safe use, disposal, and recycling of nanomaterials, with the establishment of clear regulatory frameworks and safety standards.

CONCLUSION

Nanotechnology offers transformative potential for improving water treatment, providing efficient solutions for removing a wide range of contaminants and enhancing water quality and safety. While significant advancements have been made, challenges such as cost, scalability, and environmental impact must still be addressed. Integrating nanotechnology with emerging technologies like AI and machine learning will optimize water treatment processes and enable more intelligent, adaptive systems. Ensuring the safe use of nanomaterials through continued research and the development of clear regulatory standards is crucial for the widespread adoption of these technologies. With ongoing innovation, nanotechnology can revolutionize global water treatment practices and ensure access to clean, safe water for all.

REFERENCES

Adel, M., Ahmed, M. A., Elabiad, M. A., & Mohamed, A. A. (2022). Removal of heavy metals and dyes from wastewater using graphene oxide-based nanomaterials: A critical review. *Environmental Nanotechnology, Monitoring & Management*, *18*, 100719. https://doi.org/10.1016/J.ENMM.2022.100719

Adeola, A. O., & Nomngongo, P. N. (2022). Advanced polymeric nanocomposites for water treatment applications: A holistic perspective. *Polymers*, *14*(12), 14122462. https://doi.org/10.3390/polym14122462

Al-Arjan, W. S. (2022). Zinc oxide nanoparticles and their application in adsorption of toxic dye from aqueous solution. *Polymers*, *14*(15), 14153086. https://doi.org/10.3390/polym14153086

Alavi, M., Karimi, N., & Valadbeigi, T. (2019). Antibacterial, antibiofilm, antiquorum sensing, antimotility, and antioxidant activities of green fabricated Ag, Cu, TiO_2, ZnO, and Fe_3O_4 NPs via protoparmeliopsis muralis lichen aqueous extract against multi-drug-resistant bacteria. *ACS Biomaterials Science and Engineering*, *5*(9), 4228–4243. https://doi.org/10.1021/acsbiomaterials.9b00274

Alhalili, Z. (2023). Metal oxides nanoparticles: General structural description, chemical, physical, and biological synthesis methods, role in pesticides and heavy metal removal through wastewater treatment. *Molecules*, *28*(7), 28073086. https://doi.org/10.3390/molecules28073086

Al-Mamun, M. R., Kader, S., Islam, M. S., & Khan, M. Z. H. (2019). Photocatalytic activity improvement and application of UV-TiO_2 photocatalysis in textile wastewater treatment: A review. *Journal of Environmental Chemical Engineering*, *7*(5), 103248. https://doi.org/10.1016/j.jece.2019.103248

Álvarez, P. M., Jaramillo, J., López-Piñero, F., & Plucinski, P. K. (2010). Preparation and characterization of magnetic TiO_2 nanoparticles and their utilization for the degradation of emerging pollutants in water. *Applied Catalysis B: Environmental*, *100*(1–2), 338–345. https://doi.org/10.1016/j.apcatb.2010.08.010

Aragaw, T. A., Bogale, F. M., & Aragaw, B. A. (2021). Iron-based nanoparticles in wastewater treatment: A review on synthesis methods, applications, and removal mechanisms. *Journal of Saudi Chemical Society*, *25*(8), 101280. https://doi.org/10.1016/J.JSCS.2021.101280

Arellano Vidal, C. L., & Govan, J. E. (2024). Machine learning techniques for improving nanosensors in agroenvironmental applications. *Agronomy*, *14*(2), 14020341. https://doi.org/10.3390/agronomy14020341

Aslam, M. M. A., Kuo, H. W., Den, W., Usman, M., Sultan, M., & Ashraf, H. (2021). Functionalized carbon nanotubes (CNTs) for water and wastewater treatment: Preparation to application. *Sustainability*, *13*(10), 13105717. https://doi.org/10.3390/su13105717

Athira, K., Merin, K. T., Raguram, T., & Rajni, K. S. (2020). Synthesis and characterization of Mg doped TiO_2 nanoparticles for photocatalytic applications. *Materials Today: Proceedings*, *33*, 2321–2327. https://doi.org/10.1016/j.matpr.2020.04.580

Barka, N., Bakas, I., Qourzal, S., Assabbane, A., & Ait-Ichou, Y. (2013). Degradation of phenol in water by titanium dioxide photocatalysis. *Oriental Journal of Chemistry*, *29*(3), 1055–1060. https://doi.org/10.13005/ojc/290328

Barton, L. E., Auffan, M., Olivi, L., Bottero, J. Y., & Wiesner, M. R. (2015). Heteroaggregation, transformation and fate of CeO_2 nanoparticles in wastewater treatment. *Environmental Pollution*, *203*, 122–129. https://doi.org/10.1016/J.ENVPOL.2015.03.035

Bhardwaj, A. K., Sundaram, S., Yadav, K. K., & Srivastav, A. L. (2021). An overview of silver nano-particles as promising materials for water disinfection. *Environmental Technology & Innovation*, *23*, 101721. https://doi.org/10.1016/J.ETI.2021.101721

Bhol, P., Yadav, S., Altaee, A., Saxena, M., Misra, P. K., & Samal, A. K. (n.d.). Graphene-based membranes for water and wastewater treatment: A review. *ACS Applied Nano Materials*, *4*(4), 3274–3293.

Bhutto, A. A., Baig, J. A., Uddin, S., Kazi, T. G., Sierra-Alvarez, R., Akhtar, K., Perveen, S., Afridi, H. I., Ali, H. E., Hol, A., & Samejo, S. (2023). Biosynthesis of aluminium oxide nanobiocomposite and its application for the removal of toxic metals from drinking water. *Ceramics International*, *49*(9), 14615–14623. https://doi.org/10.1016/J.CERAMINT.2023.01.052

Chircov, C., Spoială, A., Păun, C., Crăciun, L., Ficai, D., Ficai, A., Andronescu, E., & Turculeţ, S. C. (2020). Mesoporous silica platforms with potential applications in release and adsorption of active agents. *Molecules*, *25*(17), 3814. https://doi.org/10.3390/molecules25173814

Ciccarelli, D., Samanipour, S., Rapp-Wright, H., Bieber, S., Letzel, T., O'Brien, J. W., Marczylo, T., Gant, T. W., Vineis, P., & Barron, L. P. (2024). Bridging knowledge gaps in human chemical exposure via drinking water with non-target screening. *Critical Reviews in Environmental Science and Technology*, *55*(3), 190–214. https://doi.org/10.1080/10643389.2024.2396690

Danish, M. S. S., Estrella, L. L., Alemaida, I. M. A., Lisin, A., Moiseev, N., Ahmadi, M., Nazari, M., Wali, M., Zaheb, H., & Senjyu, T. (2021). Photocatalytic applications of metal oxides for sustainable environmental remediation. *Metals*, *11*(1), 1–25. https://doi.org/10.3390/met11010080

Dasgupta, N., Ranjan, S., & Ramalingam, C. (2017). Applications of nanotechnology in agriculture and water quality management. *Environmental Chemistry Letters*, *15*(4), 591–605. https://doi.org/10.1007/s10311-017-0648-9

du Plessis, A. (2022). Persistent degradation: Global water quality challenges and required actions. *One Earth*, *5*(2), 129–131. https://doi.org/10.1016/j.oneear.2022.01.005

Durán, N., Marcato, P. D., De Souza, G. I. H., Alves, O. L., & Esposito, E. (2007). Antibacterial effect of silver nanoparticles produced by fungal process on textile fabrics and their effluent treatment. *Journal of Biomedical Nanotechnology*, *3*(2), 203–208. https://doi.org/10.1166/jbn.2007.022

Eid, A. M., Fouda, A., Hassan, S. E. D., Hamza, M. F., Alharbi, N. K., Elkelish, A., Alharthi, A., & Salem, W. M. (2023). Plant-based copper oxide nanoparticles; biosynthesis, characterization, antibacterial activity, tanning wastewater treatment, and heavy metals sorption. *Catalysts*, *13*(2), 13020348. https://doi.org/10.3390/catal13020348

Elijah, O., Rahman, T. A., Leow, C. Y., Yeen, H. C., Sarijari, M. A., Aris, A., Salleh, J., & Chua, T. H. (2018). A concept paper on smart river monitoring system for sustainability in river. *International Journal of Integrated Engineering*, *10*(7), 130–139. https://doi.org/10.30880/ijie.2018.10.07.012

Fang, Z., Li, Z., Zhang, X., Pan, S., Wu, M., & Pan, B. (2021). Enhanced arsenite removal from silicate-containing water by using redox polymer-based Fe(III) oxides nanocomposite. *Water Research*, *189*, 116673. https://doi.org/10.1016/J.WATRES.2020.116673

Fouda, A., Hassan, S. E. D., Saied, E., & Hamza, M. F. (2021). Photocatalytic degradation of real textile and tannery effluent using biosynthesized magnesium oxide nanoparticles (MgO-NPs), heavy metal adsorption, phytotoxicity, and antimicrobial activity. *Journal of Environmental Chemical Engineering*, *9*(4), 105346 https://doi.org/10.1016/j.jece.2021.105346

Franco, D., Calabrese, G., Guglielmino, S. P. P., & Conoci, S. (2022). Metal-based nanoparticles: Antibacterial mechanisms and biomedical application. *Microorganisms*, *10*(9), 10091778. https://doi.org/10.3390/microorganisms10091778

Fu, L., Liao, K., Tang, B., Jiang, L., & Huang, W. (2020). Applications of graphene and its derivatives in the upstream oil and gas industry: A systematic review. *Nanomaterials*, *10*(6), 10061013. https://doi.org/10.3390/nano10061013

Fuller, R., Landrigan, P. J., Balakrishnan, K., Bathan, G., Bose-O'Reilly, S., Brauer, M., Caravanos, J., Chiles, T., Cohen, A., Corra, L., Cropper, M., Ferraro, G., Hanna, J., Hanrahan, D., Hu, H., Hunter, D., Janata, G., Kupka, R., Lanphear, B., & Yan, C. (2022). Pollution and health: A progress update. *The Lancet Planetary Health*, *6*(6), e535–e547. https://doi.org/10.1016/S2542-5196(22)00090-0

Geldasa, F. T., Kebede, M. A., Shura, M. W., & Hone, F. G. (2023). Experimental and computational study of metal oxide nanoparticles for the photocatalytic degradation of organic pollutants: A review. *RSC Advances*, *13*(27), 18404–18442. https://doi.org/10.1039/d3ra01505j

Ghasemi, Z., & Mohammadi, A. (2020). Sensitive and selective colorimetric detection of Cu (II) in water samples by thiazolylazopyrimidine-functionalized TiO_2 nanoparticles. *Spectrochimica Acta Part A: Molecular and Biomolecular Spectroscopy*, *239*, 118554. https://doi.org/10.1016/J.SAA.2020.118554

Gupta, P., Tamrakar, G., Biswas, S., Dhundhel, R. S., Soni, V. K., Das, H. S., Sahu, P., Chaturwedi, A. K., Kashyap, N. K., Akitsu, T., & Hait, M. (2024). Quantum dots as a fluorescent sensor for detecting heavy metal ions in aqueous environment: An overview. *ES Chemistry and Sustainability*, *1*, 1359. https://doi.org/10.30919/escs1359

Han, Z. Y., Huang, L. J., Qu, H. J., Wang, Y. X., Zhang, Z. J., Rong, Q. L., Sang, Z. Q., Wang, Y., Kipper, M. J., & Tang, J. G. (2021). A review of performance improvement strategies for graphene oxide-based and graphene-based membranes in water treatment. *Journal of Materials Science*, *56*(16), 9545–9574. https://doi.org/10.1007/s10853-021-05873-7

Iqbal, M. S., Aslam, A. A., Iftikhar, R., Junaid, M., Imran, S. M., Nazir, M. S., Ali, Z., Zafar, M., Kanwal, A., Othman, N. K., Pal, U., Pang, A. L., & Ahmadipour, M. (2023). The potential of functionalized graphene-based composites for removing heavy metals and organic pollutants. *Journal of Water Process Engineering*, *53*, 103809. https://doi.org/10.1016/J.JWPE.2023.103809

Jena, A., Kumar Sahoo, P., Ghosal, A., & Kumar Sahoo, N. (2022). Bioinspired synthesis of iron-based nanomaterials and nanocomposite: For environmental remediation. *Materials Today: Proceedings*, *67*, 1090–1096. https://doi.org/10.1016/J.MATPR.2022.07.058

Kahandal, A., Chaudhary, S., Methe, S., Nagwade, P., Sivaram, A., & Tagad, C. K. (2023). Galactomannan polysaccharide as a biotemplate for the synthesis of zinc oxide nanoparticles with photocatalytic, antimicrobial and anticancer applications. *International Journal of Biological Macromolecules*, *253*, 126787. https://doi.org/10.1016/J.IJBIOMAC.2023.126787

Khan, F. S. A., Mubarak, N. M., Tan, Y. H., Khalid, M., Karri, R. R., Walvekar, R., Abdullah, E. C., Nizamuddin, S., & Mazari, S. A. (2021). A comprehensive review on magnetic carbon nanotubes and carbon nanotube-based buckypaper for removal of heavy metals and dyes. *Journal of Hazardous Materials*, *413*, 125375. https://doi.org/10.1016/j.jhazmat.2021.125375

Khodakarami, M., & Bagheri, M. (2021). Recent advances in synthesis and application of polymer nanocomposites for water and wastewater treatment. *Journal of Cleaner Production*, *296*, 126404. https://doi.org/10.1016/J.JCLEPRO.2021.126404

Kolya, H., & Kang, C. W. (2023). Next-generation water treatment: Exploring the potential of biopolymer-based nanocomposites in adsorption and membrane filtration. *Polymers*, *15*(16), 15163421. https://doi.org/10.3390/polym15163421

Kr Basu, A., Barman, P., & Roy, S. (2024). Performance of mesoporous ceria supported Ru catalysts for oxidative degradation of aqueous solution of plastic monomer bisphenol A in a fixed bed flow reactor: Structure-activity insights, optimization of reaction conditions, kinetics, catalyst stability, and characterization. *Chemical Engineering Journal*, *498*, 155229. https://doi.org/10.1016/J.CEJ.2024.155229

Kumar, A., Sharma, G., Naushad, M., Al-Muhtaseb, A. H., García-Peñas, A., Mola, G. T., Si, C., & Stadler, F. J. (2020). Bio-inspired and biomaterials-based hybrid photocatalysts for environmental detoxification: A review. *Chemical Engineering Journal*, *382*, 122937. https://doi.org/10.1016/J.CEJ.2019.122937

Kyzas, G. Z., Deliyanni, E. A., & Matis, K. A. (2014). Graphene oxide and its application as an adsorbent for wastewater treatment. *Journal of Chemical Technology and Biotechnology*, *89*(2) 196–205. https://doi.org/10.1002/jctb.4220

Lalithadevi, B., Yadav, A., Pandey, A., & Adhikari, M. (2019). Iot based WSN ground water monitoring system with cloud-based monitoring as a service (Maas) and prediction using machine learning. *International Journal of Innovative Technology and Exploring Engineering*, *9*(1), 816–821. https://doi.org/10.35940/ijitee.A4316.119119

Li, S., Li, L., & Zhang, W. (2024). Nanoscale zero-valent iron (nZVI) for heavy metal wastewater treatment: A perspective. *Engineering*, *36*, 16–20. https://doi.org/10.1016/J.ENG.2023.08.012

Li, X., Zhao, C. X., & Lin, L. (2022). Plasma-based instant synthesis of functionalized gold nanoparticles for colorimetric detection of lead ions. *Chemical Engineering Science*, *260*, 117849. https://doi.org/10.1016/J.CES.2022.117849

Liu, J., Wang, S., Yang, R., Li, L., Liang, S., & Chen, L. (2022a). Bio-inspired graphene oxide-amino acid cross-linked framework membrane trigger high water permeance and high metal ions rejection. *Journal of Membrane Science*, *659*, 120745. https://doi.org/10.1016/J.MEMSCI.2022.120745

Liu, X., Sathishkumar, K., Zhang, H., Saxena, K. K., Zhang, F., Naraginti, S., Anbarasu, K., Ramya, R., Rajasekar, A., & Guo, X. (2024). Frontiers in environmental cleanup: Recent advances in remediation of emerging pollutants from soil and water. *Journal of Hazardous Materials Advances*, *16*, 100461. https://doi.org/10.1016/j.hazadv.2024.100461

Liu, Y., Wu, T., White, J. C., & Lin, D. (2021). A new strategy using nanoscale zero-valent iron to simultaneously promote remediation and safe crop production in contaminated soil. *Nature Nanotechnology*, *16*(2), 197–205. https://doi.org/10.1038/S41565-020-00803-1

Liu, Z., Wang, Q., Huang, X., & Qian, X. (2022b). Surface functionalization of graphene oxide with hyperbranched polyamide-amine and microcrystalline cellulose for efficient adsorption of heavy metal ions. *ACS Omega*, *7*(13), 10944–10954. https://doi.org/10.1021/acsomega.1c06647

Madhura, L., Singh, S., Kanchi, S., Sabela, M., Bisetty, K., & Inamuddin, A. (2019). Nanotechnology-based water quality management for wastewater treatment. *Environmental Chemistry Letters*, *17*(1), 65–121. https://doi.org/10.1007/s10311-018-0778-8

Mecha, A. C., Chollom, M. N., Babatunde, B. F., Tetteh, E. K., & Rathilal, S. (2023). Versatile silver-nanoparticle-impregnated membranes for water treatment: A review. *Membranes*, *13*(4), 13040432. https://doi.org/10.3390/membranes13040432

Mustapha, S., Ndamitso, M. M., Abdulkareem, A. S., Tijani, J. O., Shuaib, D. T., Ajala, A. O., & Mohammed, A. K. (2020). Application of TiO_2 and ZnO nanoparticles immobilized on clay in wastewater treatment: A review. *Applied Water Science*, *10*(1), 1138. https://doi.org/10.1007/s13201-019-1138-y

Naidu, R., Biswas, B., Willett, I. R., Cribb, J., Kumar Singh, B., Paul Nathanail, C., Coulon, F., Semple, K. T., Jones, K. C., Barclay, A., & John Aitken, R. (2021). Chemical pollution: A growing peril and potential catastrophic risk to humanity. *Environment International*, *156*, 106616. https://doi.org/10.1016/j.envint.2021.106616

Nguyen, M. D., Tran, H. V., Xu, S., & Lee, T. R. (2021). Fe_3O_4 nanoparticles: Structures, synthesis, magnetic properties, surface functionalization, and emerging applications. *Applied Sciences*, *11*(23), 112311301. https://doi.org/10.3390/app112311301

Nikić, J., Tubić, A., Watson, M., Maletić, S., Šolić, M., Majkić, T., & Agbaba, J. (2019). Arsenic removal from water by green synthesized magnetic nanoparticles. *Water*, *11*(12), 1122520. https://doi.org/10.3390/w11122520

Noorjannah Ibrahim, S., Asnawi, A. L., Abdul Malik, N., Mohd Azmin, N. F., Jusoh, A. Z., & Mohd Isa, F. N. (2018). Web based water turbidity monitoring and automated filtration system: IoT application in water management. *International Journal of Electrical and Computer Engineering*, *8*(4), 2503–2511. https://doi.org/10.11591/ijece.v8i4

Palani, G., Trilaksana, H., Sujatha, R. M., Kannan, K., Rajendran, S., Korniejenko, K., Nykiel, M., & Uthayakumar, M. (2023). Silver nanoparticles for waste water management. *Molecules*, *28*(8), 28083520. https://doi.org/10.3390/molecules28083520

Papanicolaou, G. C., & Portan, D. V. (2015). Carbon and titanium dioxide nanotube polymer composite manufacturing - characterization and interphase modeling. In Beaumont, P.W.R., Soutis, C., & Hodzic, A. (eds) *Structural Integrity and Durability of Advanced Composites: Innovative Modelling Methods and Intelligent Design* (pp. 735–761). Elsevier, Amsterdam, Netherlands. https://doi.org/10.1016/B978-0-08-100137-0.00027-4

Qu, X., Brame, J., Li, Q., & Alvarez, P. J. J. (2013). Nanotechnology for a safe and sustainable water supply: Enabling integrated water treatment and reuse. *Accounts of Chemical Research*, *46*(3), 834–843. https://doi.org/10.1021/ar300029v

Ramirez Arenas, L., Le Coustumer, P., Ramseier Gentile, S., Zimmermann, S., & Stoll, S. (2023). Removal efficiency and adsorption mechanisms of CeO_2 nanoparticles onto granular activated carbon used in drinking water treatment plants. *Science of the Total Environment*, *856*, 159261. https://doi.org/10.1016/J.SCITOTENV.2022.159261

Ray, S. S., Iroegbu, A. O. C., & Bordado, J. C. (2020). Polymer-based membranes and composites for safe, potable, and usable water: A survey of recent advances. *Chemistry Africa, 3*(3), 593–608. https://doi.org/10.1007/s42250-020-00166-z

Raza, S., Ghasali, E., Orooji, Y., Lin, H., Karaman, C., Dragoi, E. N., & Erk, N. (2023). Two dimensional (2D) materials and biomaterials for water desalination; structure, properties, and recent advances. *Environmental Research, 219*, 114998. https://doi.org/10.1016/J.ENVRES.2022.114998

Recillas, S., García, A., González, E., Casals, E., Puntes, V., Sánchez, A., & Font, X. (2011). Use of CeO_2, TiO_2 and Fe_3O_4 nanoparticles for the removal of lead from water. Toxicity of nanoparticles and derived compounds. *Desalination, 277*(1–3), 213–220. https://doi.org/10.1016/j.desal.2011.04.036

Rottman, J., Platt, L. C., Sierra-Alvarez, R., & Shadman, F. (2013). Removal of TiO_2 nanoparticles by porous media: Effect of filtration media and water chemistry. *Chemical Engineering Journal, 217*, 212–220. https://doi.org/10.1016/j.cej.2012.11.117

Saha, I., Bhattacharya, S., Mukhopadhyay, A., Chattopadhyay, D., Ghosh, U. C., & Chatterjee, D. (2013). Role of nanotechnology in water treatment and purification: Potential applications and implications. *International Journal of Chemical Science and Technology, 3*(3), 59–64. https://www.urpjournals.com

Sahu, A., Dosi, R., Kwiatkowski, C., Schmal, S., & Poler, J. C. (2023). Advanced polymeric nanocomposite membranes for water and wastewater treatment: A comprehensive review. *Polymers, 15*(3), 15030540. https://doi.org/10.3390/polym15030540

Saied, E., Eid, A. M., Hassan, S. E. D., Salem, S. S., Radwan, A. A., Halawa, M., Saleh, F. M., Saad, H. A., Saied, E. M., & Fouda, A. (2021). The catalytic activity of biosynthesized magnesium oxide nanoparticles (MgO-NPs) for inhibiting the growth of pathogenic microbes, tanning effluent treatment, and chromium ion removal. Catalysts, *11*(7), 11070821. https://doi.org/10.3390/catal11070821

Sibhatu, A. K., Weldegebrieal, G. K., Sagadevan, S., Tran, N. N., & Hessel, V. (2022). Photocatalytic activity of CuO nanoparticles for organic and inorganic pollutants removal in wastewater remediation. *Chemosphere, 300*, 134623. https://doi.org/10.1016/J.CHEMOSPHERE.2022.134623

Sicard, C., Glen, C., Aubie, B., Wallace, D., Jahanshahi-Anbuhi, S., Pennings, K., Daigger, G. T., Pelton, R., Brennan, J. D., & Filipe, C. D. M. (2015). Tools for water quality monitoring and mapping using paper-based sensors and cell phones. *Water Research, 70*, 360–369. https://doi.org/10.1016/J.WATRES.2014.12.005

Spoială, A., Ilie, C. I., Ficai, D., Ficai, A., & Andronescu, E. (2021). Chitosan-based nanocomposite polymeric membranes for water purification—a review. *Materials, 14*(9), 14092091. https://doi.org/10.3390/ma14092091

Sv, K., Rao, K., & Sunitha, V. (2021). Recent advances in effective remediation of groundwater fluoride. In *Advances in Chemical Science & Biotechnology for Water Purification, Energy Production and Stress Management.* KROS Publications, Andhra Pradesh, India.

Tan, M., Qiu, G., & Ting, Y. P. (2015). Effects of ZnO nanoparticles on wastewater treatment and their removal behavior in a membrane bioreactor. *Bioresource Technology, 185*, 125–133. https://doi.org/10.1016/j.biortech.2015.02.094

Theron, J., Walker, J. A., & Cloete, T. E. (2008). Nanotechnology and water treatment: Applications and emerging opportunities. *Critical Reviews in Microbiology, 34*(1), 43–69. https://doi.org/10.1080/10408410701710442

Tripathy, J., Mishra, A., Pandey, M., Thakur, R. R., Chand, S., Rout, P. R., & Shahid, M. K. (2024). Advances in nanoparticles and nanocomposites for water and wastewater treatment: A review. *Water, 16*(11), 16111481. https://doi.org/10.3390/w16111481

Ucar, A., Findik, M., Gubbuk, I. H., Kocak, N., & Bingol, H. (2017). Catalytic degradation of organic dye using reduced graphene oxide–polyoxometalate nanocomposite. *Materials Chemistry and Physics*, *196*, 21–28. https://doi.org/10.1016/j.matchemphys.2017.04.047

Vikesland, P. J. (2018). Nanosensors for water quality monitoring. *Nature Nanotechnology*, *13*(8), 651–660. https://doi.org/10.1038/s41565-018-0209-9

Wang, F., Xiang, L., Sze-Yin Leung, K., Elsner, M., Zhang, Y., Guo, Y., Pan, B., Sun, H., An, T., Ying, G., Brooks, B. W., Hou, D., Helbling, D. E., Sun, J., Qiu, H., Vogel, T. M., Zhang, W., Gao, Y., Simpson, M. J., & Tiedje, J. M. (2024). Emerging contaminants: A one health perspective. *Innovation*, *5*(4), 100612. https://doi.org/10.1016/j.xinn.2024.100612

Wang, J., Wang, J., Zhou, P., Tao, H., Wang, X., & Wu, Y. (2020). Oligonucleotide-induced regulation of the oxidase-mimicking activity of octahedral Mn_3O_4 nanoparticles for colorimetric detection of heavy metals. *Microchimica Acta*, *187*(2), 2. https://doi.org/10.1007/s00604-019-4069-2

Wang, P., Li, L., Pang, X., Zhang, Y., Zhang, Y., Dong, W. F., & Yan, R. (2021). Chitosan-based carbon nanoparticles as a heavy metal indicator and for wastewater treatment. *RSC Advances*, *11*(20), 12015–12021. https://doi.org/10.1039/d1ra00692d

Yan, X. W., Bigdeli, F., Abbasi-Azad, M., Wang, S. J., & Morsali, A. (2024a). Selective separation of CO_2/CH_4 gases by metal-organic framework-based composites. *Coordination Chemistry Reviews*, *520*, 216126. https://doi.org/10.1016/J.CCR.2024.216126

Yan, Z., Ouyang, J., Wu, B., Liu, C., Wang, H., Wang, A., & Li, Z. (2024b). Nonmetallic modified zero-valent iron for remediating halogenated organic compounds and heavy metals: A comprehensive review. *Environmental Science and Ecotechnology*, *21*, 100417. https://doi.org/10.1016/J.ESE.2024.100417

Yin, Z., Cui, C., Chen, H., Duoni, Y. X., & Qian, W. (2020). The application of carbon nanotube/graphene-based nanomaterials in wastewater treatment. *Small*, *16*(15), 201902301. https://doi.org/10.1002/smll.201902301

Zhou, J., Wu, Z., Jin, C., & Zhang, J. X. J. (2024). Machine learning assisted dual-functional nanophotonic sensor for organic pollutant detection and degradation in water. *Clean Water*, *7*(1), 4. https://doi.org/10.1038/s41545-023-00292-4

Zou, Y., Wang, X., Khan, A., Wang, P., Liu, Y., Alsaedi, A., Hayat, T., & Wang, X. (2016). Environmental remediation and application of nanoscale zero-valent iron and its composites for the removal of heavy metal ions: A review. *Environmental Science and Technology*, *50*(14), 7290–7304. https://doi.org/10.1021/ACS.EST.6B01897

Zulkifli, C. Z., Garfan, S., Talal, M., Alamoodi, A. H., Alamleh, A., Ahmaro, I. Y. Y., Sulaiman, S., Ibrahim, A. B., Zaidan, B. B., Ismail, A. R., Albahri, O. S., Albahri, A. S., Soon, C. F., Harun, N. H., & Chiang, H. H. (2022). IoT-based water monitoring systems: A systematic review. *Water*, *14*(22), 14223621. https://doi.org/10.3390/w14223621

Nanotech in Wastewater Treatment for Quality and Safety

10

Nishtha Naudiyal and
Rajeev Kumar Srivastava

INTRODUCTION

Water, often celebrated as the "soup of life," is one of Earth's most vital resources, essential for sustaining all forms of life. Yet, the quality of this invaluable resource is increasingly at risk, posing significant challenges in the 21st century. Issues such as food security, biodiversity loss, climate change and climate resilience are deeply intertwined with the availability and quality of water resources (UNESCO, 2009). These challenges not only disrupt water availability but also degrade water quality, putting freshwater systems under tremendous pressure (Huntington, 2006; Oki and Kanae, 2006). Human activities contribute significantly to water pollution through synthetic chemicals and natural geogenic pollutants, which tend to exacerbate water quality issues (Schwarzenbach et al., 2010).

In addition, an increasing trend in the quantity of emerging pollutants (EPs), which include pharmaceuticals, personal care products (PCPs), pesticides, herbicides,

 DOI: 10.1201/9781003613183-10

industrial chemicals, microplastics and endocrine-disrupting compounds (EDCs), poses significant ecological and health risks (Bracamontes-Ruelas et al., 2022). Unlike conventional pollutants, EPs persist in trace amounts but have substantial long-term impacts. Traditional wastewater treatment systems effectively remove organic matter, nutrients and pathogens, but they struggle with trace-level pollutants and are ill-equipped to handle the complexity and persistence of EPs (Zhou et al., 2019). Furthermore, the sludge generated by these processes often contains concentrated levels of EPs, creating additional environmental risks. Advanced treatment methods show promise but are hindered by high energy demands and operational costs (Bhat et al., 2022).

WHY NANOTECHNOLOGY?

Nanotechnology offers a transformative solution to the above-mentioned limitations, with its unique properties enabling more effective wastewater treatment. Nanomaterials, defined as materials smaller than 100 nm in at least one dimension, exhibit properties such as high surface area, tunable pore size, reactivity and electrostatic properties, making them highly efficient while reducing costs and energy requirements (Chhabra and Kumar, 2020).

Despite their promise, nanotechnology-enabled solutions require careful consideration of safety concerns and environmental impacts. The production, application and disposal of nanomaterials must be managed to prevent unintended consequences. Further research and investment are essential to scale up these technologies, integrate them into existing wastewater systems and ensure their sustainability (Bhat et al., 2022; Kyzas and Mitropoulos, 2021). In conclusion, nanotechnology presents a groundbreaking opportunity to overcome the limitations of traditional wastewater treatment systems. The chapter further discusses the various categories, applications and major technological advancements in the field of nanotechnology for the wastewater treatment.

CATEGORIES OF NANOMATERIALS IN WASTEWATER TREATMENT

Nanomaterials are becoming more and more acknowledged as useful wastewater treatment solutions because of their intricate atomic and molecular structures, which provide them a variety of chemical and physical characteristics. They fall into a number of groups, including metal-oxide, polymeric, carbon-based and nanocomposite materials, each of which has unique benefits for eliminating pollutants from wastewater (Ghosh et al., 2022).

METAL OXIDE NANOPARTICLES

The potential of metal oxide nanoparticles (MONPs) to cure a variety of pollutants has attracted a lot of interest in wastewater treatment. By means of processes including photocatalysis, adsorption and antibacterial properties, these nanoparticles effectively remove colors, heavy metals and organic pollutants. Because of their enormous surface area and nanoscale size, which improve their catalytic and adsorption capabilities, MONPs are very desirable. According to research, many MONP types, including TiO_2, ZnO, MgO and FeO_4/FeO_3, may effectively handle difficult water treatment problems (Abdelbasir and Shalan, 2019; Singh et al., 2019; Yang et al., 2013). Additionally, MONPs can be combined with other substances, such as carbon-based and polymeric nanomaterials, to improve the effectiveness of pollution removal (Vilardi et al., 2018). For instance, polymer-MONP composites can combine the high adsorption capacity of MONPs with the mechanical stability and flexibility of polymers, optimizing the removal of diverse pollutants (Naseem and Durrani, 2021).

CARBON-BASED NANOMATERIALS

Carbon-based nanomaterials including graphene, graphene oxide, single- and double-walled carbon nanotubes (CNTs) have remarkable adsorptive and antibacterial qualities and are frequently utilized in composites with polymers or MONPs. CNTs and graphene-based nanomaterials are excellent in removing organic pollutants from wastewater because of their large surface area, mesoporous structure, chemical stability and adjustable features, but fullerenes are still underutilized. They can withstand harsh wastewater conditions, such as high salinity, acidity and temperature. Their adsorption effectiveness is because of the hydrophilic and hydrophobic surface characteristics as well as atomic-level structural interactions. The goal of developments in nanotechnological advancement in carbon-based nanomaterials is in the recombination and structural modification methods in order to improve adsorption efficiency and reduce the cost of these carbon-based nanomaterials (Gopinath et al., 2021; Thines et al., 2017; Yin et al., 2020).

POLYMERIC NANOMATERIALS

Polymeric nanomaterials (PNMs) are becoming promising solutions in wastewater treatment as they tend to use the combined qualities of polymers and nanoparticles. PNMs are known to effectively remove organic and inorganic pollutants.

Microorganisms make high-molecular-weight polymers called extracellular polymeric substances (EPSs), which are made up of biomolecules such as polysaccharides, proteins, nucleic acids and lipids and are found in microbial aggregates like biofilms and activated sludge. The adsorption and conversion of pollutants are facilitated by EPSs, and therefore, they are used for treating wastewater. MONPs interact with EPSs through processes like ion exchange, complexation and hydrophobic contacts. This interaction is affected by a number of variables, viz., oxidation–reduction potential, ionic strength and pH. Also, MONPs that are efficiently immobilized and transformed by EPSs lower their toxicity and environmental impact (Huangfu et al., 2019; Khodakarami and Bagheri, 2021).

NANOCOMPOSITES

Nanocomposites (NCs) are very useful materials because of their adaptability, environmental sustainability, economic nature and customizability using various procedures depending on their utility in different wastewater treatment processes. Polymer-NCs made by adding MONPs, PNMs and polymer matrices are found to have improved permeability, thermal and mechanical stability and antifouling properties. For instance, NC membranes created using iron oxide nanoparticles and carboxylate graphene oxide can significantly remove dye from wastewater as this combination of nanoparticles results in their increased surface area and active sites (Guo et al., 2017); poly(ε-caprolactone)/poly(ethylene oxide) NCs created using gold nanoparticles have outstanding catalytic capability for 4-nitrophenol degradation (Wang et al., 2017). Metal oxide NCs have adjustable surface characteristics, which increases their adsorption and photocatalytic properties significantly. Clay-based NCs show great promise as pollutant adsorbent because of their large surface areas and thermal stability (Ayalew, 2022; Joshi et al., 2020; Nasir et al., 2019) (Table 10.1).

APPLICATIONS OF NANOTECHNOLOGY IN WASTEWATER TREATMENT

Nanomaterials efficiently remove/degrade pollutants, microbes and heavy metals present in wastewater (Figure 10.1). Table 10.2 shows the removal of pollutants from wastewater using different types of nanomaterials.

TABLE 10.1 Types of nanomaterials, their advantages and limitations

S.NO.	*NANOMATERIALS*	*ADVANTAGES*	*LIMITATIONS*
1.	Metal oxide nanoparticles (MONPs)	Substantial surface area, chemical stability and capacity to function at low concentrations	Potential toxicity, aggregation and limited reusability
2.	Carbon-based nanomaterials	Readily available and inexpensive to produce	High production costs for some types (e.g., graphene) and potential environmental impact
3.	Polymeric nanomaterials (PNMs)	Special functional qualities, adjustable properties and the ability to integrate polymers	Limited thermal stability and potential leaching of polymer components
4.	Nanocomposites (NCs)	Exceptional pollutant removal capabilities, flexibility, low space needs and chemical and mechanical durability	Complex synthesis, possible environmental toxicity and challenges in large-scale production

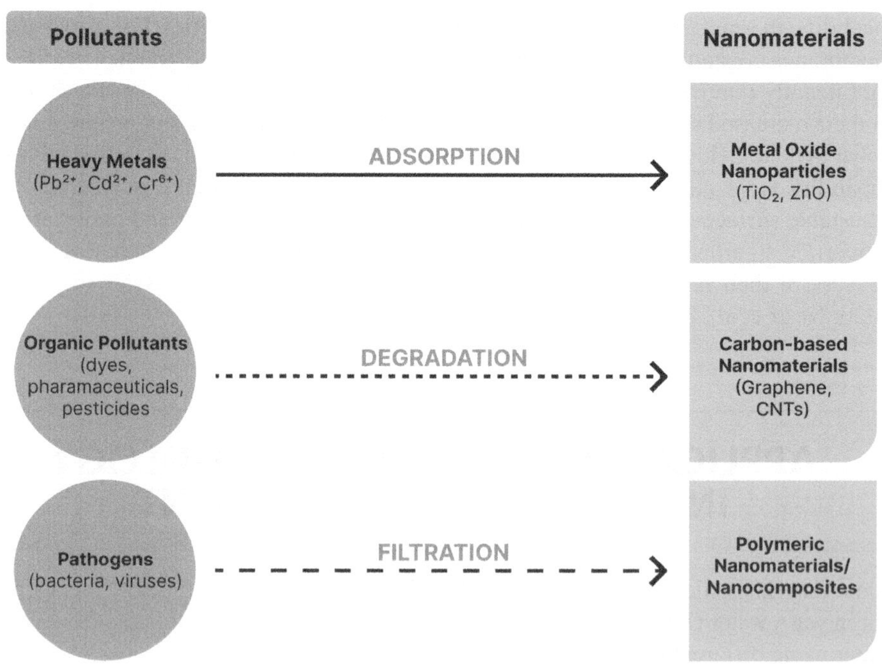

FIGURE 10.1 Pollutant separation mechanisms through various nanomaterials.

TABLE 10.2 Nanomaterial-based pollutant removal

S.NO.	POLLUTANT REMOVED	NANOMATERIAL USED	REFERENCES
1.	Methylene blue and phenol	Cobalt-doped MoS_2	Iqbal et al. (2025)
2.	Caffeine	MnO_2/MXene/chitosan nanocomposite	Arshad et al. (2025)
3.	Cd^{2+} and Pb^{2+} ions	Nickel-oxide-infused orange peel nanobiocomposite	Njimou et al. (2025)
4.	Crystal violet and malachite green oxalate	$NiAg_2O$-infused chitosan nanocomposites	Shabnum et al. (2025)
5.	chlorothiazide, ranitidine, azo blue, methylene blue dye	Chitosan/nitrogen-doped graphene quantum dot	Amari et al. (2021)
6.	Cr^{6+} ions	TiO_2, ZnO and CdS	Joshi and Shrivastava (2011)
7.	Zn^{2+} ions	GO, SiO_2 and SnO_2 nanomaterials	Ahmed et al. (2019)
8.	Cr^{6+} ions	ZrO_2/Fe_3O_4/chitosan nanomaterials	Kumar et al. (2016)
9.	Rose bengal and methylene blue dye	Fe_2O_3 nanoparticles	Dutta et al. (2014)
10.	Rhodamine B dye	ZnO nanomaterials	Slimani et al. (2024)
11.	Malachite green dye	ZnO nanomaterials	Saikia et al. (2015)
12.	Bisphenol A	Cu_2FeSnS_4 nanomaterial	Kong et al. (2019)
13.	Congo red dye	Zinc peroxide nanomaterial	Chawla et al. (2017)
14.	Remazol Red B and Rhodamine B dyes	TiO_2 nanomaterial	El-Dossoki et al. (2021)
15.	Organic pollutants	ZnO nanomaterials	Yang et al. (2023)
16.	Antibiotic	Nano zero-valent iron	Liu et al. (2022)

REMOVAL OF HEAVY METALS

Heavy metals such as Pb^{2+}, Cd^{2+}, Zn^{2+}, Ni^{2+}, Cr^{6+} and Hg^{2+} in water are one of the worst wastewater pollutants, which may be extremely harmful to both human and animal health and are poisonous, non-biodegradable and detrimental to ecosystems. These ions are often released by industrial, municipal, agricultural and residential wastes. Nanomaterials assist in removing heavy metals from polluted wastewater, which are often immobilized on carriers or organic membranes and used in membrane separators, thus safeguarding water quality. Heavy metals can be isolated using a variety of nanomaterials, such as MONPs, nanotubes, magnetic nanoparticles, nanofilter membranes and nanosensors because of their special structural characteristics, which include excellent selectivity and adsorption capacity. Nanomaterials are very effective adsorbents that can effectively remove heavy metal ions even at lower concentrations (Jang et al., 2017; Liu et al., 2019; Vidu et al., 2020).

DEGRADATION OF ORGANIC POLLUTANTS

Industrial and agricultural wastewater (which frequently contains hazardous and non-biodegradable organic pollutants from textiles, pharmaceuticals, agriculture fields, etc.), which when released untreated poses serious environmental threats. Nanomaterials are used as adsorbents, catalysts and photocatalysts to remove these organic pollutants. Out of the three, the most environmentally sound and energy-efficient method for breaking down organic pollutants is "photocatalysis," which is based on sophisticated oxidation processes and uses sunlight to breakdown organic pollutants into smaller, less-toxic substances. For instance, titanium dioxide (TiO_2) is a commonly used photocatalyst to degrade organic pollutants because of its exceptional qualities under ultraviolet (UV) irradiation, strong oxidation resistance, long-term stability, low cost and low toxicity. Apart from photocatalytic degradation, TiO_2 membranes have several other functions, which include disinfection of pathogen and pollutant separation. Also, TiO_2 nanofibers have shown the ability to remove pharmaceuticals (80%), phenol (70%), 4-nitrophenol (50%) and paraquat herbicide (50%) (Kumar, 2021; Pasini et al., 2021).

DISINFECTION AND PATHOGEN REMOVAL

Millions of people die each year from infectious diseases caused by waterborne microbes. Nanotechnology provides viable advances for disinfection and microbial control in wastewater treatment because of its superior antibacterial qualities. One such technique is nanofiltration (NF) membranes composed of materials like carbon nanotubes and bioactive nanoparticles. Advanced oxidation processes (AOPs) that use TiO_2 or ZnO under UV light to produce reactive oxygen species like hydroxyl radicals (•OH) is another main technique used in nanotechnology-based disinfection. These techniques have demonstrated efficacy against fungus, bacteria and viruses. The demand for alternate antimicrobial methods has increased globally due to the emergence of multidrug-resistant (MDR) pathogens, and therefore, nanoparticles are potential options as they have special properties, which allow them to break through biofilms and interfere with essential molecular functions in pathogens. Silver, zinc, gold and copper nanoparticles are utilized in a variety of sectors, viz., food packaging, pharmaceuticals and agriculture. Particularly useful in combating the rising issue of bacterial resistance are silver and zinc nanoparticles, which are routinely used to eradicate resistant infections (Jain et al., 2021; Prasse and Tenes, 2010; Verma et al., 2023).

NUTRIENT RECOVERY

For wastewater treatment to be sustainable, phosphorus and nitrate from polluted water must be recovered and reused. Conventional nutrient recovery techniques concentrate on removing solids-bound or precipitating phosphorus as brushite or struvite, which frequently produces slow-release, low-quality fertilizers. These techniques, however, leave behind dissolved nutrients that might damage the ecosystem, encourage toxic algal blooms and deteriorate the overall water quality, making it not safe for human and animal consumption. A promising answer to this problem is provided by hybrid ion exchange nanotechnology (HIX-Nano), originally developed for the removal of arsenic, which makes it possible to remove and recover phosphorus, and after desorption from the saturated medium, produces a beneficial nutrient solution. This method offers a sustainable and economical solution to the world's phosphorus problems in wastewater treatment. HIX-Nanotechnology has also demonstrated promise in the removal of nitrates from wastewater and is cost-effective due to its regenerable nature. Its value is further increased as it turns the recovered nutrients into marketable, high-value byproducts like non-fertilizer products or liquid N-P-K fertilizers. This creative method is a sustainable and significant solution for wastewater nutrient treatment globally (Ownby et al., 2021; Vaneeckhaute and Weinberg, 2017).

EMERGING POLLUTANTS

For the scientific community and those involved in wastewater treatment, the emergence of new pollutants also called emergent pollutants (EPs), which include hormones, antibiotics, illegal medications, EDCs, PCPs, pesticides, surfactants and microplastics; in water systems, it is a serious problem as conventional wastewater treatment systems are unable to remove these pollutants even when they are in trace levels. To tackle this issue, sophisticated methods have been created, emphasizing adsorption-oriented procedures using novel adsorbents such as metal–organic frameworks, chitosan, nanocellulose and cyclodextrin polymers. AOPs and biological-based technologies also exhibit potential because of their effectiveness and environmental friendliness. CNTs have shown great promise in cleaning up EPs. Under ideal circumstances, their vast surface area, developed pore structure and photocatalytic properties of nanomaterials allow for the adsorption and degradation of pharmaceuticals and PCPs and EDCs, with removal rates of up to 95%. Nanomaterials provide scalable potential for wastewater treatment by transforming EPs into innocuous byproducts like CO_2 and H_2O when they are included in NC membranes or utilized as photocatalysts (Kurwadkar et al., 2019; Morin et al., 2022).

NANOTECHNOLOGICAL PROCESSES FOR WASTEWATER TREATMENT

Three primary categories in which nanotechnological wastewater treatment methods can be distinguished are: (a) pollution prevention, (b) sensing and detection, and (c) treatment and remediation (Watlington, 2005).

ADVANCED OXIDATION PROCESSES

For wastewater to properly breakdown organic pollutants, advanced treatment techniques are necessary. AOPs have become more popular for treating wastewater in the last 20 years. AOPs are very effective because they produce reactive hydroxyl radicals (•OH), which break down contaminants quickly and without discrimination. In contrast to traditional techniques, AOPs can work well even on persistent organic pollutants and generate less harmful intermediates. Numerous AOPs have been effectively used, such as Fenton reactions (Fe^{2+}/H_2O_2), UV-based processes and ozonation (O_3). Nanomaterials such as nanoparticles, nanomembranes and nanotubes have been used in recent developments to greatly enhance adsorption and catalysis. The sustainability of certain AOPs can be further increased by using photocatalysis. Heavy metals, organic and other pollutants can be effectively removed by nanomaterials like TiO_2, FeO_2, cerium oxide, magnetic chitosan and other NCs. By making it possible to create sophisticated catalysts with increased efficiency and lower energy requirements, nanotechnology with AOPs has the potential of more innovation (Bethi et al., 2016; Hodges et al., 2018; Kurian, 2021).

NANOFILTRATION MEMBRANES

NF are a "fourth class" of pressure-driven membranes that were initially presented by Dr. Peter Eriksson in 1984 where the advantages of reverse osmosis (RO) and ultrafiltration (UF) are combined in order to remove heavy metals, organic pollutants, pesticides, bacteria and viruses. This approach is notable for its ease of use, dependability, minimal energy usage and great removal efficiency to ensure that public drinking water sources are safe. NF membrane, also referred to as "membrane softening," uses less energy than lime or zeolite softening and runs at lower pressures than RO, which lowers operating expenses and therefore is well-suited for treating wastewater. They are adaptable and work via both solution-diffusion and pore-size flow (convective) processes. NF membranes can withstand high temperatures since they are composed of materials like ceramics, polyamide, cellulose acetate and chemically resistant polymers. NF is a key technique for water treatment because of its compositional and design flexibility, which offers further uses (Abdel-Fatah, 2018; Chorawalaa and Mehta, 2015; Vidu et al., 2020; Zekic ́et al., 2018).

NANOSCALE ZERO-VALENT IRON AND MAGNETIC NANOPARTICLES

Nanoscale zero-valent iron (nZVI) is synthesized from iron (II) and iron (III) using sodium borohydride, producing particles 10–100 nm in size with a zero-valent iron core and a mixed-valence oxide shell. It has large surface area, high reactivity and dual adsorption-reduction capabilities that make it an excellent material for wastewater treatment. nZVI effectively breaks down pollutants such as chlorinated hydrocarbons, nitroaromatics, organic dyes and heavy metals into less harmful byproducts. Also, it outperforms granular iron, enabling the reduction of inorganic anions. Bimetallic variants of nZVI enhance its redox properties, enabling the detoxification of persistent pollutants (Gebre, 2023; Kumar, 2023; Li et al., 2014).

Magnetic nanoparticles are highly effective nanoadsorbents that are emerging as a promising solution for wastewater treatment, particularly in removing both organic and inorganic pollutants due to their unique properties, including dynamic morphology, superparamagnetism, high saturation magnetization and optimal size. These nanoparticles are highly efficient in removing pollutants such as heavy metals, dyes, pesticides and pharmaceuticals, achieving removal rates of up to 98% within minutes. Functionalizing their surfaces with inorganic compounds, small organics or polymers further enhances their adsorption capacity and redox reaction efficiency. However, challenges such as scalability, regeneration methods and optimization of production protocols must be addressed to facilitate commercial applications (Ali et al., 2017; Gao, 2019; Shukla et al., 2021).

MONITORING, DETECTION AND NANOSENSORS

Monitoring water quality remains critical for wastewater treatment due to low concentrations of pollutants and failure of traditional methods to detect emerging pathogens. Nanosensors offer transformative solutions with superior sensitivity, selectivity and response times. These sensors incorporate three core components: signal transduction technologies, recognition agents and nanomaterials, which enhance detection efficiency through their optical, electrochemical and magnetic properties, enabling multiplex target detection. This field is advancing rapidly with nanomaterials like nanotubes, nanowires and nanocomposites offering customized properties. Enhanced bionanosensors now incorporate fluorescent, electrochemical and piezoelectric technologies for unmatched precision, revolutionizing water monitoring by exceeding conventional detection capabilities (Falciola et al., 2020; Kumar, 2023).

ENVIRONMENTAL AND HUMAN SAFETY CONSIDERATIONS WITH NANOTECHNOLOGY IN WASTEWATER TREATMENT

As with any revolutionary technology, the environmental and human health implications of nanotechnology necessitate thorough examination. Nanotechnology introduces three primary safety concerns: deliberate human exposure, accidental exposure and environmental harm (Roco et al., 2011). The environmental harm caused by nanotechnology includes accumulation of nanomaterials in soil, challenging safe disposal of nanomaterials and traditional nanomaterial synthesis (Lens et al., 2013; Osman, 2019).

To address these environmental harms caused by nanomaterials, a balanced approach that integrates robust nano-safety frameworks, collaborative research efforts and green synthesis methods is essential. A systematic framework known as "nano-safety" has been proposed, which aims to evaluate risks and implement safeguards to protect both the environment and human health (Roco et al., 2011; Zekic et al., 2018). Establishing research priorities and creating new safety standards are urgent steps to ensure the safe integration of nanotechnology and wastewater treatment into society for which collaborations among academia, regulatory agencies, independent institutions and industry are vital (Osman, 2019; Rathod et al., 2024). The shift toward green synthesis represents a transformative approach to nanoparticle production, supporting the sustainable development of nanotechnology while minimizing its environmental footprint. This approach not only reduces reliance on hazardous chemicals but also aligns with zero-effluent discharge goals, promoting a circular economy (Gautam et al., 2019).

By addressing the risks associated with nanomaterials while harnessing their benefits, nanotechnology can contribute to a cleaner, safer and more sustainable future.

FUTURE OUTLOOK OF NANOTECHNOLOGY IN WASTEWATER TREATMENT

Future developments in nanotechnology are essential to advance sustainable wastewater treatment and address the limitations of traditional methods. Improvements in MONPs, carbon-based nanomaterials and PNMs can enhance wastewater treatment efficiency.

Research into biogenic synthesis pathways offers eco-friendly solutions using natural biomolecules like proteins and flavonoids. These green approaches reduce pollution, align with sustainability goals and show promise in removing pollutants. However, most studies use synthetic solutions, underscoring the need for real-world applications. A lifecycle approach is vital to manage environmental and health risks, while addressing nanotoxicology's early-stage knowledge gaps for safe and effective implementation.

CONCLUSION

Nanotechnology has emerged as a transformative approach to addressing critical challenges in wastewater treatment. Leveraging the unique properties of nanomaterials, related technologies have demonstrated remarkable efficiency in tackling complex wastewater treatment issues. Additionally, nanomaterial-based sensors enhance water quality monitoring with precise, real-time detection of pollutants, improving overall water quality and safety.

Despite these advancements, scaling nanotechnology for widespread use requires addressing associated environmental and safety concerns. These challenges necessitate robust "nano-safety" frameworks and transition toward green nanotechnology as it aligns with the principles of environmental stewardship, reduce secondary pollution and support circular economy practices.

At last, for nanotechnology to realize its full potential, continued research is essential to optimize nanomaterials for large-scale applications, enhance economic feasibility and ensure sustainable integration into existing systems. By responsibly advancing nanotechnological innovations, we can redefine global wastewater treatment methods and contribute significantly in achieving sustainable development goals. Nanotechnology offers a path to cleaner water, resilience against pollution and sustainable resource management, ensuring ecological balance and human well-being for future generations.

REFERENCES

Abdelbasir, S. M. and A. E. Shalan. 2019. An overview of nanomaterials for industrial wastewater treatment. *Korean Journal of Chemical Engineering*, 36: 1209–1225.

Abdel-Fatah, M. A. 2018. Nanofiltration systems and applications in wastewater treatment. *Ain Shams Engineering Journal*, 9(4): 3077–3092.

Ahmed, M., M. Elektorowicz, and S. W. Hasan. 2019. GO, SiO_2, and SnO_2 nanomaterials as highly efficient adsorbents for Zn^{2+} from industrial wastewater—A second stage treatment to electrically enhanced membrane bioreactor. *Journal of Water Process Engineering*, 31: 100815.

Ali, I., C. Peng, I. Naz, et al. 2017. Phytogenic magnetic nanoparticles for wastewater treatment: A review. *RSC Advances*, 7(64): 40158–40178.

Amari, A., N. Elboughdiri, D. Ghernaout, R. H. Lajimi, A. M. Alshahrani, M. A. Tahoon, and F. B. Rebah. 2021. Multifunctional crosslinked chitosan/nitrogen-doped graphene quantum dot for wastewater treatment. *Ain Shams Engineering Journal*, 12(4): 4007–4014.

Arshad, F., M. S. Ali, L. Ali, and L. Zou. 2025. Catalytic MnO_2/MXene/chitosan nanocomposite membrane in removing caffeine from wastewater. *Separation and Purification Technology*, 361: 131504.

Ayalew, A. A. 2022. A critical review on clay-based nanocomposite particles for application of wastewater treatment. *Water Science and Technology*, 85(10): 3002–3022.

Bethi, B., S. H. Sonawane, B. A. Bhanvase, and S. P. Gumfekar. 2016. Nanomaterials-based advanced oxidation processes for wastewater treatment: A review. *Chemical Engineering and Processing-Process Intensification*, 109: 178–189.

Bhat, S. A., F. Sher, M. Hameed, et al. 2022. Sustainable nanotechnology based wastewater treatment strategies: Achievements, challenges and future perspectives. *Chemosphere*, 288: 132606.

Bracamontes-Ruelas, A. R., L. A. Ordaz-Díaz, A. M. Bailón-Salas, J. C. Ríos-Saucedo, Y. Reyes-Vidal, and L. Reynoso-Cuevas. 2022. Emerging pollutants in wastewater, advanced oxidation processes as an alternative treatment and perspectives. *Processes,* 10(5):1041.

Chawla, S., H. Uppal, M. Yadav, N. Bahadur, and N. Singh. 2017. Zinc peroxide nanomaterial as an adsorbent for removal of Congo red dye from waste water. *Ecotoxicology and Environmental Safety*, 135: 68–74.

Chhabra, H., and M. Kumar. 2020. Size and shape dependent equation of state for nanomaterials with application to bulk materials. *Journal of Physics and Chemistry of Solids,* 139: 109308.

Chorawalaa, K. K., and M. J. Mehta. 2015. Applications of nanotechnology in wastewater treatment. *International Journal of Innovative and Emerging Research in Engineering*, 2(1): 21–26.

Dutta, A. K., S. K. Maji, and B. Adhikary. 2014. γ-Fe_2O_3 nanoparticles: an easily recoverable effective photo-catalyst for the degradation of rose Bengal and methylene blue dyes in the waste-water treatment plant. *Materials Research Bulletin*, 49: 28–34.

El-Dossoki, F. I., T. M. Atwee, A. M. Hamada, and A. A. El-Bindary. 2021. Photocatalytic degradation of Remazol Red B and Rhodamine B dyes using TiO_2 nanomaterial: Estimation of the effective operating parameters. *Desalination and Water Treatment*, 233: 319–330.

Falciola, L., V. Pifferi, and A. Testolin. 2020. Detection methods of wastewater contaminants: state of the art and role of nanotechnology. In Barbara, B., S. F. Francesca, R. Ilenia, and S. Rajandrea (eds), *Nanomaterials for the Detection and Removal of Wastewater Pollutants* (pp. 47–68). Elsevier, Amsterdam, Netherlands.

Gao, F. 2019. An overview of surface-functionalized magnetic nanoparticles: Preparation and application for wastewater treatment. *ChemistrySelect*, 4(22): 6805–6811.

Gautam, P. K., A. Singh, K. Misra, A. K. Sahoo, and S. K. Samanta. 2019. Synthesis and applications of biogenic nanomaterials in drinking and wastewater treatment. *Journal of Environmental Management*, 231: 734–748.

Gebre, S. H. 2023. Nanoscale zero-valent iron for remediation of toxicants and wastewater treatment. *Environmental Technology Reviews*, 12(1): 390–419.

Ghosh, N., S. Das, G. Biswas, and P. K. Haldar. 2022. Review on some metal oxide nanoparticles as effective adsorbent in wastewater treatment. *Water Science and Technology*, 85(12): 3370–3395.

Gopinath, K. P., D. V. N. Vo, D. Gnana Prakash, A. AdithyaJoseph, S. Viswanathan, and J. Arun. 2021. Environmental applications of carbon-based materials: A review. *Environmental Chemistry Letters*, 19: 557–582.

Guo, Z., C. Xie, P. Zhang, et al. 2017. Toxicity and transformation of graphene oxide and reduced graphene oxide in bacteria biofilm. *Science of the Total Environment*, 580: 1300–1308.

Hodges, B. C., E. L. Cates, and J. H. Kim. 2018. Challenges and prospects of advanced oxidation water treatment processes using catalytic nanomaterials. *Nature Nanotechnology*, 13(8): 642–650.

Huangfu, X., Y. Xu, C. Liu, et al. 2019. A review on the interactions between engineered nanoparticles with extracellular and intracellular polymeric substances from wastewater treatment aggregates. *Chemosphere*, 219: 766–783.

Huntington, T. G. 2006. Evidence for intensification of the global water cycle: Review and synthesis. *Journal of Hydrology,* 319: 83–95.

Iqbal, T., A. Basit, A. M. Elgorban, et al. 2025. Cobalt doped MoS_2: A photoactivated nanomaterial for removal of methylene blue and phenol. *Journal of Physics and Chemistry of Solids*, 197: 112422.

Jain, K., A. S. Patel, V. P. Pardhi, and S. J. S. Flora. 2021. Nanotechnology in wastewater management: A new paradigm towards wastewater treatment. *Molecules*, 26(6): 1797.

Jang, H. J., M. H. Lee, E. J. Lee, X. Yang, and I. C. Kong. 2017. Toxicity evaluation of metals and metal-oxide nanoparticles based on the absorbance, chlorophyll content and cell count of chlorella vulgaris. *Clean Technology*, 23: 27–33.

Joshi, K. M., and V. S. Shrivastava. 2011. Photocatalytic degradation of Chromium (VI) from wastewater using nanomaterials like TiO_2, ZnO, and CdS. *Applied Nanoscience*, 1(3): 147–155.

Joshi, P., K. Gupta, R. Gusain, and O. P. Khatri. 2020. Metal oxide nanocomposites for wastewater treatment. In Raneesh, B. and P. M. Visakh (eds), *Metal Oxide Nanocomposites: Synthesis and Applications* (pp. 361–397). Wiley, New York.

Khodakarami, M. and M. Bagheri. 2021. Recent advances in synthesis and application of polymer nanocomposites for water and wastewater treatment. *Journal of Cleaner Production*, 296: 126404.

Kong, L., G. Fang, Y. Chen, et al. 2019. Efficient activation of persulfate decomposition by Cu_2FeSnS_4 nanomaterial for bisphenol A degradation: Kinetics, performance and mechanism studies. *Applied Catalysis B: Environmental*, 253: 278–285.

Kumar, A., C. Guo, G. Sharma, et al. 2016. Magnetically recoverable ZrO_2/Fe_3O_4/chitosan nanomaterials for enhanced sunlight driven photoreduction of carcinogenic Cr (VI) and dechlorination & mineralization of 4-chlorophenol from simulated waste water. *RSC Advances*, 6(16): 13251–13263.

Kumar, B. 2021. Green synthesis of gold, silver, and iron nanoparticles for the degradation of organic pollutants in wastewater. *Journal of Composites Science*, 5(8): 219.

Kumar, S. 2023. Smart and innovative nanotechnology applications for water purification. *Hybrid Advances*, 3: 100044.

Kurian, M. 2021. Advanced oxidation processes and nanomaterials-a review. *Cleaner Engineering and Technology*, 2: 100090.

Kurwadkar, S., T. V. Hoang, K. Malwade, S. R. Kanel, W. F. Harper, and G. Struckhoff. 2019. Application of carbon nanotubes for removal of emerging contaminants of concern in engineered water and wastewater treatment systems. *Nanotechnology for Environmental Engineering*, 4: 1–16.

Kyzas, G. Z. and A. C. Mitropoulos. 2021. Nanomaterials and nanotechnology in wastewater treatment. *Nanomaterials*, 11(6): 1539.

Lens, P. N. L., J. Virkutye, V. Jegatheesan, S. H. Kim, and S. Al-Abed. 2013. *Nanotechnology for Water and Wastewater Treatment*. IWA Publishing, London.

Li, S., W. Wang, Y. Liu, and W. X. Zhang. 2014. Zero-valent iron nanoparticles (nZVI) for the treatment of smelting wastewater: A pilot-scale demonstration. *Chemical Engineering Journal*, 254: 115–123.

Liu, L., X. B. Luo, L. Ding, and S. L. Luo. 2019. Application of nanotechnology in the removal of heavy metal from water. In Luo, X. and F. Deng (eds), *Nanomaterials for the Removal of Pollutants and Resource Reutilization* (pp. 83–147). Elsevier, Amsterdam, Netherlands.

Liu, W., R. Xia, X. Ding, et al. 2022. Impacts of nano-zero-valent iron on antibiotic removal by anaerobic membrane bioreactor for swine wastewater treatment. *Journal of Membrane Science*, 659: 120762.

Morin-Crini, N., E. Lichtfouse, M. Fourmentin, et al. 2022. Removal of emerging contaminants from wastewater using advanced treatments. A review. *Environmental Chemistry Letters*, 20(2): 1333–1375.

Naseem, T. and T. Durrani. 2021. The role of some important metal oxide nanoparticles for wastewater and antibacterial applications: A review. *Environmental Chemistry and Ecotoxicology*, 3: 59–75.

Nasir, A., F. Masood, T. Yasin, and A. Hameed. 2019. Progress in polymeric nanocomposite membranes for wastewater treatment: Preparation, properties and applications. *Journal of Industrial and Engineering Chemistry*, 79: 29–40.

Njimou, J. R., V. Fai, M. T. Sieugaing, et al. 2025. Hydrothermal synthesis of a nickel-oxide-infused orange peel nanobiocomposite for enhanced heavy metal removal from mining wastewater. *Hybrid Advances*, 9: 100392.

Oki, T. and S. Kanae. 2006. Global hydrological cycles and world water resources. *Science*, 313: 1068–72.

Osman, E. M. 2019. Environmental and health safety considerations of nanotechnology, nano safety. *Biomedical Journal of Scientific & Technical Research*, 19(4): 14501–14515.

Ownby, M., D. A. Desrosiers, and C. Vaneeckhaute. 2021. Phosphorus removal and recovery from wastewater via hybrid ion exchange nanotechnology: A study on sustainable regeneration chemistries. *Clean Water*, 4(1): 6.

Pasini, S. M., A. Valerio, G. Yin, et al. 2021. An overview on nanostructured TiO_2–containing fibers for photocatalytic degradation of organic pollutants in wastewater treatment. *Journal of Water Process Engineering*, 40: 101827.

Prasse, C., and T. Ternes. 2010. Removal of organic and inorganic pollutants and pathogens from wastewater and drinking water using nanoparticles–a review. In Frimmel, F., and R. Niessner, (eds) *Nanoparticles in the Water Cycle: Properties, Analysis and Environmental Relevance* (pp. 55–79). Springer, Berlin.

Rathod, S., S. Preetam, C. Pandey, and S. P. Bera. 2024. Exploring synthesis and applications of green nanoparticles and the role of nanotechnology in wastewater treatment. *Biotechnology Reports*, 41: e00830.

Roco, M. C., M. C. Hersam, C. A. Mirkin, et al. 2011. Nanotechnology environmental, health, and safety issues. In Mihail, C. R., C. A. Mirkin, and M. C. Hersam (eds), *Nanotechnology Research Directions for Societal Needs in 2020: Retrospective and Outlook* (pp. 159–220). Springer, Dordrecht.

Saikia, L., D. Bhuyan, M. Saikia, B. Malakar, D. K. Dutta, and P. Sengupta. 2015. Photocatalytic performance of ZnO nanomaterials for self sensitized degradation of malachite green dye under solar light. *Applied Catalysis A: General*, 490: 42–49.

Schwarzenbach, R. P., T. Egli, T. B. Hofstetter, U. VonGunten, and B. Wehrli. 2010. Global water pollution and human health. *Annual Review of Environment and Resources*, 35(1): 109–136.

Shabnum, S. S., R. Siranjeevi, R. Susmitha, et al. 2025. Enhanced photocatalytic degradation of crystal violet and malachite green oxalate dyes by $NiAg_2O$ infused chitosan nanocomposites. *International Journal of Biological Macromolecules*, 286: 138365.

Shukla, S., R. Khan, and A. Daverey. 2021. Synthesis and characterization of magnetic nanoparticles, and their applications in wastewater treatment: A review. *Environmental Technology & Innovation*, 24: 101924.

Singh, S., V. Kumar, R. Romero, K. Sharma, and J. Singh. 2019. Applications of nanoparticles in wastewater treatment. In Prasad, R., V. Kumar, M. Kumar, and D. Choudhary (eds), *Nanobiotechnology in Bioformulations* (pp. 395–418). Springer, Cham.

Slimani, Y., A. Khan, M. Nawaz, M. K. Hossain, and A. Thakur. 2024. Efficient rhodamine B dye photocatalytic degradation in aqueous media using novel ZnO nanomaterials co-doped with Ce and Dy. *Spectrochimica Acta Part A: Molecular and Biomolecular Spectroscopy*, 321: 124725.

Thines, R. K., N. M. Mubarak, S. Nizamuddin, J. N. Sahu, E. C. Abdullah, and P. Ganesan. 2017. Application potential of carbon nanomaterials in water and wastewater treatment: A review. *Journal of the Taiwan Institute of Chemical Engineers*, 72: 116–133.

UNESCO. 2009. *The United Nations World Water Development Report 3: Water in a Changing World.* UNESCO/Berghahn Books, Paris/New York.

Vaneeckhaute, C., and E. Weinberg. 2017. Nutrient recovery from wastewaters as high-valued hydroponic fertilizer and non-fertilizer products using hybrid ion exchange nanotechnology. *Communications in Agricultural and Applied Biological Sciences*, 352: 102.

Verma, D. K., A. Sharma, L. Awasthi, et al. 2023. Recent development and importance of nanoparticles in disinfection and pathogen control. In Bachheti, R.K., A. Bachheti, and A. Husen, (eds), *Nanomaterials for environmental and agricultural sectors* (pp. 83–106). Springer Nature, Singapore.

Vidu, R., E. Matei, A. M. Predescu, et al. 2020. Removal of heavy metals from wastewaters: A challenge from current treatment methods to nanotechnology applications. *Toxics*, 8(4): 101.

Vilardi, G., J. M. Ochando-Pulido, M. Stoller, N. Verdone, and L. DiPalma. 2018. Fenton oxidation and chromium recovery from tannery wastewater by means of iron-based coated biomass as heterogeneous catalyst in fixed-bed columns. *Chemical Engineering Journal*, 351: 1–11.

Wang, L., Q. Yang, Y. Cui, et al. 2017. Highly stable and biocompatible dendrimer-encapsulated gold nanoparticle catalysts for the reduction of 4-nitrophenol. *New Journal of Chemistry*, 41(16): 8399–8406.

Watlington, K. 2005. *Emerging Nanotechnologies for Site Remediation and Wastewater Treatment.* National Network for Environmental Management Studies Fellow, North Carolina State University, Washington, DC.

Yang, Y., J. Peng, H. Tao, et al. 2023. Synergistic effect of exposed facets and surface defects of ZnO nanomaterials for photocatalytic ozonation of organic pollutants. *Environmental Science: Nano*, 10(7): 1897–1906.

Yang, Y., C. Zhang, and Z. Hu. 2013. Impact of metallic and metal oxide nanoparticles on wastewater treatment and anaerobic digestion. *Environmental Science: Processes & Impacts is a Monthly*, 15: 39–48.

Yin, Z., C. Cui, H. Chen, Duoni, X. Yu, and W. Qian. 2020. The application of carbon nanotube/graphene-based nanomaterials in wastewater treatment. *Small*, 16(15): 1902301.

Zekic, E., Ž. Vukovic, and I. Halkijevic. 2018. Application of nanotechnology in wastewater treatment. *Građevinar*, 70(04): 315–323.

Zhou, Y., J. Meng, M. Zhang, et al. 2019. Which type of pollutants needs to be controlled with priority in wastewater treatment plants: Traditional or emerging pollutants?. *Environment International*, 131: 104982.

11 Nanomaterials as Adsorbents for Heavy Metal Decontamination of Wastewater

Inayat M. Khan, Asma Shakeel, Syed Andleeba, Aneesa Manzoor, Rehana Shakeel, and M. H. Chesti

INTRODUCTION

Water represents an indispensable resource and a critical component for life on Earth. Earth is often designated as the "blue planet" in debt to the fact that approximately 70% of its surface is covered by water. Out of this, 95% is saline water, leaving only 2.5% as fresh water. Of this restricted freshwater resource, about 68.9% is confined to glaciers, ice caps, and snow, whereas 30.8% prevails as groundwater. A mere 0.3% of groundwater is readily accessible for human consumption (Dhingra, 2024). Clean and safe water is indispensable for the well-being and economic development of a society (Jain et al., 2021). However, a significant decline in water quality with emerging global concerns has been marked in response to escalating global population, amplified urban growth, industrial expansion, and intensive agricultural activities (Singha & Mishrab, 2020). In the present era, about 1.2 billion populations are devoid of accessibility to fresh water

DOI: 10.1201/9781003613183-11

globally, while 2.6 billion cope with the challenge of meeting basic sanitation requirements. Millions of deaths have been attributed to unsafe and polluted water, especially children in debt of numerous waterborne diseases (Kumar et al., 2014). This scary snapshot necessitates the advent of strategies to address freshwater scarcity and encompasses the reclamation and reuse of wastewater. The application of reclaimed water is predominantly employed for non-drinking engagements encircling groundwater replenishment, agriculture, landscape irrigation, firefighting, vehicle washing, construction purposes, and cooling systems in heat-based power facilities. On a global scale, treated wastewater imparts an instrumental part in enhancing farm productivity, apart from streamlining the livelihoods of countless smallholder agrarians (Kesari et al., 2021).

The proficient employment of wastewater reclamation is subject to specific criteria to guarantee its maximum benefits. However, the selection of the most suitable approach and materials for treating wastewater is imperative to balance the efficiency and cost involved therein. The prime factors considered for the selection process encircle the effectiveness of the approach, reusability of materials, accessibility, affordability, and environmental sustainability (Jangid & Inbaraj, 2021). Established wastewater treatment approaches encompassing standard filtration, biosand filtration, distillation, reverse osmosis, flocculation-coagulation, bio-sand filtration, and standard filtration are quite inadequate in the remediation of heavy metals from wastewater completely. The adsorption approach outshines most of the available technologies on account of its cost-effectiveness, exceptional efficiency, and operational simplicity. Adsorption entails the employment of various adsorbent materials such as activated carbon, zeolites, waste biomaterials, and polymers. Undoubtedly, conventional adsorbents are widely employed given the adsorption of noxious pollutants; however, the limited adsorption efficiency often restricts their use for treating waste water (Sadegh et al., 2017). Consequently, to address these inadequacies, intensive research efforts have been directed at exploring advanced adsorbents with improved performance. Nanostructure materials characterized by exceptional surface area have proven to be significantly more efficient in contrast to conventional adsorbents. These materials exhibit rapid adsorption rates in hand with amplified removal efficiencies, illustrating their edge in modern wastewater treatment applications (Dil et al., 2017).

Nanotechnology entangles the engineering of matter at a scale of atoms and molecules to establish improved structures with enhanced attributes entailing improved electronic, optical, mechanical, conductive, and magnetic characteristics (Jain, 2019). With the inception of advanced technical interventions, the scientific domain increasingly embraces interventions that are more affordable, cleaner, and safer in contrast to conventional ones. Given the rapid depletion of the natural resource base, the financial implications of these technologies are also considered (Roco, 2011). Nanotechnology offers a solution to these technical gaps. Nanotechnology is clearer, cleaner, and more affordable than traditional mass production approaches and intensive machinery. Additionally, the nanotechnology approach aligns with the realm of being applied across all areas of life. These areas encompass nanomaterial science, nanoelectronics, and nanomedicine and integrate into various aspects of chemistry, as well as physical and biological sciences. Consequently, it is reasonable to project that nanotechnological solutions would manifest as an essential aspect of research for upcoming cohorts

(Malik et al., 2023). Nanotechnology is acknowledged for its significant potential in furnishing perks in diverse sectors, such as developing drugs, purification of water, and information and communication interventions in hand with the fabrication of lighter and stronger materials. This field entails the material formation and manipulation at a nanoscale, accomplished by either assembling individual atoms or refining and reducing the bulk materials (Benelmekki, 2015). Nanotechnology goes beyond gradual advancements to prevailing techniques by offering paradigm shifting and ground-breaking interventions. These innovations might provide an instantaneous way out to benefit the society, ecosphere, and the overall planet. Key expanses where nanotechnology engraves its indelible imprints on account of smart material development and allied toolkits encompass resource management, energy, environmental protection, and health services (Pokrajac et al., 2021).

The term "nano" finds its roots to the Greek works connoting to "dwarf," and signifies an extremely tiny scale. A nanometre is equal to one-billionth of a metre (10^{-9} m) (Zhang et al., 2024). Materials exhibiting dimensions of 1–100 nm are generally classified as nanomaterials. Such materials exhibit notable and unique attributes comprising exceptional adsorption proficiencies, an elevated surface-to-volume ratio, enhanced catalytic activity, and increased surface reactivity. Nanostructures furnish exceptional perks for creating efficient catalysts and redox-active media concerning the water purification in debt of their nanoscale dimensions, extensive surface area, and effortless functionalization. These characteristics craft nanomaterials as highly effective adsorbents in removing a diverse array of contaminants from wastewater, encompassing organic compounds, heavy metals, and inorganic compounds, biological toxins, dyes, and pathogens accountable for diseases like cholera and typhoid (Lu et al., 2016).

ADSORPTION PHENOMENON

The adsorption phenomenon represents a surface-based aspect experiencing adsorbate accumulation on the adsorbent surface. The accretion or deposition of adsorbate molecules onto the adsorbent surface is facilitated by the intermolecular forces of attraction in response to the interplay of adsorbent–adsorbate molecules. In bulk materials, atomic bonding encircling ionic, covalent, or metallic is quenched by neighbouring atoms. The atomic entities residing on the surface of adsorbent material lack complete bonding with surrounding atoms, enabling them to attract the adsorbate molecules (Wang et al., 2012). The type of bonding formed during adsorption relies on the nature of the substances involved. Adsorption can be characterized as either physisorption, encompassing weak van der Waals forces for adherence to the surface of adsorbents, or chemisorption, featuring stronger interplay like covalent bonding or electrostatic attraction (Sadegh et al., 2017). Figure 11.1 illustrates various underlying principles involved in the course of adsorption using various adsorbents. The crosstalk between adsorbate and adsorbent is imperative in determining the effectiveness of the adsorption phenomena.

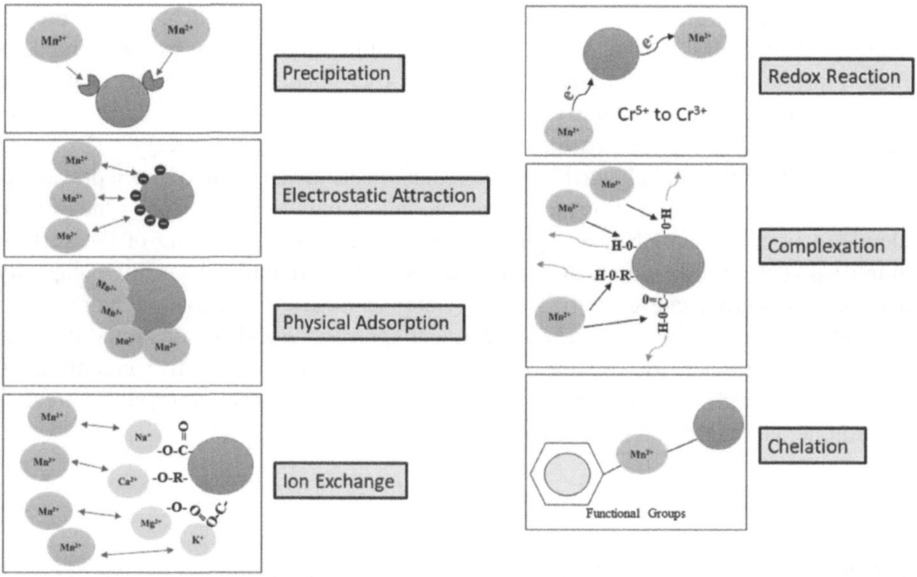

FIGURE 11.1 Mechanistic overview of adsorbent–adsorbate interactions.

A large surface area pertaining to the respective adsorbents amplifies the adsorption capacity and reflects the adsorbent selection as a crucial footstep for optimizing the proficiency, adaptability, and economic effectiveness of the employed approach (Pandey et al., 2022). While traditional adsorbents, such as zeolites, activated alumina, activated carbon, silica gels, and ion exchange resins, play their part in pollutant removal, they do exhibit certain challenges. These challenges encompass the expensive nature of materials, the need for pre-treatment, disposal, and regeneration issues. And these often pose issues related to disposal and regeneration. Alternate adsorbents encircling magnetic materials, natural clays, and nanoadsorbents are capturing attention in debt of their proficiency in removing environmental contaminants (Pandey, 2017).

NANOADSORBENTS FOR WASTEWATER TREATMENT

Nanoadsorbents have emerged as the most appropriate resources for treating wastewater in debt of their easy application and impending potential to yield an array of adsorbents (Basheer, 2018). They are recognized for their exceptional surface area, enormous adsorption sites, requirement of low modification temperatures, configurable pores, truncated intraparticle distance, and improved surface chemistry (Khin et al., 2012). Nanoadsorbents exhibit incomparable morphologies that impart exceptional proficiencies and functional properties. The implicit stability and balance amid the energies from polar charge, surface area, and agile deformation are accounted for these morphological characteristics of nanoadsorbent materials. Given these unparalleled attributes, nanoadsorbents have engraved their promise and substitution for conventional adsorbents employed for various wastewater treatment processes (El-Sayed, 2020). The properties and behaviour of nanoadsorbents are elucidated by the prevalence and characteristics of active surface spots, encompassing the quantity and type of prevalent functional groups. Key attributes encompass atomic positioning of the surface, amplified surface area, augmented chemical reactivity, remarkable sorptive capacity, and elevated binding energy at the surface (Gatoo et al., 2014). The inception of nanoadsorbents has transformed wastewater treatment approaches and demonstrated the most appropriate approach for treating organic, inorganic, and radioactive contaminants (Jiang et al., 2018). The description of various nanoadsorbents employed for treating environmental contaminants is portrayed in Table 11.1.

TABLE 11.1 Performance of various nanoadsorbents in treating environmental contaminants

NANOADSORBENT USED	*CONTAMINANT*	*REMOVAL EFFICIENCY/ MAXIMUM ADSORPTION CAPACITY*	*REFERENCES*
Carbon dots modified mesoporous organosilica	2,4-Dichlorophenol	99.70 mg/g	Wang et al. (2015)
Iron oxide nanoparticles	Lead	36.00%	Keshvardoostchokami et al. (2017)
Multi-walled carbon nanotubules	Copper	93.00%	Gupta et al. (2017)
Polyaniline/ZnO nanocomposite	Chromium	346.18 mg/g	Ahmad and Hasan (2017)
$CoFe_2O_4$ nanozyme	Methylene Blue	91.20%	Wu et al. (2018)
Polyhydroxylbutyrate-functionalized carbon nanotubes	Chromium	98.19%	Bankole et al. (2019)
Starch–silver NPs from Cocoa pods	Rhodamine B	70.00%	Azeez et al. (2020)
Polydopamine/metal organic framework thin film nanocomposite	Cadmium	99.20%	He et al. (2020)

(*Continued*)

TABLE 11.1 (*Continued*) Performance of various nanoadsorbents in treating environmental contaminants

NANOADSORBENT USED	*CONTAMINANT*	*REMOVAL EFFICIENCY/ MAXIMUM ADSORPTION CAPACITY*	*REFERENCES*
Silver nanoparticle modified palm kernel	Phenol	90.30%	Aremu et al. (2020)
Shell-activated carbon			
Copper nanozyme	Methyl Orange	93.00%	Geng et al. (2021)
Fluorescent chitosan-based hydrogel	Chromium	228.2 mg/g	Luo et al. (2021)
g-C_3N_4 nanosheet-carbon dot-FeOCl nanocomposites	Tetracycline hydrochloride	99%	Asadzadeh-Khaneghah et al. (2021)
Silver oxide (Ag_2O) nanoparticles	Sulfamethoxazole	98.93%	El Messaoudi et al. (2023)
Carbon @ nano zero-valent iron composite	Lead	98.70%	Yang et al. (2023)
Chitosan-Titanium Dioxide	Cadmium	133.33 mg/g	Abdulrasool et al. (2024)
Chitosan-functionalized MnO_2 nanoadsorbent	Lead	94.40%	Singh et al. (2024)

Carbon Nanotubes

Carbon nanomaterials comprise a captivating subject of research and a versatile tool for diverse domains including the adsorption process. These reflections might be accredited to the prevalence of unique structural and electronic features affiliated with these carbon nanotubes. These nanomaterials exhibit exceptional effectiveness in wastewater treatment on account of their ability to adsorb diverse contaminants, possession of extensive surface area, rapid adsorption kinetics, and specificity for certain compounds such as aromatics. Among carbon nanomaterials, carbon nanotubes possess an edge over other forms encasing fibres, carbon beads, and nanoporous carbon materials. Carbon nanotubes are primarily composed of graphene sheets rolled into cylindrical shapes and possess diameters as small as 1 nm. The porous structure of carbon nanotubes accounts for their incomparable characteristics encircling amplified specific surface area, apart from an excellent adsorption capacity for an array of pollutants. To further augment their functionality, carbon nanotubes are pooled with metals or other materials to enhance their surface area, mechanical strength, optical behaviour, and electrical properties (Khin et al., 2012).

Silver Nanomaterials

Silver nanomaterials are predominantly employed in the colloidal form and have demonstrated substantial proficiency as antimicrobial agents, specifically against naturally occurring microorganisms (Reidy et al., 2013). These nanomaterials disrupt the essential cellular functions in microbes and direct to the death of respective organisms. Silver nanomaterials facilitate cell lysis by adhering to the cell membrane and eliminating pathogens thereof. Silver nanomaterials find their place in community water systems and hospitals concerning water purification, and they have been employed as an alternative to chlorine in the filtration process. The antimicrobial action of silver nanomaterials operates by triggering oxidative stress, which disrupts the critical cellular organizations in pathogenic microbes. Owing to their cost-effective nature and sound anti-bacterial properties, silver nanomaterials are considered a highly promising alternative for water disinfection (Quang et al., 2013).

Recently, Polyethersulfone (PES) microfiltration membranes have been integrated with chemically synthesized silver nanomaterials. This amalgam intends to inhibit microbial activity across the membrane and promote the effectiveness of water purification systems (Ferreira et al., 2015).

Iron Nanomaterials

Iron nanomaterials have acquired prominent applications in wastewater reclamation due to their excellent proficiency and versatility. The treatment mechanism affiliated with iron nanomaterials primarily relies on the reductive dehalogenation reactions in which contaminants are reduced and converted to non-toxic forms. Iron nanomaterials are economically effective and adhere to the pollutants to constitute hydroxides that play their part as flocculants and facilitate the remediation of both inorganic compounds and organic contaminants (Klačanová et al., 2013). Iron oxide nanomaterials characterized by extensive surface area and exceptional reactivity act as solid adsorbents to bind with the solid and colloidal impurities. Elemental iron in conjunction with metal catalysts embracing palladium, nickel, or platinum constitutes bimetallic nanomaterials that heighten the kinetics of redox reactions (Zhang & Elliott, 2006). These adsorbents are remarked for their strong sorption capacities, ease of use, adaptability, and swift accessibility, making them a preference in wastewater treatment. The predominant iron oxide forms employed include magnetite, maghemite, and nonmagnetic hematite. The distinctive chemical, physical, and magnetic attributes of iron oxide nanomaterials pay for the effectiveness and cost-efficiency of these materials in contrast to conventional ones (Babel & Kurniawan, 2003). In addition, the iron oxide nanomaterials exhibit photocatalytic properties due to absorbance of visible wavelengths and generate electron–hole pairs, which eventually facilitate the removal of water contaminants. Mechanisms encompassing the selective adsorption, electrostatic interplay, surface interactions, and ligand binding further back the removal potential of these contaminants from wastewater (Yang & Yin, 2017).

Titanium Oxide Nanomaterials

The photocatalytic degradation approach has been acknowledged to treat wastewater effectively due to the disintegration of contaminants to intermediate products exhibiting low molecular weights. These intermediate compounds are further converted into reusable forms encircling carbon dioxide, water, and anions like chloride, nitrate, and phosphates. Metal oxides and sulphide semiconductors comprise the prominent group of photocatalysts employed in wastewater treatment. Titanium oxide (TiO_2) has been extensively explored owing to its remarkable attributes entailing photocatalytic efficiency, affordability, photostability, and chemical and biological durability (Oghyanous, 2022). TiO_2 represents one of the most efficient photocatalysts on account of its non-toxic, stable, cost-efficient, and accessible nature. The broad-spectrum activity and low selectivity affiliated with titanium oxide allow it to degrade an extensive array of noxious pollutants (Ohsaka et al., 2008). The photocatalytic performance of titanium oxide has been marked to enhance in response to metal doping, with silver being the most prominent and explored element.

Nano Zero-Valent Iron (nZVI) as Adsorbents

Zero-valent nanoparticles have garnered considerable buzz for their outstanding part in wastewater treatment in debt of their peculiar characteristics. Among these, nano zero-valent iron stands as the prime and widely explored adsorbent owing to its exceptional adsorption capacity, precipitation capability, and proficiency of acting as a reducing entity for contaminants that exhibit vulnerability to the redox reactions (Nassar, 2012). nZVI is instrumental in addressing a wide array of contaminants prevalent in wastewater. Despite having countless advantages, nZVI adsorbents possess certain constraints and the prime being its tendency to aggregate which aftermaths complicates its separation from wastewater system. To overcome this constraint and augment its performance in wastewater treatment, certain modification approaches have been introduced. These interventions encompass doping, surface coating of metals, encapsulation, conjugation, and emulsification and are directed at improving the stability and reactivity of nZVI (Lu et al., 2016).

Carbon quantum dots (CQDs)

Carbon quantum dots (CQDs) have emerged as a promising category of nanomaterials in debt of their exceptional properties such as multi-colour fluorescence, excellent photostability, biocompatibility, and ease of surface modification (Rossi et al., 2022). The mechanisms backing the adsorption process via CQD application encompass surface adsorption in debt of high large surface area in hand with the predominance of surface functional moieties. These functional groups exhibit a healthy interplay with contaminants via chemical adsorption or electrostatic interaction, leading to the strong adherence of adsorbate molecules and their subsequent removal from wastewater. In addition, the hydrophobic interaction of CQDs with non-polar contaminants accounts for the contaminant removal from wastewater. This interplay facilitates the partitioning of aqueous phase and hydrophobic surface of quantum carbon dots. Chelation and coordination of various pollutants, specifically heavy metal ions, comprise another profound mechanism. The contaminants form stable complexes with the functional moieties prevalent on CQD surface. These functional groups act as ligands to claw the metal ions and thereby promote the adsorption capacity of CQDs. Photocatalytic properties of CQDs assist in the accrual of reactive oxygen species, which assist in pollutant degradation by oxidation process and pay to the comprehensive performance of wastewater treatment processes (Agbogo et al., 2024).

Nanozymes in Wastewater Treatment

Nanozymes connote to the nanomaterials exhibiting an enzyme-mimicking activity and emerge as the budding substitute for natural enzymes in wastewater reclamation. Nanozymes are documented to exhibit peroxidase, oxidase, superoxide dismutase, and catalase resembling activity. In addition, nanozymes possess an elevated stability in contrast to the natural enzymes, which eventually aids them to enact at a diverse temperature and pH range values. The remediation of organic pollutants with nanozymes is predominantly accomplished using the degradation process (Liu et al., 2022). However, the degradation processes involved are intricate and are accompanied by certain intermediate reactions, producing different intermediate products. In the case of catalytic processes, nanozymes facilitate the catalysis of pollutant degradation and constitute certain intermediates, which eventually produce degraded products. The degradation of contaminants using nanozymes is illustrated in Figure 11.2. The illustration represents

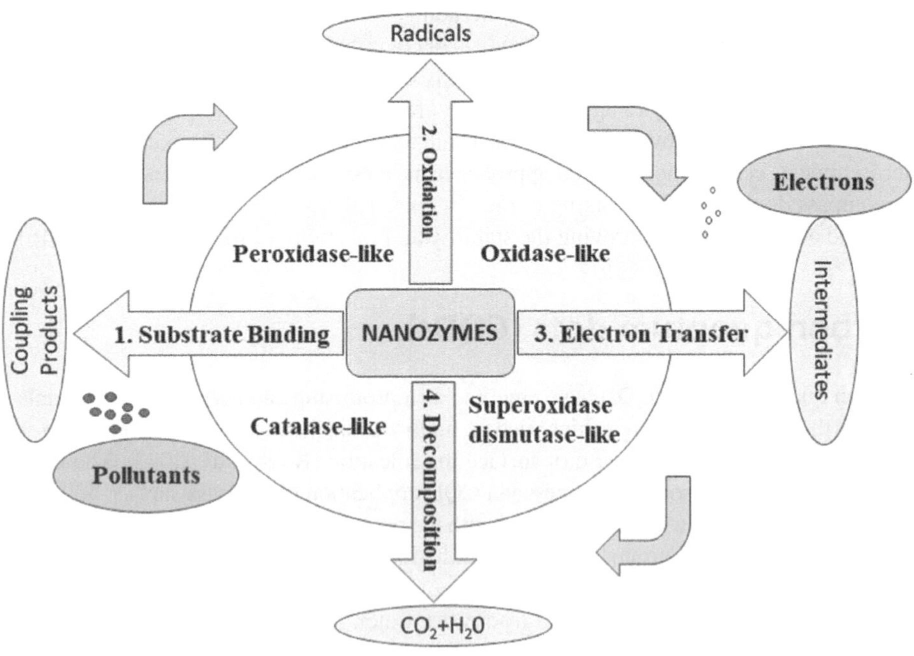

FIGURE 11.2 Nanozyme-mediated degradation process of pollutants.

that the nanozymes adhere to pollutants and form coupling products (Wang et al., 2019), which are subsequently subjected to oxidation process to form free radicals, followed by electron transfer reactions (Tang et al., 2018).

SYNTHESIS OF NANOADSORBENTS

The synthesis of nanosorbent materials is accomplished using two prominent approaches: (a) Bottom-up approach and (b) top-down approach. The suitability of the method is contingent on the specific advantages each offers. The top-down approach is a conventional approach involving a breakdown of large materials into nanoscale particles. This is achieved with interventions such as acid oxidation, electrochemical or sonochemical exfoliation, micromechanical approaches, energy-intensive ball milling, and reactive milling. Structures possessing long-range order on macroscopic scale are prime contenders of the top-down approach. Undoubtedly, the top-down approach is readily accessible, yet it falls short in furnishing the nanoparticles with complex sizes and shapes. This approach is afflicted with the challenge of producing particles of the right size and form in debt of enhanced surface dislocation, polydispersions, aggregate formation, and clustering (Khan et al., 2022). Nevertheless, it continues to play a crucial part in nanoparticle synthesis (Yadav et al., 2012). In parallel, the bottom-up technique entangles the development of nanosorbent materials through integrations of atoms or molecules. The bottom-up approach is accomplished through the employment of atomic layer deposition, sol-gel processes, unzipping carbon nanotubes, arch discharge, graphene oxide reduction, and physical/chemical vapour deposition (Saleh, 2021). The bottom-up approach stands as the advanced intervention and delivers a precise control over structure of nanosorbent materials. Consequently, the approach ensures the delivery of nanostructures exhibiting a meagre volume of defects, improved homogeneity in chemical composition, and the desired shape and size of the crystal structures (Kumar and Kumar, 2014). In addition, this approach has the potential to deliver less waste and hence be more cost-effective. However, the prominent drawbacks afflicted with this approach encircle the need for chemical purification during the nanofabrication process, cost-intensive production, and difficulty in producing at a large scale. Despite these confinements, it outshines other approaches. Apart from chemical and physical preparation approaches, nanoparticle synthesis is performed in line with the green approaches utilizing biological processes. Such an approach entangles the exploitation of microorganisms encircling fungi, bacteria, algae, and actinomycetes, plant extracts, enzymes, and bio-reduction methods (Nadaroglu et al., 2017). The synthetic pathways commonly acknowledged for nanoadsorbent synthesis encompass (a) co-precipitation, (b) sol-gel, (c) reduction, and (d) hot-injection methods. Figure 11.3 illustrates the synthesis pathways for nanoadsorbents.

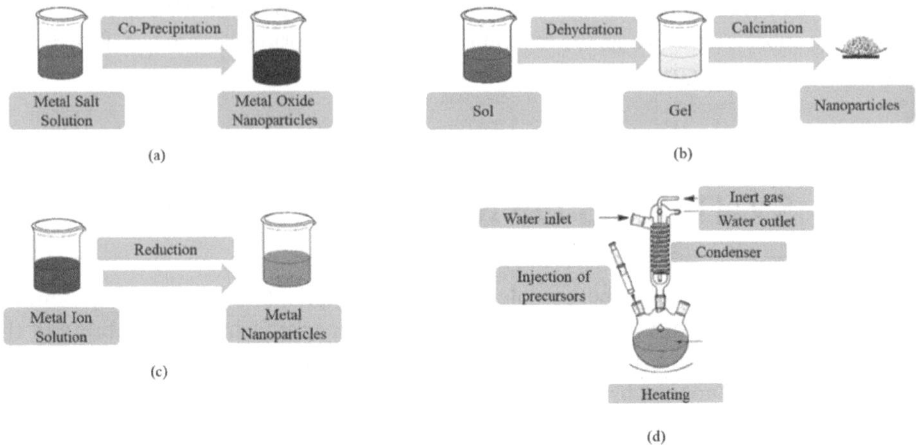

FIGURE 11.3 Schematic representation of synthesis pathways for nanoadsorbents.

Green Synthesis of Nanoadsorbents

Green synthesis connotes an economically effective and environmentally friendly tactic in nanostructured material synthesis that possesses tunable morphologies, structures, and particle size distribution (Khan et al. 2019). The green approach engages organisms entangling bacteria, plants, fungi, etc. to synthesize nanomaterials and has manifested as an escalating area of research owing to their reduced toxicity or non-toxic feature, environmental-friendly attribute, economic feasibility, and sustainability (Goutam et al. 2018). The employment of plant-assisted nanoadsorbent synthesis demands the exploitation of those plants that possess the impending part in heavy metal accretion and detoxification (Iravani, 2011; Goutam et al., 2018). The acquisition of plant extract could be accomplished using distinct plant parts such as leaves, stems, flowers, and seeds. The extract entails biomolecules encircling amino acids, proteins, vitamins, enzymes, flavonoids, terpenoids, phenolic acids, and alkaloids, which perform as reducing or capping agents for nanomaterial production. Phenolics and terpenoids are organic polymers with strong antioxidant properties that act as stabilizing drivers for nanomaterial synthesis. Flavonoids encasing flavonol, anthocyanin, flavanone, chalcone, flavone, and isoflavonoid possess the proficiency of reducing metal ions during nanoparticle synthesis (Goutam et al., 2020). However, the challenges in the extraction of raw materials, enduring synthesis period, and homogeneity in particle size of the products persist, which eventually affect the quality of ultimate produce and reaction time (Ying et al., 2022). In addition, the nanoparticle synthesis during the bioreduction of plant extracts is governed by different physical and chemical parameters entailing temperature, pH, etc. (Iravani, 2011). The low pH values are acknowledged to reduce the nucleation rate of metal ions, which enhances the particle agglomeration and results in the development of large-sized nanoparticles, while the higher pH values result in the development of nanoparticles with small dimensions (Thanh et al., 2014).

CHALLENGES IN NM APPLICATION

Nanotechnology has introduced significant strides in wastewater treatment processes for contaminant removal and is afflicted by several challenges that need to be addressed for effective remediation of environmental pollutants. One major issue is the tiring recovery and separation of nanoadsorbents after the treatment process on account of their small size (Lu et al., 2016). Fouling and shrinkage aspects further complicate the engagement of nanofibres in treating wastewaters. These fibres often exhibit poor mechanical characteristics and might deform after their employment in wastewater treatment (Qayoom et al., 2022). Agglomeration constitutes another prominent issue with nanoscale particles. The agglomeration of respective particles prevails due to the development of electrostatic interplay, increased surface energy, or simple physical entanglement (Asghar and Mushtaq, 2023). Moreover, the potential toxicity of residual nanoadsorbents remains a concern. The requisition of enormous quantities of nanomaterials for the accomplishment of treatment processes within a reasonable time frame may restrict their utilization potential. The above-mentioned aspects necessitate that future research is directed to advance the comprehension of these challenges and ensure the efficiency and safety of nanoadsorbent materials for wastewater reclamation. The advanced interventions in nanosorbent technology intend to address the prime constraints such as scalability challenges, poor selectivity, and reduced stability. Functionalization of nanoadsorbents with specific chemic moieties including thiol, amine, and carboxyl groups has promoted their selectivity for specific contaminants, while biomimetic tactics comprising molecularly imprinted polymers have further enhanced the specificity through biological recognition mimicking (Chen et al., 2016). Stability issues have been addressed through the inception of core–shell structures possessing robust coatings of carbon or silica and the employment of cross-linked nanoadsorbents to sustain the environmental extremes (Satyam and Patra, 2024). Recovery and Reusability issues have been managed through the integration of magnetic properties with nanoadsorbents, which ensures the swift recovery and reuse via external magnetic fields. Magnetic nanoadsorbents comprise the prime constituents of water treatment systems, specifically for the elimination of organic contaminants (Sahebi et al., 2019). Magnetic separation of nanoadsorbents is accomplished by layering ligands onto magnetic nanoparticles and moulds them into the adsorbents having specific attraction to target contaminants (Perez et al., 2019). The regeneration of nanoadsorbents through ion-exchange process, cleaning agents, and magnetic forces has crafted their promise in commercial viability and economic effectiveness (Tripathy et al., 2024). The green synthesis of nanoadsorbents using plant extracts facilitates the adoption of sustainable synthesis approaches, reduces the operational costs and environmental imprints, and tackles the scalability issues affiliated with nanoadsorbents (Phouthavong et al., 2022). The fouling issue associated with nanoparticles has been mitigated through the functionalization of the membrane surfaces with amine and carboxylic acid groups, which enhance the hydrophilicity of the membranes (Enfrin et al., 2021). Physical cleaning is usually performed to alleviate reversible fouling and to dislodge the particles accrued on the membrane surfaces. The application of a two-step process has been acknowledged to address this issue in which forward flushing is employed to rinse off the foulants conveyed by a backwashing step (Jang et al., 2018). The gas scouring technique entailing the injection of gas into the feed stream has been recognized as a cost-effective solution to optimize the membrane proficiency in resisting particulate fouling such as bentonite, silica, and water-suspended solids (Kim et al., 2020). Furthermore, the nanoparticles are finely engineered with antifouling coatings, such as PEG and zwitterion coatings, to boost the proficient application of nanoadsorbent materials (Bevilacqua et al., 2021).

CONCLUSION AND FUTURE RESEARCH DIRECTIONS

Adsorption stands as the most proficient approach for the removal of environmental pollutants with minimal or no production of harmful by-products. Nanomaterials showcase a promising potential with regard to wastewater treatment, owing it to the cost-effective, seamless operation, eco-friendly, and time-efficient nature. Nanoadsorbents exhibit an exceptional efficiency in pollutant removal and offer an environmentally sustainable alternative to conventional adsorbent materials. However, some prominent issues encircling potential environmental toxicity, regeneration challenges, and scalability need to be resolved in future.

The research directions need to be aligned at developing the biodegradable materials to minimize the ill environmental imprints and enhance the performance and long-term viability of nanoadsorbents in wastewater treatment. In addition, the cost-effective regeneration approaches without comprising the adsorption efficiency need to be validated. Exploring the hybrid adsorbents that combine nanomaterials with polymers or natural adsorbents could reinforce their adsorption characteristics. Furthermore, the engineering and tailoring of nanoadsorbents for remediating target specific pollutants such as heavy metals, emerging contaminants encircling pharmaceuticals, and microplastics is imperative. It is formidable to appraise the environmental imprints affiliated with nanoadsorbents, entangling the potential toxicity to terrestrial and aquatic life forms. By addressing these gaps, nanoadsorbents might deliver a proficient and sustainable preference for wastewater treatment in the upcoming epoch.

REFERENCES

Abdulrasool, M. M., Kamari, A., Wong, S. T. S., Abdullah, N. N. A., Bakar, N. A., Ahmad, W. H. W., & Sutapa, I. W. (2024). Chitosan-titanium dioxide nanoadsorbent for Cd (II), Cu (II), Ni (II) and Pb (II) ions: Adsorption isotherm and characterization studies. *IOP Conference Series: Earth and Environmental Science*, 1425(1), 012029.

Agbogo, U. V., Rifore, B. S., Arum, C. T., Iorver, P. D., Mathew, J., & Tanko, S. A. (2024). Carbon quantum dots for wastewater treatment: Present progress and future prospects. *Fudma Journal of Sciences*, 8(1), 93–102.

Ahmad, R., & Hasan, I. (2017). Efficient remediation of an aquatic environment contaminated by Cr (VI) and 2, 4-dinitrophenol by XG-g-polyaniline@ ZnO nanocomposite. *Journal of Chemical & Engineering Data*, 62(5), 1594–1607.

Aremu, M. O., Arinkoola, A. O., Olowonyo, I. A., & Salam, K. K. (2020). Improved phenol sequestration from aqueous solution using silver nanoparticle modified palm kernel shell activated carbon. *Heliyon*, 6(7), e04492.

Asadzadeh-Khaneghah, S., Habibi-Yangjeh, A., Seifzadeh, D., Chand, H., & Krishnan, V. (2021). Visible-light-activated g-C_3N_4 nanosheet/carbon dot/FeOCl nanocomposites: Photodegradation of dye pollutants and tetracycline hydrochloride. *Colloids and Surfaces A: Physicochemical and Engineering Aspects*, 617, 126424.

Asghar, F., & Mushtaq, A. (2023). The future of nanomaterial in wastewater treatment: A review. *International Journal of Chemical and Biochemical Science*, 23, 150–157.

Azeez, L., Lateef, A., Adejumo, A. L., Adeleke, J. T., Adetoro, R. O., & Mustapha, Z. (2020). Adsorption behaviour of rhodamine B on hen feather and corn starch functionalized with green synthesized silver nanoparticles (AgNPs) mediated with cocoa pods extracts. *Chemistry Africa*, 3(1), 237–250.

Babel, S., & Kurniawan, T. A. (2003). Low-cost adsorbents for heavy metals uptake from contaminated water: A review. *Journal of Hazardous Materials*, 97(1–3), 219–243.

Bankole, M. T., Abdulkareem, A. S., Mohammed, I. A., Ochigbo, S. S., Tijani, J. O., Abubakre, O. K., & Roos, W. D. (2019). Selected heavy metals removal from electroplating wastewater by purified and polyhydroxylbutyrate functionalized carbon nanotubes adsorbents. *Scientific Reports*, 9(1), 4475.

Basheer, A. A. (2018). New generation nano-adsorbents for the removal of emerging contaminants in water. *Journal of Molecular Liquids*, 261, 583–593.

Benelmekki, M. (2015). An introduction to nanoparticles and nanotechnology. *Designing Hybrid Nanoparticles*, 2015, 1–14.

Bevilacqua, P., Nuzzo, S., Torino, E., Condorelli, G., Salvatore, M., & Grimaldi, A. M. (2021). Antifouling strategies of nanoparticles for diagnostic and therapeutic application: A systematic review of the literature. *Nanomaterials*, 11(3), 780.

Chen, L., Wang, X., Lu, W., Wu, X., & Li, J. (2016). Molecular imprinting: Perspectives and applications. *Chemical Society Reviews*, 45(8), 2137–2211.

Dhingra, N. (2024). Application of nanoadsorbents for lead decontamination in water. In: Kumar, N., and Jha, A. K. (eds), *Lead Toxicity Mitigation: Sustainable Nexus Approaches* (pp. 169–182). Cham: Springer Nature Switzerland.

Dil, E. A., Ghaedi, M., & Asfaram, A. (2017). The performance of nanorods material as adsorbent for removal of azo dyes and heavy metal ions: Application of ultrasound wave, optimization and modeling. *Ultrasonics Sonochemistry*, 34, 792–802.

Enfrin, M., Wang, J., Merenda, A., Dumée, L. F., & Lee, J. (2021). Mitigation of membrane fouling by nano/microplastics via surface chemistry control. *Journal of Membrane Science*, 633, 119379.

El Messaoudi, N., El Mouden, A., Fernine, Y., El Khomri, M., Bouich, A., Faska, N., & Lacherai, A. (2023). Green synthesis of Ag_2O nanoparticles using Punica granatum leaf extract for sulfamethoxazole antibiotic adsorption: Characterization, experimental study, modeling, and DFT calculation. *Environmental Science and Pollution Research*, 30(34), 81352–81369.

El-Sayed, M. E. (2020). Nanoadsorbents for water and wastewater remediation. *Science of the Total Environment*, 739, 139903.

Ferreira, A. M., Roque, É. B., da Fonseca, F. V., & Borges, C. P. (2015). High flux microfiltration membranes with silver nanoparticles for water disinfection. *Desalination and Water Treatment*, 56(13), 3590–3598.

Gatoo, M. A., Naseem, S., Arfat, M. Y., Mahmood Dar, A., Qasim, K., & Zubair, S. (2014). Physicochemical properties of nanomaterials: Implication in associated toxic manifestations. *BioMed Research International*, 2014(1), 498420.

Geng, X., Xie, X., Liang, Y., Li, Z., Yang, K., Tao, J., & Wang, Z. (2021). Facile fabrication of a novel copper nanozyme for efficient dye degradation. *ACS Omega*, 6(9), 6284–6291.

Goutam, S. P., Saxena, G., Roy, D., Yadav, A. K., & Bharagava, R. N. (2020). Green synthesis of nanoparticles and their applications in water and wastewater treatment. In: Saxena, G., and Bharagava, R. (eds), *Bioremediation of Industrial Waste for Environmental Safety: Volume I: Industrial Waste and Its Management* (pp. 349–379). Springer, Singapore.

Goutam, S. P., Saxena, G., Singh, V., Yadav, A. K., Bharagava, R. N., & Thapa, K. B. (2018). Green synthesis of TiO_2 nanoparticles using leaf extract of *Jatropha curcas* L. for photocatalytic degradation of tannery wastewater. *Chemical Engineering Journal*, 336, 386–396.

Gupta, V. K., Agarwal, S., Bharti, A. K., & Sadegh, H. (2017). Adsorption mechanism of functionalized multi-walled carbon nanotubes for advanced Cu (II) removal. *Journal of Molecular Liquids*, 230, 667–673.

He, M., Wang, L., Lv, Y., Wang, X., Zhu, J., Zhang, Y., & Liu, T. (2020). Novel polydopamine/metal organic framework thin film nanocomposite forward osmosis membrane for salt rejection and heavy metal removal. *Chemical Engineering Journal*, 389, 124452.

Iravani, S. (2011). Green synthesis of metal nanoparticles using plants. *Green Chemistry*, 13(10), 2638–2650.

Jain, K. (2019). Nanohybrids of dendrimers and carbon nanotubes: A benefaction or forfeit in drug delivery?. *Nanoscience & Nanotechnology-Asia*, 9(1), 21–29.

Jain, K., Patel, A. S., Pardhi, V. P., & Flora, S. J. S. (2021). Nanotechnology in wastewater management: A new paradigm towards wastewater treatment. *Molecules*, 26(6), 1797.

Jang, H., Charfi, A., Lee, J., Park, B., Jang, J., Kim, H., & Kim, J. (2018). Combined effect of periodic backwashing and forward flushing on fouling mitigation in a pressurized UF membrane process for high turbid surface water treatment. *Desalination and Water Treatment*, 101, 24–30.

Jangid, P., & Inbaraj, M. P. (2021). Applications of nanomaterials in wastewater treatment. *Materials Today: Proceedings*, 43, 2877–2881.

Jiang, Y., Liu, Z., Zeng, G., Liu, Y., Shao, B., Li, Z., & He, Q. (2018). Polyaniline-based adsorbents for removal of hexavalent chromium from aqueous solution: A mini review. *Environmental Science and Pollution Research*, 25, 6158–6174.

Kesari, K. K., Soni, R., Jamal, Q. M. S., Tripathi, P., Lal, J. A., Jha, N. K., & Ruokolainen, J. (2021). Wastewater treatment and reuse: A review of its applications and health implications. *Water, Air, & Soil Pollution*, 232, 1–28.

Keshvardoostchokami, M., Babaei, L., Zamani, A. A., Parizanganeh, A. H., & Piri, F. (2017). Synthesized chitosan/iron oxide nanocomposite and shrimp shell in removal of nickel, cadmium and lead from aqueous solution. *Global Journal of Environmental Science and Management*, 3(3), 267–278.

Khan, I., Saeed, K., & Khan, I. (2019). Nanoparticles: Properties, applications and toxicities. *Arabian Journal of Chemistry*, 12(7), 908–931.

Khan, Y., Sadia, H., Ali Shah, S. Z., Khan, M. N., Shah, A. A., Ullah, N., & Khan, M. I. (2022). Classification, synthetic, and characterization approaches to nanoparticles, and their applications in various fields of nanotechnology: A review. *Catalysts*, 12(11), 1386.

Khin, M. M., Nair, A. S., Babu, V. J., Murugan, R., & Ramakrishna, S. (2012). A review on nanomaterials for environmental remediation. *Energy & Environmental Science*, 5(8), 8075–8109.

Kim, Y., Li, S., & Ghaffour, N. (2020). Evaluation of different cleaning strategies for different types of forward osmosis membrane fouling and scaling. *Journal of Membrane Science*, 596, 117731.

Klačanová, K., Fodran, P., Šimon, P., Rapta, P., Boča, R., Jorík, V., & Čaplovič, L. (2013). Formation of Fe (0)-nanoparticles via reduction of Fe (II) compounds by amino acids and their subsequent oxidation to iron oxides. *Journal of Chemistry*, 2013(1), 961629.

Kumar, N., & Kumar, R. (2014). Introduction. *Nanotechnology and Nanomaterials in the Treatment of Life-Threatening Diseases* (pp. 1–51). William Andrew Publishing, Norwich, NY. doi:10.1016/b978-0-323-26433-4.00001-4.

Kumar, S., Ahlawat, W., Bhanjana, G., Heydarifard, S., Nazhad, M. M., & Dilbaghi, N. (2014). Nanotechnology-based water treatment strategies. *Journal of Nanoscience and Nanotechnology*, 14(2), 1838–1858.

Liu, Q., Zhu, X., Zhong, L., Zhang, S., Luo, X., Liu, Q., & Lu, Y. (2022). Recent advances in the applications of nanozymes for the efficient detection/removal of organic pollutants: A review. *Environmental Science: Nano*, 9(4), 1212–1235.

Luo, Q., Huang, X., Luo, Y., Yuan, H., Ren, T., Li, X., & Wu, Y. (2021). Fluorescent chitosan-based hydrogel incorporating titanate and cellulose nanofibers modified with carbon dots for adsorption and detection of Cr (VI). *Chemical Engineering Journal*, 407, 127050.

Malik, S., Muhammad, K., & Waheed, Y. (2023). Nanotechnology: A revolution in modern industry. *Molecules*, 28(2), 661.

Nadaroglu, H., Güngör, A. A., & Ince, S. (2017). Synthesis of nanoparticles by green synthesis method. *International Journal of Innovative Research and Reviews*, 1(1), 6–9.

Nassar, N. N. (2012). Iron oxide nanoadsorbents for removal of various pollutants from wastewater: An overview. In: Bhatnagar, A. (ed), *Application of Adsorbents for Water Pollution Control* (pp. 81–118). The Netherlands: Bentham Science Publishers.

Oghyanous, F. A. (2022). Nanoparticles in wastewater treatment. *Water Conservation: Inevitable Strategy*, 107, 107–119.

Ohsaka, T., Shinozaki, K., Tsuruta, K., & Hirano, K. (2008). Photo-electrochemical degradation of some chlorinated organic compounds on *n*-TiO_2 electrode. *Chemosphere*, 73(8), 1279–1283.

Pandey, P., Khan, F., Agarwal, S., & Singh, M. (2022). Nano adsorbents in wastewater treatment: A new paradigm in wastewater management. *Letters in Applied NanoBioScience*, 12, 125.

Pandey, S. (2017). A comprehensive review on recent developments in bentonite-based materials used as adsorbents for wastewater treatment. *Journal of Molecular Liquids*, 241, 1091–1113.

Perez, T., Pasquini, D., de Faria Lima, A., Rosa, E. V., Sousa, M. H., Cerqueira, D. A., & de Morais, L. C. (2019). Efficient removal of lead ions from water by magnetic nanosorbents based on manganese ferrite nanoparticles capped with thin layers of modified biopolymers. *Journal of Environmental Chemical Engineering*, 7(1), 102892.

Phouthavong, V., Yan, R., Nijpanich, S., Hagio, T., Ichino, R., Kong, L., & Li, L. (2022). Magnetic adsorbents for wastewater treatment: Advancements in their synthesis methods. *Materials*, 15(3), 1053.

Pokrajac, L., Abbas, A., Chrzanowski, W., Dias, G. M., Eggleton, B. J., Maguire, S., & Mitra, S. (2021). Nanotechnology for a sustainable future: Addressing global challenges with the international network4sustainable nanotechnology. *ACS Nano*, 15(12), 18608–18623.

Qayoom, A., Ashraf, I., Rashid, A., Ayoub, M., & Kumar, D. (2022). *Synthesis and Characterization of NiS and Zn Doped NiS Quantum Dots and Their Application in Methylene Blue Degradation in Aqueous Solution*. Research Square Platform LLC, Durham, NC.

Quang, D. V., Sarawade, P. B., Jeon, S. J., Kim, S. H., Kim, J. K., Chai, Y. G., & Kim, H. T. (2013). Effective water disinfection using silver nanoparticle containing silica beads. *Applied Surface Science*, 266, 280–287.

Reidy, B., Haase, A., Luch, A., Dawson, K. A., & Lynch, I. (2013). Mechanisms of silver nanoparticle release, transformation and toxicity: A critical review of current knowledge and recommendations for future studies and applications. *Materials*, 6(6), 2295–2350.

Roco, M. C. (2011). The long view of nanotechnology development: The national nanotechnology initiative at 10 years. *Journal of Nanoparticle Research*, 13, 427–445.

Rossi, B. L., Andrade, C. M., Therézio, E. M., Ramos, R. J., Vasconcelos, L. G., Terezo, A. J., & De Siqueira, A. B. (2022). Carbon quantum dots: An environmentally friendly and valued approach to sludge disposal. *Frontiers in Chemistry*, 10, 858323.

Sadegh, H., Ali, G. A., Gupta, V. K., Makhlouf, A. S. H., Shahryari-Ghoshekandi, R., Nadagouda, M. N., & Megiel, E. (2017). The role of nanomaterials as effective adsorbents and their applications in wastewater treatment. *Journal of Nanostructure in Chemistry*, 7, 1–14.

Sahebi, S., Sheikhi, M., & Ramavandi, B. (2019). A new biomimetic aquaporin thin film composite membrane for forward osmosis: Characterization and performance assessment. *Desalination and Water Treatment*, 148, 42–50.

Saleh, T. A. (2021). Protocols for synthesis of nanomaterials, polymers, and green materials as adsorbents for water treatment technologies. *Environmental Technology & Innovation*, 24, 101821.

Satyam, S., & Patra, S. (2024). Innovations and challenges in adsorption-based wastewater remediation: a comprehensive review. *Heliyon*, 10(9), e29573.

Singh, V., Singh, N., Rai, S. N., Chaturvedi, V. K., Singh, S. K., Kumar, A., & Mishra, V. (2024). Chitosan functionalized recyclable and eco-friendly nanoadsorbent for Pb (II) adsorption from water. *Nanocomposites*, 10(1), 214–226.

Singha, I., & Mishrab, P. K. (2020). Nano-membrane filtration a novel application of nanotechnology for waste water treatment. *Materials Today: Proceedings*, 29, 327–332.

Tang, L., Liu, Y., Wang, J., Zeng, G., Deng, Y., Dong, H., & Peng, B. (2018). Enhanced activation process of persulfate by mesoporous carbon for degradation of aqueous organic pollutants: Electron transfer mechanism. *Applied Catalysis B: Environmental*, 231, 1–10.

Thanh, N. T., Maclean, N., & Mahiddine, S. (2014). Mechanisms of nucleation and growth of nanoparticles in solution. *Chemical Reviews*, 114(15), 7610–7630.

Tripathy, J., Mishra, A., Pandey, M., Thakur, R. R., Chand, S., Rout, P. R., & Shahid, M. K. (2024). Advances in nanoparticles and nanocomposites for water and wastewater treatment: A review. *Water*, 16(11), 1481.

Wang, L., Cheng, C., Tapas, S., Lei, J., Matsuoka, M., Zhang, J., & Zhang, F. (2015). Carbon dots modified mesoporous organosilica as an adsorbent for the removal of 2, 4-dichlorophenol and heavy metal ions. *Journal of Materials Chemistry A*, 3(25), 13357–13364.

Wang, J., Huang, R., Qi, W., Su, R., Binks, B. P., & He, Z. (2019). Construction of a bioinspired laccase-mimicking nanozyme for the degradation and detection of phenolic pollutants. *Applied Catalysis B: Environmental*, 254, 452–462.

Wang, L., Li, J., Jiang, Q., & Zhao, L. (2012). Water-soluble Fe_3O_4 nanoparticles with high solubility for removal of heavy-metal ions from waste water. *Dalton Transactions*, 41(15), 4544–4551.

Wu, L., Wan, G., Hu, N., He, Z., Shi, S., Suo, Y., & Wang, G. (2018). Synthesis of porous $CoFe_2O_4$ and its application as a peroxidase mimetic for colorimetric detection of H_2O_2 and organic pollutant degradation. *Nanomaterials*, 8(7), 451.

Yadav, T. P., Yadav, R. M., & Singh, D. P. (2012). Mechanical milling: A top down approach for the synthesis of nanomaterials and nanocomposites. *Nanoscience and Nanotechnology*, 2(3), 22–48.

Yang, J. C., & Yin, X. B. (2017). $CoFe_2O_4$@ MIL-100 (Fe) hybrid magnetic nanoparticles exhibit fast and selective adsorption of arsenic with high adsorption capacity. *Scientific Reports*, 7(1), 40955.

Yang, L., Jin, X., Lin, Q., Owens, G., & Chen, Z. (2023). Enhanced adsorption and reduction of Pb (II) and Zn (II) from mining wastewater by carbon@ nano-zero-valent iron (C@ nZVI) derived from biosynthesis. *Separation and Purification Technology*, 311, 123249.

Ying, S., Guan, Z., Ofoegbu, P. C., Clubb, P., Rico, C., He, F., & Hong, J. (2022). Green synthesis of nanoparticles: Current developments and limitations. *Environmental Technology & Innovation*, 26, 102336.

Zhang, W. X., & Elliott, D. W. (2006). Applications of iron nanoparticles for groundwater remediation. *Remediation Journal: The Journal of Environmental Cleanup Costs, Technologies & Techniques*, 16(2), 7–21.

Zhang, X., Zhang, K., Shi, Y., Xiang, H., Yang, W., & Zhao, F. (2024). Surface engineering of multifunctional nanostructured adsorbents for enhanced wastewater treatment: A review. *Science of The Total Environment*, 920, 170951.

12 Unleashing Blockchain's Potential in Water Governance

Toward Transparent and Equitable Resource Management

Reyana A, Mathu T, and Esther Daniel

INTRODUCTION

Water Resource Management is essential to human life sustainability and economic growth. As the population grows, the challenges in water governance are increasing nowadays (Radmehr et al., 2022). Also, there are water issues due to pollution, scarcity, and inequitable distribution. With the recent developments in blockchain technology, a new pursuit has emerged in the governance and management of water resources.

DOI: 10.1201/9781003613183-12

This exploration delves into the untapped potential of blockchain in water governance, aiming to pave the way for transparent and equitable resource management (Minhas et al., 2020).

The urgent requirement for sustainable and efficient governance structures emphasizes tackling water-related concerns. Traditional systems frequently face data opacity, bureaucratic inefficiency, and a lack of diversity (Tonkin et al., 2021). Blockchain technology integration acts as a change catalyst, providing decentralized, transparent, and secure processes that have the potential to revolutionize the landscape of water resource management (Xia et al., 2022). This study elucidates the complex interaction between blockchain technology and water governance. A future in which blockchain enables traceable and accountable water use and promotes a fairer distribution of this valuable resource (Sandhu et al., 2020). This chapter provides policymakers, industry stakeholders, and the global community with the knowledge they need to fully realize blockchain's potential in achieving transparent and equitable water resource management through case studies, regulatory considerations, and forward-looking analysis.

UNDERSTANDING WATER GOVERNANCE

Definition and Importance

Water governance encompasses the processes, structures, policies, and practices that guide decision-making and implementing actions regarding water resources. Managing water fairly and responsibly enables everyone to get enough and healthy water resources in the future. It's about balancing the needs of people, businesses, and nature to make sure water is used wisely and doesn't run out. In a broader sense, water governance extends beyond government agencies and includes the active participation of various stakeholders, such as communities, non-governmental organizations, businesses, and individuals (Jiménez et al., 2020). It emphasizes collaboration, transparency, and inclusivity in decision-making processes, recognizing that water resource management is a complex and interconnected challenge that requires collective efforts. Healthy aquatic ecosystems depend on well-managed water resources, and sound governance practices help prevent pollution, habitat degradation, and biodiversity loss (Pahl-Wostl et al., 2020). With climate change causing more frequent and severe water-related events, adaptive governance structures are necessary to mitigate the impacts, manage water scarcity, and ensure the resilience of communities and ecosystems (Di Vaio et al., 2021). Effective governance becomes a tool for preventing conflicts in regions where water resources are shared among multiple users or countries (Babuna et al., 2023). Clear policies and cooperative frameworks can help manage transboundary water resources and promote collaboration rather than competition (Koebele et al., 2023).

Current Challenges in Water Governance

Water governance faces a variety of complex challenges, reflecting the intricate nature of managing a vital and shared resource. These challenges encompass a range of issues, from scarcity and pollution to governance structures and global water justice (Woodhouse & Muller, 2017; Mishra et al., 2021; Millington & Scheba, 2021; Koop et al., 2022). These are:

Water Scarcity: Rapid population growth, urbanization, and increased industrialization contribute to rising water demand, exacerbating scarcity issues in many regions. Water governance (Woodhouse & Muller, 2017).

Climate Change: Climate change exacerbates water scarcity challenges through altered precipitation patterns, extreme weather events, and shifts in hydrological cycles (Pahl-Wostl et al., 2020).

Water Quality and Pollution: Runoff from agriculture and industrial activities introduces pollutants into water bodies, affecting water quality and posing risks to ecosystems and public health.

Wastewater Management: Inadequate treatment and disposal of wastewater contribute to contamination of water sources, impacting both surface and groundwater quality.

Inadequate Infrastructure: Inadequate water infrastructure leads to water losses, inefficient distribution, and increased susceptibility to disruptions.

Lack of Access: The access to safe drinking water and sanitation is lacking largely due to inadequate infrastructure, especially in rural and underserved urban areas (Jiménez et al., 2020).

Corruption and Absence of Transparency: Corruption in water-related institutions and a lack of transparency can impede fair and equitable water governance (Mutschinski & Coles, 2021).

Transboundary Water Management: Shared water resources among different countries can lead to geopolitical tensions and conflicts over access, usage, and management (Makaya et al., 2020).

Lack of Cooperation: Limited cooperation and coordination among neighboring countries on transboundary water issues impede effective and sustainable management.

Addressing these challenges requires effective water governance that balances policy effectiveness, operational efficiency, and stakeholder trust and engagement, as shown in Figure 12.1.

Need for Transparency and Equity

Transparency and equity in water governance are paramount in sustaining life, supporting ecosystems, and fostering socioeconomic development. Achieving transparency and equity in water governance is essential for addressing the challenges associated

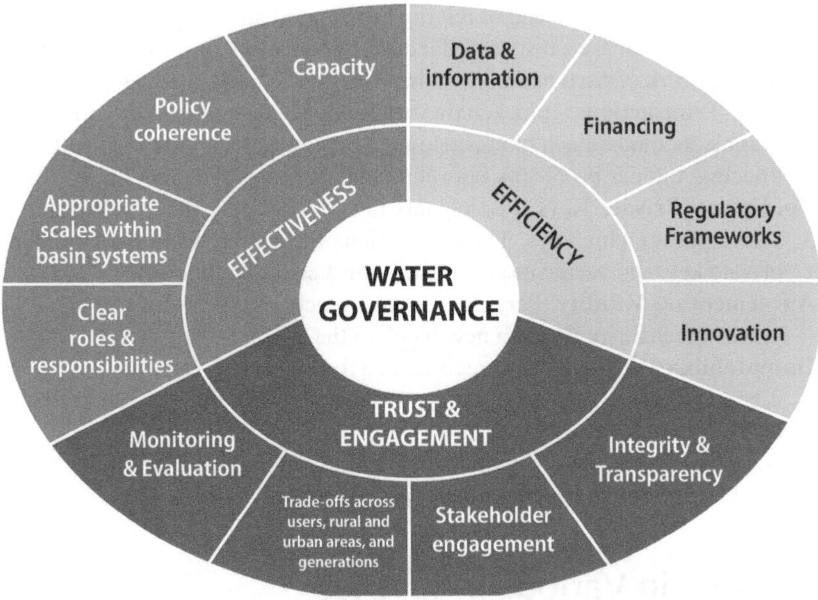

FIGURE 12.1 Strategies for effective water governance. (Romano & Akhmouch, 2019.)

with water scarcity, pollution, and unequal access (Salmoral et al., 2020). Transparency in decision-making processes builds accountability, allowing stakeholders to understand how water-related decisions are made, implemented, and evaluated. Transparent governance fosters trust among stakeholders, including communities, businesses, and governmental bodies, which is crucial for effective collaboration in water management. Equitable water governance addresses the needs of marginalized and vulnerable communities, preventing the concentration of water resources (Lieberherr et al., 2022). Transparency facilitates the active participation of local people and those involved in water governance, ensuring that diverse perspectives and needs are considered.

BLOCKCHAIN TECHNOLOGY OVERVIEW

Basics of Blockchain

Blockchain is a shared digital record-keeping system that runs cryptocurrencies like Bitcoin and can be used for many other purposes beyond digital money. The key concepts and features of blockchain are:

Distributed Ledger: Instead of relying on a central authority or database, water governance can use a distributed ledger, like blockchain, which is maintained by a network of participants working together (Anandan & Deepak, 2022).

Secure Transactions: Blockchain transactions are protected with cryptographic methods. Each block includes a previous block's hash value, forming a secure, tamper-proof link between them (Lesavre et al., 2020).

Public and Private Keys: Participants in the blockchain have public and private cryptographic keys. The transactions are signed using the confidential private key, and verification is done using a shared public key.

Agreement on Validity: Participants must reach an agreement through a consensus mechanism to add a new block to the chain.

Immutability: After a block is recorded on the blockchain, altering or removing it becomes extremely challenging. This ensures the integrity and reliability of the transaction history.

Smart Contracts: Smart contracts help enforce the terms directly without the need for intermediaries (Jaradat et al., 2022).

Blockchain in Various Industries

Blockchain technology can potentially transform a wide range of industries by introducing transparency, security, and efficiency into various processes (Kaur & Gupta, 2024). These are:

Financial Services: Streamlining financial processes such as settlements, asset transfers, and compliance through self-executing smart contracts.

Supply Chain Management: Blockchain boosts traceability with transparent records, reducing fraud and ensuring authenticity.

Healthcare: Blockchain guarantees the security and integrity of patient records, enabling secure sharing of health information among authorized parties.

Government and Public Services: Blockchain provides a transparent, secure identity management system that can enhance identity verification, mitigate identity theft, and streamline government services.

Applications of Blockchain in Water Governance

Blockchain technology offers several applications in enhancing the traceability of water sources, which include:

Origin Tracking (Zhang, 2023)
Supply Chain Transparency (Xia et al., 2022)
Verification of Water Quality (Vladucu et al., 2024)
Preventing Contamination and Fraud (Ramprasath et al., 2024)
IoT Integration for Real-time Monitoring (Sitanggang et al., 2023)

Smart Contracts for Compliance (Anglen, n.d.)
Water Rights and Ownership (Miller, 2021)
Decentralized Access to Information (Xia et al., 2022)
Collaboration Across Stakeholders (Kaushik, 2019)
Smart Contracts for Water Agreements

Smart contracts in water governance enable various improvements in resource management. They automate water allocations based on predefined conditions, such as weather patterns (Li & Kassem, 2021), and facilitate peer-to-peer water trading without intermediaries. Real-time monitoring through IoT integration allows dynamic allocation adjustments (Nguyen et al., 2019). Smart contracts can implement dynamic pricing, ensure regulation compliance, and create decentralized water markets, reducing costs. They also support contingency planning for emergencies, streamline administrative tasks (Oberhauser, 2019), and provide transparent billing. They also support multi-party agreements, enhancing stakeholder collaboration (Gurgun & Koc, 2022).

Blockchain for Water Quality Monitoring

A water quality monitoring system detects high levels of contaminants in a river. The data is documented on the blockchain, and a smart contract automatically notifies relevant authorities and triggers water treatment processes. The data collected is securely documented on the blockchain, creating a continuous, tamper-resistant record that ensures the integrity of historical data (Zeng et al., 2023; Alharbi et al., 2021). Furthermore, blockchain supports decentralized environmental monitoring systems for ecosystems and water bodies (Dogo et al., 2019), enabling dynamic responses and enhanced decision-making for sustainable water resource management and safety. Blockchain technology enhances water quality monitoring by improving data transparency, accuracy, and security, as shown in Figure 12.2. When integrated with IoT devices like water quality sensors, the parameters such as pH, turbidity, and chemical concentrations are monitored (Xia et al., 2022).

Success Case Studies

The success case studies are given as follows:

a. **Smart Water Grids in California, USA**
   Santa Monica, California, has implemented a blockchain platform to monitor and manage urban water usage. In partnership with Sweet Bridge, the system uses smart water meters linked to the blockchain to securely record and verify water consumption, enhancing data accuracy and transparency.
b. **Agricultural Water Management in California, USA**
   In collaboration with the California State Government, IBM developed a blockchain platform to manage water rights. This platform facilitates the water trading process among farmers, promoting a transparent and secure system for optimizing agricultural water usage and ensuring equitable distribution.

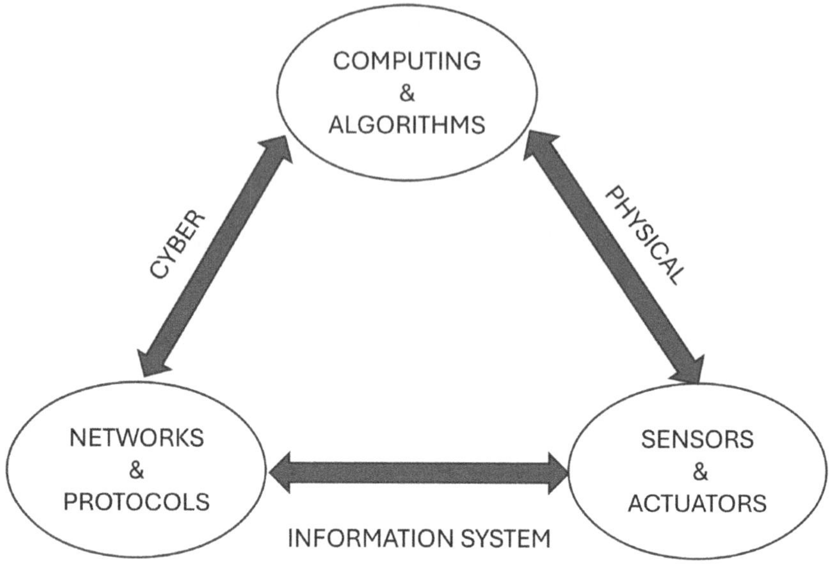

FIGURE 12.2 Blockchain for water quality monitoring.

c. **Transboundary Water Management in the Middle East**
The World Bank is exploring blockchain to improve trust and transparency in transboundary water management in the Middle East. Blockchain's decentralized structure can secure data sharing and transactions between countries sharing water resources.

d. **Decentralized Water Management in Kenya**
Hydro-Blockchain, a project in Kenya, aims to decentralize water management by allowing communities to manage and monitor water sources autonomously, ensuring data integrity and transparent water transactions.

From the above case studies, we come to know how Blockchain technology is being leveraged globally to enhance water management systems by improving transparency, security, and efficiency. Applications range from urban water monitoring and agricultural water trading to transboundary resource management and community-led water systems. These innovations address challenges like data accuracy, equitable distribution, and trust in shared resources.

REGULATORY AND ETHICAL CONSIDERATIONS

Regulatory and ethical considerations are key to water governance, ensuring water resources are managed sustainably and fairly. They are crucial in tackling issues such as water scarcity, pollution, and access to clean water. Here are some key aspects of regulatory and ethical considerations in water governance:

a. **Water Rights and Allocation**:

Fair and sustainable water distribution among users (agriculture, industry, households).

Ethical Focus: Equity for marginalized communities and future generations.

b. **Water Quality Standards**:

Regulations to prevent contamination and ensure safe drinking water.

Ethical Focus: Protecting ecosystems and vulnerable populations.

c. **Transboundary Water Management**:

Cooperation in shared water resources across regions/countries.

Ethical Focus: Equitable use, conflict prevention, and ecosystem protection.

d. **Access to Water and Basic Services**:

Ensuring universal, affordable access to clean water.

Ethical Focus: Recognizing water as a human right and common good.

e. **Climate Change Adaptation**:

Strategies for managing climate-related impacts on water resources.

Ethical Focus: Protecting vulnerable populations and ecosystems.

f. **Community Engagement and Participation**:

Inclusive decision-making in water governance.

Ethical Focus: Representation of diverse stakeholders.

g. **Enforcement and Compliance**:

Strengthening regulatory agencies for law enforcement.

Ethical Focus: Accountability for polluters and consistent implementation.

h. **Data Transparency and Information Sharing**:

Open access to water-related data for governance and decision-making.

Ethical Focus: Preventing data misuse and ensuring public awareness.

While all aspects are important, Transboundary Water Management has a broader geopolitical impact that directly influences water allocation, quality, enforcement, and climate adaptation. Poor transboundary management can worsen water crises, leading to social, economic, and political instability.

LEGAL FRAMEWORK FOR BLOCKCHAIN IN WATER GOVERNANCE

Effective governance ensures equitable access, sustainable use, and conflict prevention among riparian (water-sharing) nations. It requires strong legal frameworks, particularly for transboundary water management. Blockchain, an emerging technology, offers transparency, accountability, and efficiency in managing shared water resources. Various mechanisms, such as social planning, water markets, game theory, and

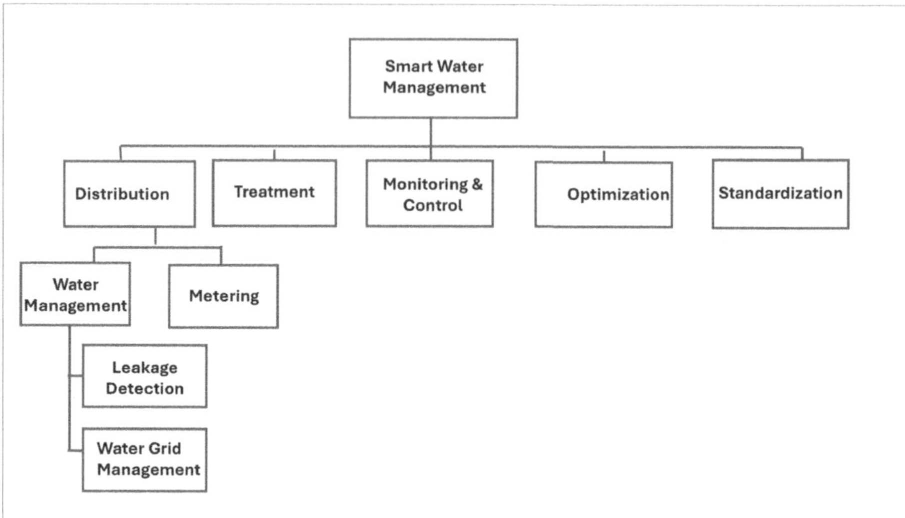

FIGURE 12.3 Legal framework for blockchain.

negotiation analyses, resolve transboundary water conflicts. Blockchain can complement these approaches by offering a transparent and tamper-proof system for recording agreements, monitoring compliance, and facilitating negotiations, thus contributing to more effective conflict resolution. A general overview of the potential legal framework considerations for blockchain in water governance is shown in Figure 12.3.

a. **Distribution**:

   Smart Metering and IoT Regulations: Blockchain can enhance transparency in smart water metering systems. Legal frameworks should regulate IoT devices and smart meters to ensure compliance with data protection laws and technical standards.

   Regulation of Tokenized Assets: If blockchain is used for tokenizing water rights, regulations must define their legal status, ownership, and transferability to avoid fraud and ensure accountability.

b. **Treatment**:

   Environmental Regulations: Water treatment facilities must comply with environmental laws that regulate pollution levels, chemical usage, and energy consumption. Blockchain can track and verify compliance with these regulations in real time.

   Digital Signatures and Authentication: Digital signatures ensure that once water quality reports or treatment logs are logged on the blockchain, alterations become formidable, thus reducing fraud and human errors.

c. **Monitoring and Control**:

   Data Protection and Privacy laws protect sensitive information from misuse. Blockchain transactions are transparent and traceable, raising concerns about data privacy in water governance. Legal frameworks must comply with data protection laws to ensure the security of sensitive water-related data.

Cybersecurity Regulations: Blockchain systems must be protected from cyber threats through strong cybersecurity regulations to ensure the integrity of water infrastructure.

d. **Optimization**:

Smart Contracts and Legal Validity: Though Smart contracts offer automation in water governance, the legal systems must recognize their validity and enforceability. Regulations need to adapt to accommodate this new technology.

Cross-Border Transactions: In transboundary water management, legal frameworks must address the recognition and enforcement of blockchain transactions and agreements across borders.

The above points can be related to optimization because they improve water resource management's efficiency, automation, and scalability.

e. **Data Standardization**:

Interoperability Standards ensures blockchain compatibility across water governance systems, enabling seamless data exchange.

Contract Dispute Resolution: Blockchain-based water contracts should have clear conflict resolution mechanisms between the agreed parties.

ENVIRONMENTAL IMPACT OF BLOCKCHAIN

Blockchain technology, especially in networks that use Proof of Work (PoW) like Bitcoin, has sparked concerns about its environmental impact because it consumes a lot of energy and produces a large carbon footprint. Bitcoin mining alone consumes an estimated 120–150 terawatt-hours (TWh) of electricity every year, which is similar to the energy consumption of entire countries like Argentina or Norway, and largely relies on non-renewable energy sources in many regions (Bitcoin Energy Consumption Index, 2023). This energy-intensive process contributes to greenhouse gas emissions and generates electronic waste from obsolete mining hardware.

Mitigative Strategies

However, efforts are underway to mitigate these impacts, such as transitioning to Proof of Stake (PoS) mechanisms—exemplified by Ethereum's 2022 upgrade—and adopting renewable energy sources for mining operations. While these advancements offer hope for reducing blockchain's environmental footprint, achieving sustainability remains a critical challenge for the industry.

FUTURE OUTLOOK

Emerging trends in blockchain technology for water governance are shaping the future of how societies manage and allocate water resources. The integration of blockchain with latest technologies such as Artificial Intelligence (AI) and Internet of Things (IoT) is increasing and is a remarkable development (Xia et al., 2022; Zeng et al., 2023). This convergence helps monitor water quality, predictive modeling, and data analytics to provide a holistic approach to water management. Another trend involves exploring decentralized autonomous organizations (DAOs) in water governance, empowering communities to participate in decision-making and promoting a more inclusive and transparent approach. As blockchain gains wider acceptance, interoperability standards are emerging, enabling seamless collaboration on a global scale for managing transboundary water resources.

Furthermore, there is an emergent focus on developing privacy-preserving solutions within blockchain frameworks to address data security and confidentiality concerns. Pilot projects and research initiatives actively explore blockchain's scalability and applicability in diverse water management contexts, emphasizing the need for interdisciplinary collaboration and innovative solutions to address water-related challenges in an ever-changing world (Satilmisoglu et al., 2024). There are several notable pilot projects and emerging technologies in this field. Few of them are:

Atlantis DAO is piloting a decentralized water network in rural India using blockchain tokens to incentivize water collection and distribution (Ekshian & Emily, 2024).

IBM has integrated blockchain with IoT sensors in California's water systems to track groundwater usage for sustainability(Wolfson, 2019)

Utilizing advanced analytics and AI algorithms to analyze large datasets related to water resources, predictive modeling, machine learning, and intelligent decision-making can help optimize water allocation, identify patterns, and predict water-related challenges. Implement smart water infrastructure technologies, such as sensor networks, IoT devices, and real-time monitoring systems. This enables efficient water distribution, early detection of leaks, and improved overall infrastructure management. Introducing decentralized water treatment solutions, including small-scale and modular treatment facilities. These systems can be deployed closer to the point of use, reducing transmission losses and providing tailored solutions for specific communities.

Advanced treatment methods, including membrane technologies and advanced oxidation processes, can enhance the quality of recycled wastewater for non-drinking uses like agricultural irrigation or industrial processes. Implementing green infrastructure and natural water purification methods shall improve water quality and enhance ecosystem services, contributing to sustainable water management. Citizen science initiatives aim to encourage active participation of local communities in water governance. Crowdsourced data collection, raising awareness, and engaging communities can help build a sense of shared responsibility for water resources. Introducing water banking systems that allow users to buy, sell, or lease water rights. This encourages efficient water use, provides economic incentives for conservation, and facilitates the transfer of water allocations based on market demand.

Exploring innovations in desalination technologies, with a focus on distributed desalination units. Small-scale, solar-powered desalination facilities can provide freshwater in coastal areas without the large energy and environmental footprint associated with traditional desalination plants. Leveraging cloud computing technologies to create centralized water data management, analysis, and collaboration platforms. Cloud-based systems can facilitate real-time sharing of information among stakeholders and streamline decision-making processes. Developing water governance strategies that incorporate resilience planning to address the impacts of climate change on water availability, extreme weather events, and shifting precipitation patterns. Utilizing robotic and drone technologies for efficient water monitoring, inspection of infrastructure, and data collection in hard-to-reach or hazardous areas. These technologies can enhance the speed and accuracy of data collection and infrastructure maintenance. Addressing the water-energy nexus by implementing integrated solutions that optimize the use of water and energy resources. This includes co-locating water and energy infrastructure, exploring energy-efficient water technologies, and considering the dual impacts of water and energy policies. These innovations can revolutionize water governance, enhancing its resilience, efficiency, and sustainability in response to evolving challenges. Implementing these innovations often demands cooperation between government agencies, private sector organizations, academia, and local communities.

CONCLUSION

Blockchain application in water governance shows great potential for managing and allocating water resources. This cutting-edge technology offers numerous benefits that address key challenges in the water sector. Blockchain's transparency and immutability provide a reliable and verifiable record of water-related data, fostering trust among stakeholders such as governments, communities, and businesses. By offering a secure and transparent platform, blockchain enhances the management of water resources. Additionally, automation streamlines processes like water agreements and compliance checks, reducing bureaucratic delays, cutting down paperwork, and boosting overall efficiency. Blockchain also empowers local communities and individuals by giving them more control over their water-related data. Transparent and accessible information strengthen public participation in decision-making, promoting greater community involvement in water governance. Furthermore, standardized protocols and transparent systems support cross-border cooperation, especially when managing transboundary water resources.

REFERENCES

Alharbi, N., Althagafi, A., Alshomrani, O., Almotiry, A., & Alhazmi, S. (2021). A blockchain based secure IoT solution for water quality management. In *2021 International Congress of Advanced Technology and Engineering (ICOTEN)* (pp. 1–8). IEEE, New York.

Anandan, R., & Deepak, B. S. (2022). An overview of blockchain technology: Fundamental theories and concepts. In Goundar, S., Suseendran, G. & Anandan, R, (Eds), *The Convergence of Artificial Intelligence and Blockchain Technologies: Challenges and Opportunities* (pp. 1–22). World Scientific Publishing Co. Pte. Ltd., London.

Anglen, J. (n.d.). *Smart Contracts in Supply Chain Management: Enhancing Transparency and Efficiency.* https://www.rapidinnovation.io/post/smart-contracts-in-supply-chain-management-enhancing-transparency-and-efficiency?utm_source=chatgpt.com#9-use-case-6-quality-control-and-compliance

Babuna, P., Yang, X., Tulcan, R. X. S., Dehui, B., Takase, M., Guba, B. Y., & Li, M. (2023). Modeling water inequality and water security: The role of water governance. *Journal of Environmental Management*, 326, 116815.

Bitcoin Energy Consumption Index. (2023). https://digiconomist.net/bitcoin-energy consumption

Di Vaio, A., Trujillo, L., D'Amore, G., & Palladino, R. (2021). Water governance models for meeting sustainable development goals: A structured literature review. *Utilities Policy*, 72, 101255.

Dogo, E. M., Salami, A. F., Nwulu, N. I., & Aigbavboa, C. O. (2019). Blockchain and internet of things-based technologies for intelligent water management system. In Al-Turjman, F. (Eds), *Artificial Intelligence in IoT* (pp. 129–150). Springer, Cham.

Ekshian, R. B., & Emily, E. (2024). *How is Web3 Helping Solve the Water Crisis in Rural India?* https://cryptoforinnovation.org/how-are-web3-solutions-addressing-rural-indias-water-crisis/?utm_source=chatgpt.com

Gurgun, A. P., & Koc, K. (2022). Administrative risks challenging the adoption of smart contracts in construction projects. *Engineering, Construction and Architectural Management*, 29(2), 989–1015.

Jaradat, A., Ali, O., & AlAhmad, A. (2022). Blockchain technology: A fundamental overview. In Muthu, S. S. (Eds), *Blockchain Technologies for Sustainability* (pp. 1–24). Springer, Singapore.

Jiménez, A., Saikia, P., Giné, R., Avello, P., Leten, J., Liss Lymer, B., & Ward, R. (2020). Unpacking water governance: A framework for practitioners. *Water*, 12(3), 827.

Kaur, M., & Gupta, S. (2024). Blockchain technology: An overview with notable features and challenges. *Artificial Intelligence, Blockchain, Computing and Security*, 2, 186–193.

Kaushik, A. K. (2019). *The Development of Smart Water Markets Using Blockchain Technology.* New America's India-U.S. Fellows. https://www.newamerica.org/fellows/reports/anthology-working-papers-new-americas-us-india-fellows/the-development-of-smart-water-markets-using-blockchain-technology-aditya-k-kaushik/

Koebele, E. A., Méndez-Barrientos, L. E., Nadeau, N., & Gerlak, A. K. (2023). Beyond engagement: Enhancing equity in collaborative water governance. *Wiley Interdisciplinary Reviews: Water*, 11, e1687.

Koop, S. H. A., Grison, C., Eisenreich, S. J., Hofman, J., & van Leeuwen, K. (2022). Integrated water resources management in cities in the world: Global solutions. *Sustainable Cities and Society*, 86, 104137.

Lesavre, L., Varin, P., & Yaga, D. (2020). *Blockchain Networks: Token Design and Management Overview (No. NIST Internal or Interagency Report (NISTIR) 8301 (Withdrawn)).* National Institute of Standards and Technology, Gaithersburg, MD.

Li, J., & Kassem, M. (2021). Applications of distributed ledger technology (DLT) and blockchain-enabled smart contracts in construction. *Automation in Construction*, 132, 103955.

Lieberherr, E., Hüesker, F., & Pakizer, K. (2022). Rethinking urban water governance and infrastructure in Europe: Challenges and opportunities of regionalization and organizational autonomy. In Thomas, B., Francisco, S. P., & Megan, F. (Eds), *Routledge Handbook of Urban Water Governance* (pp. 272–283). Routledge, London.

Makaya, E., Rohse, M., Day, R., Vogel, C., Mehta, L., McEwen, L., & Van Loon, A. F. (2020). Water governance challenges in rural South Africa: Exploring institutional coordination in drought management. *Water Policy*, 22(4), 519–540.

Miller, B. (2021). *Application of Blockchain Capabilities to the Management of Water Rights and Water Markets*. Information Systems Water and Catchment Group Department of Environment Land Water and Planning Victoria, Australia, Victoria, Australia.

Millington, N., & Scheba, S. (2021). Day zero and the infrastructures of climate change: Water governance, inequality, and infrastructural politics in Cape Town's water crisis. *International Journal of Urban and Regional Research*, 45(1), 116–132.

Minhas, P. S., Ramos, T. B., Ben-Gal, A., & Pereira, L. S. (2020). Coping with salinity in irrigated agriculture: Crop evapotranspiration and water management issues. *Agricultural Water Management*, 227, 105832.

Mishra, B. K., Kumar, P., Saraswat, C., Chakraborty, S., & Gautam, A. (2021). Water security in a changing environment: Concept, challenges and solutions. *Water*, 13(4), 490.

Mutschinski, K., & Coles, N. A. (2021). The African water vision 2025: Its influence on water governance in the development of Africa's water sector, with an emphasis on rural communities in Kenya: A review. *Water Policy*, 23(4), 838–861.

Nguyen, T. Q., Das, A. K., & Tran, L. T. (2019, May). NEO smart contract for drought-based insurance. In *2019 IEEE Canadian Conference of Electrical and Computer Engineering (CCECE)* (pp. 1–4). IEEE, New York.

Oberhauser, D. (2019). Blockchain for environmental governance: Can smart contracts reinforce payments for ecosystem services in Namibia?. *Frontiers in Blockchain*, 2, 21.

Pahl-Wostl, C., Knieper, C., Lukat, E., Meergans, F., Schoderer, M., Schütze, N., & Vidaurre, R. (2020). Enhancing the capacity of water governance to deal with complex management challenges: A framework of analysis. *Environmental Science & Policy*, 107, 23–35.

Radmehr, A., Bozorg-Haddad, O., & Loáiciga, H. A. (2022). Developing strategies for agricultural water management of large irrigation and drainage networks with fuzzy MCDM. *Water Resources Management*, 36(13), 4885–4912.

Ramprasath, M., Bellam, K., Chandrasekar, T., Praveen Kumar, B., Thilagavathy, A., & Elangovan, G. (2024). An innovative water control system based on blockchain technologies. *Digital Twin and Blockchain for Smart Cities*, 2024, 277–296. https://onlinelibrary.wiley.com/doi/abs/10.1002/9781394303564.ch13

Romano, O., & Akhmouch, A. (2019). Water governance in cities: Current trends and future challenges. *Water*, 11(3), 500.

Salmoral, G., Zegarra, E., Vázquez-Rowe, I., González, F., Del Castillo, L., Saravia, G. R., & Knox, J. W. (2020). Water-related challenges in nexus governance for sustainable development: Insights from the city of Arequipa, Peru. *Science of the Total Environment*, 747, 141114.

Sandhu, G., Wood, M. O., Rus, H. A., & Weber, O. (2020). Bulk water pricing policies and strategies for sustainable water management: The case of Ontario, Canada. In Thomas, W., Northrop, S.-M., & Sherif, G. (Eds), *Environmental Policy: An Economic Perspective* (pp. 71–88). John Wiley & Sons Ltd, Hoboken, NJ.

Satilmisoglu, T. K., Sermet, Y., Kurt, M., & Demir, I. (2024). Blockchain opportunities for water resources management: A comprehensive review. *Sustainability*, 16(6), 2403.

Sitanggang, R., Ruttenberg, F., & Singh, A. (2023). *Blockchain and IoT for Drinking Water in G20 Countries: A Game-Changing Opportunity*. https://t20ind.org/wp-content/uploads/2023/06/T20_PolicyBrief_TF6_Blockchain-IOT-Water.pdf.

Tonkin, Z., Yen, J., Lyon, J., Kitchingman, A., Koehn, J. D., Koster, W. M., & Todd, C. (2021). Linking flow attributes to recruitment to inform water management for an Australian freshwater fish with an equilibrium life-history strategy. *Science of the Total Environment*, 752, 141863.

Vladucu, M., Wu, H., Medina, J., Salehin, K., & Blockchains, Z. D. (2024). Blockchain on sustainable environmental measures: A review. *Blockchains*, 2(3), 334–365. Retrieved January 25, 2025, from https://www.mdpi.com/2813-5288/2/3/16.

Wolfson, R. (2019). *IBM Pilots Blockchain and IoT Sensor Solution To Track Sustainable Groundwater Usage In California*. https://www.forbes.com/sites/rachelwolfson/2019/02/08/ibm-pilots-blockchain-and-iot-sensor-solution-to-track-sustainable-groundwater-usage-in-california/?utm_source=chatgpt.com

Woodhouse, P., & Muller, M. (2017). Water governance—An historical perspective on current debates. *World Development*, 92, 225–241.

Xia, W., Chen, X., & Song, C. (2022). A framework of blockchain technology in intelligent water management. *Frontiers in Environmental Science*, 10, 909606. https://doi.org/10.3389/FENVS.2022.909606/FULL

Zeng, H., Dhiman, G., Sharma, A., Sharma, A., & Tselykh, A. (2023). An IoT and blockchain-based approach for the smart water management system in agriculture. *Expert Systems*, 40(4), e12892. https://doi.org/10.1111/EXSY.12892

Zhang, J. (2023). Application of blockchain technology in water management and new energy economic development. *Water Supply*, 23(10), 4227–4241.

13

Heavy Metal Contamination in Urban Road Dust

Pollution Indices from Dhaka City

Umma Salma Hashe, Marziat Rahman, Md Nasir Ahmmed Patoary, Abdullah Al Nayeem, Kazi Khaled Mahmud, Mohammad Moniruzzaman, and Ahmad Kamruzzaman Majumder

INTRODUCTION

Dust found on the road is made up of solid, tiny particles formed by any breakdown of materials, and it is considered an indication of metal pollution in the environment from atmospheric fallout. It has emerged as a considerable issue owing to its prevalence in urban environments (Suryawanshi et al., 2016). Road dust results from intricate interactions between gaseous, liquid, and solid substances from natural and man-made sources.

DOI: 10.1201/9781003613183-13

It contains several constituents, including organic and inorganic substances, metal and nonmetal particles, pollen, spores, and mould, which can be resuspended by transportation activity and wind (Rajaram et al., 2014). Road dust accumulates along impervious roadways in densely populated urban settlements and widespread human activities, serving as an abundance of various lethal heavy metals that typically originate from both man-made activities like industrial emissions, vehicular traffic, and construction work and domestic activities, as well as environmental geochemical processes (Mathur et al., 2016). Research shows that household activities, vehicle emissions, tyre wear, auto repair shops, the use of fossil fuels, industrial processes, power generation, and the weathering of building materials are all human-induced; other sources may increase contamination in road particles (Ahmed and Ishiga, 2006). Heavy metals have the potential to penetrate living organisms' cells, such as animals and plants; eventually, they can be passed in the food chain, causing their health degradation (Adriano, 2001). The sources of heavy metals are constant throughout the year; however, the amount of heavy metals in dust might vary from winter to summer (Men et al., 2018). An urban area comprises residential, urban green, commercial, and industrial areas where vehicular movement, human activity, land use, and population distribution all contribute to the area's dynamics. Heavy metals distribution and their known sources concerning various land uses are important factors in determining pollution status (Trujillo-González et al., 2016). As a result, land use distribution and seasonal fluctuation may be deliberated to determine heavy metal concentration distributions and ecological risk factor assessments. Due to the lengthy lifecycle and non-biodegradable nature of heavy metals, road dust containing these metals has grown to be a significant issue in urban areas. The presence of pollutants in the urban environment for an extended period, particularly as the human body absorbs them by skin absorption, inhalation, and ingestion, causes illness (Abrahams, 2002). Heavy metal-laden polluted dust causes various malignancies; damage to the liver and kidneys; and cardiovascular, pulmonary, skin, and neurological problems (WHO, 2007). Children are more exposed to this contamination than adults, as they have weakened immune systems and repetitive finger and mouthing of non-food things (Rajaram et al., 2014). Surface discharge of road dust carrying heavy metals with water exposes nearby aquatic life at risk. Dissolved heavy metals and dust particles can create serious biological and environmental risks for the surrounding metropolitan region (Wijaya et al., 2012). Many studies on heavy metals in dust samples have been carried out all over the world (Diganta et al., 2020). However, data on heavy metal pollution, particularly road dust in Bangladesh, is still lacking. Bangladesh's capital, Dhaka, is the sixth-largest megacity, with a population of 21 million people (UN, 2018). Due to constant human-induced activities that damage the ecosystem in various ways, this high population density uses significant resources and generates many pollutants (Selonen et al., 2014). Dhaka City has seen rapid urbanization and industrialization in recent decades, notably mega projects like mass rapid transit and bus rapid transit, as well as extensive but unplanned residential or housing construction and unregulated increases in new vehicle inclusion on roads (Iqbal et al., 2019), putting an extra burden on the urban atmospheric environment. In Dhaka city, air pollution has become a major concern because of the above-mentioned causes, owing to significant amounts of resuspended dust. Road dust contributes to air pollution. Domestic garbage dumping, traffic emissions, industrialization, and urbanization are the root cause of the increase of Dhaka City's roadway dust. According to Ahmed and Ishiga (2006), in the past 10 years, count of motorized and non-motorized vehicles in Dhaka City has increased multiple times. According to the Bangladesh Road Transport Authority (BRTA), the government office during the year 2016 reports a number of 93,934 registered motorized vehicles in 2015, up from 72,377 in 2011 (BRTA, 2016). Based on the findings of Chowdhury et al. (2000), the total number of motorized vehicles registered under Dhaka Metropolitan increased from 66,703 in the year 1986 to 93,934 in the year 2015. Electrically charged vehicles contribute to particulate matter and contain elevated levels of heavy metals, which are components of road dust (Chowdhury et al., 2000). A minimal amount of research is available on internet sources to evaluate heavy metal distribution in road dust based on temporal variation, zone differences, traffic density variation, and the eight heavy metals ratio assessment with comparison to neutral areas. Furthermore, the geochemical pollution ratings of heavy metals have not been considered (Kabir et al., 2021). Despite extensive research on heavy metal contamination in urban environments, several gaps persist in understanding the pollution indices and ecological risks associated with road dust in Dhaka City, Bangladesh. Previous studies, such as those carried out by Li et al. (2008) and Li et al. (2013), have primarily focused on industrial and commercial zones in other regions, such as Northeast China and Hong Kong, where urbanization and industrial activities differ significantly from Dhaka's context. These studies provide valuable insights into contamination levels and distribution patterns; however, they lack region-specific data that reflect Dhaka's unique environmental and socioeconomic conditions. Furthermore, while studies like Li et al. (2001) emphasize the contamination levels regarding dust and urban soils in Hong Kong, they do not comprehensively address seasonal variations in contamination levels, which are crucial for understanding pollution dynamics in Dhaka's distinct climatic conditions. The current body of literature lacks a detailed comparative analysis of heavy metal contamination across different urban zones (industrial, residential, and commercial) within Dhaka, particularly considering the influence of monsoon and dry seasons. In addition, previous research conducted by Tchounwou et al. (2012) has discussed the detrimental effects that heavy metals have on the environment on a global scale. However, these studies do not make use of indices that provide localized risk evaluations. These indices include the Enrichment Factor (EF), the Geo-Accumulation Index (Igeo), the Pollution Load Index (PLI), and the Contamination Factor (CF). These indices are crucial for establishing baseline data, identifying pollution sources, and formulating targeted mitigation strategies specific to Dhaka's urban landscape.

Therefore, this research attempted (a) to estimate the quantities of eight carcinogenic heavy metals (Cd Fe, Cu, Cr, Pb, Zn, Mn, and Ni) in roadway dust based on seasonal change in three Dhaka City zones: residential, industrial, and commercial. (b) To assess potential pollution indices using Igeo, EF, PLI, CF, Potential Ecological Risk Index (PERI), and Ecological Risk Factor (ER).

METHODOLOGY

Study Area

Ten sampling sites (Figure 13.1) in Dhaka, Bangladesh, were chosen based on nearby human-induced activities. Nine sites representing various community activities, including commercial (zone with heavy vehicular traffic), industrial (zone with moderate vehicular traffic), and residential (zone with minimal vehicular traffic) sections of Dhaka city, as well as Ramna Park, were chosen as a control location to see the differences. Tejgaon, Karwan Bazar, and Old Dhaka are surrounded by a diverse range of industries. These three locations are used to depict an industrial zone. Dhaka's traffic-congested Mirpur-10, Dhanmondi-27, and Mohakhali are included in the sampling to reflect a commercial heavy traffic area. And the residential areas are represented by Bashundhara, Gulshan, and Baridhara. These are surrounded by prestigious educational institutions, offices, a world-renowned medical centre, and popular cafes, among other things.

Samples

A total of 57 road-dust samples had been collected from 10 sites across Dhaka City. Among these, 54 were taken in the rainy season (August 2021) and 3 were taken in the winter season (2022) from Ramna Park as controlled place samples to help us understand the differences in traffic homogeneity and frequency of density across different land-use categories (residential, commercial, and industrial areas, respectively). From each location, three samples have been collected by maintaining a 100 feet distance from one to another. A plastic spoon and a soft paintbrush were used to gather around 300 g of dust. Following that, the samples were placed inside a plastic bag and labelled accordingly. They were then placed in a plastic storage box with a clip-on lid and handle until they were sent to the lab to be prepared for analysis (Figure 13.2).

All the samples were left to dry naturally in the lab until they all had the same amount of wetness. After that, a nylon mesh measuring 1.0 mm (16 US mesh) was utilized for the purpose of filtering out any larger objects, dirt, stones, and other similar substances. This particular sieve was manufactured by W.S. Tyler Inc. in the United States. Street dust particles smaller than 100 microns in diameter are more easily resuspended and ingestible during breathing (rather than bigger particles that travel mainly by "saltation" and "creep"), hence this measurement was chosen for elemental analysis within this study (Ferreira-Baptista and Miguel, 2005). The moisture content of the samples was then determined by weighing them in Moisture Analysers (EM 120-HR – Precisa Gravimetrics). It helped to maintain homogeneity as well as further sample analysis. Final Samples were stored in a Ziplock bag for further analysis.

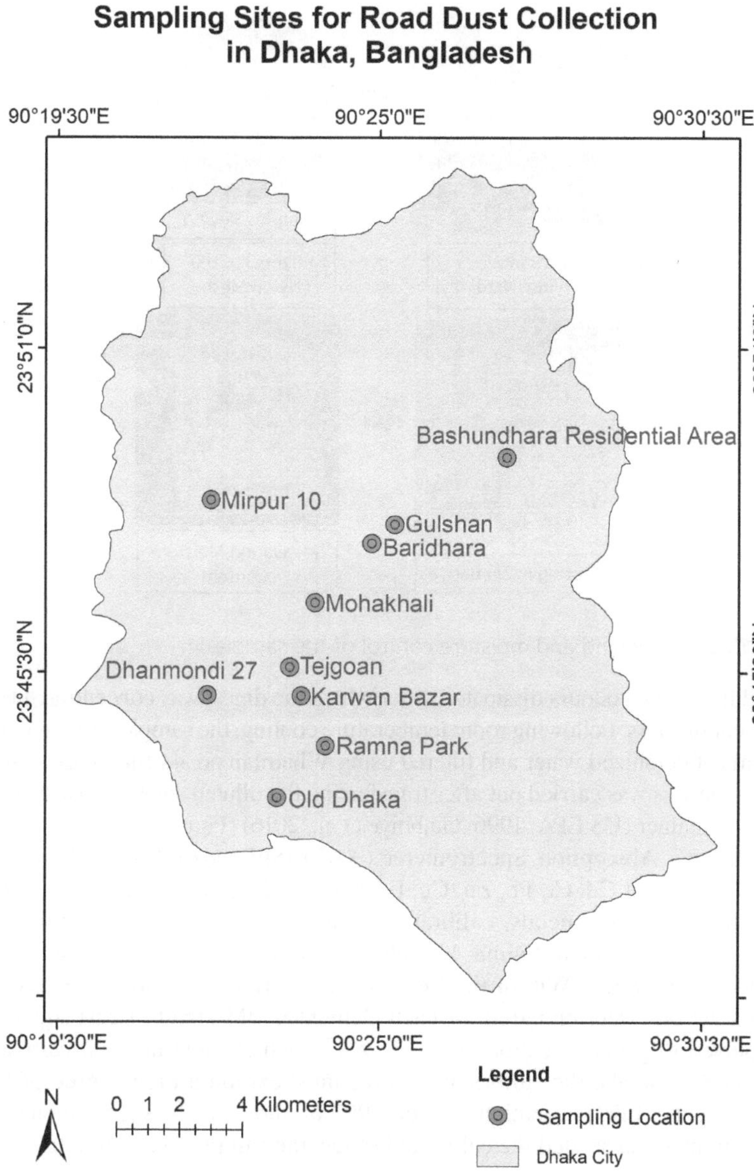

FIGURE 13.1 Sites for collecting road dust samples in Dhaka, Bangladesh.

To begin digestion, a 50 mL glass beaker was filled with 1 g of dust sample. The samples, along with empty ones, were given 15 mL of aqua regia (HNO_3: HCl (v/v) = 1:3) via pipette (Gajbhiye et al., 2016). The samples with aqua regia were kept the entire night for self-digestion before transferring them to the hotplate. Samples were heated on hot plates until they were acid digested, which took four temperature changes: first, 120°C; then 160°C; then 180°C after 40 minutes; and finally, the ambient temperature was adjusted

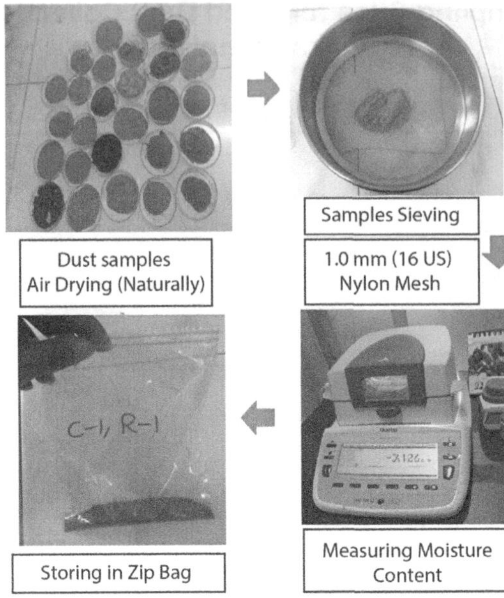

FIGURE 13.2 Screening and moisture control of the sample.

as the reddish-brown odours dissipated, after which the digest was concentrated to around 1 mL or near dryness. Following room temperature cooling, the samples were reconstituted with 50 mL of deionized water and filtered using Whatman no. 44 filter paper. Additional elemental analysis was carried out after transferring the diluted solution to a 100 mL polycarbonate container (US EPA, 1996; Gajbhiye et al., 2016) (Figure 13.3).

An Atomic Absorption Spectrometer (AAS) (Shimadzu-AA-7000) was used to identify the levels of Cd, Cr, Pb, Zn, Cu, Fe, Mn, and Ni in the samples. Utilizing standard solutions of stock metals, calibration standards were prepared for instrumental analysis (Fluka Analytical, Sigma-Aldrich, Germany, 1,000 mg/L) to assess the samples under examination. Within the linear concentration range for measurement, five standard solutions were prepared for each element in 2M HNO_3. Least squares fitting was employed to generate calibration curves for each element under investigation. To measure heavy metals, the calibration curves must exhibit a high degree of linearity, with coefficients of determination (R^2) of 0.999 or above. If the spectrometer readings for a certain metal exceeded the calibrated range, the samples were diluted.

Precision, Accuracy, and Quality Control

The analytical processes utilized blank samples for quality control, allowing for the analysis of contamination, accuracy, and bias to minimize the root cause of the error. Preparation and analysis of the blank sample preceded measurement of the dust samples. No detectable metals were found in the blank samples. The presented values are the means of all extractions and assays performed on three independent samples (n=3). For every measurement,

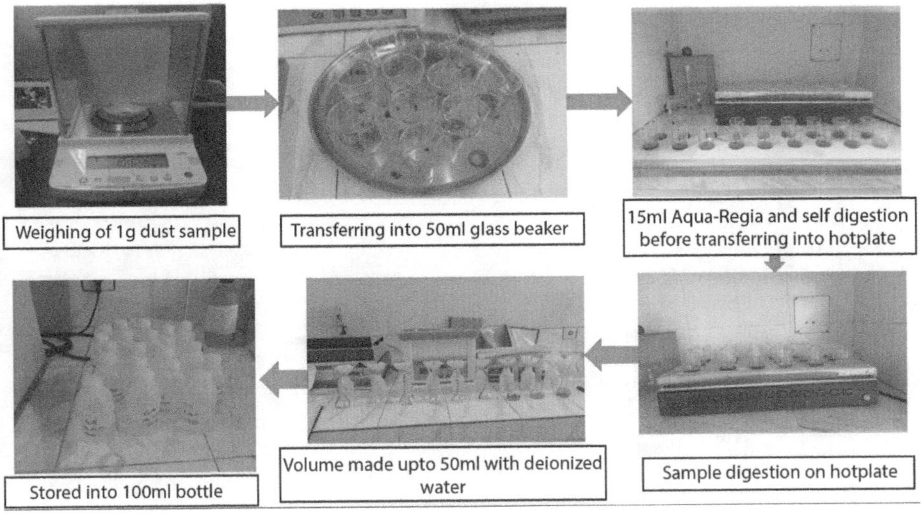

FIGURE 13.3 Preparing the sample for analysis.

the average of three consecutive replicates was utilized, provided that the RSD was below 10%. All laboratory work and facilities were given by the Bangladesh Council of Scientific and Industrial Research (BCSIR), Dhaka's Soil and Environment Laboratory.

Calculations for Heavy Metal Pollution Level

Numerous pollution indices were utilized in this study. These included the Enrichment Factor (EF) (Ergin et al., 1991; Gajbhiye et al., 2016), the Geo-Accumulation Index (Igeo) (Müller, 1969; Atiemo et al., 2011), the Pollution Load Index (Tomlinson et al., 1980; Jorfi et al., 2017), the Contamination Factor (CF) (Hakanson, 1980; Sobhanardakani et al., 2016), and Potential Ecological Risk Index (PERI) (Hakanson, 1980). For the purpose of this descriptive statistical analysis, we utilized Microsoft Excel 2016 with IBM SPSS V20 and Google Earth Pro. Various charts, tables, and diagrams were created to enhance understanding of the data.

FINDINGS AND REMARKS

Accumulation of Heavy Metals

Concentrations of Cd, Cr, Cu, Fe, Mn, Ni, Pb, and Zn found in samples of road dust have been compared to the normal levels in Ramna Park, which is considered a clean and unaffected area.

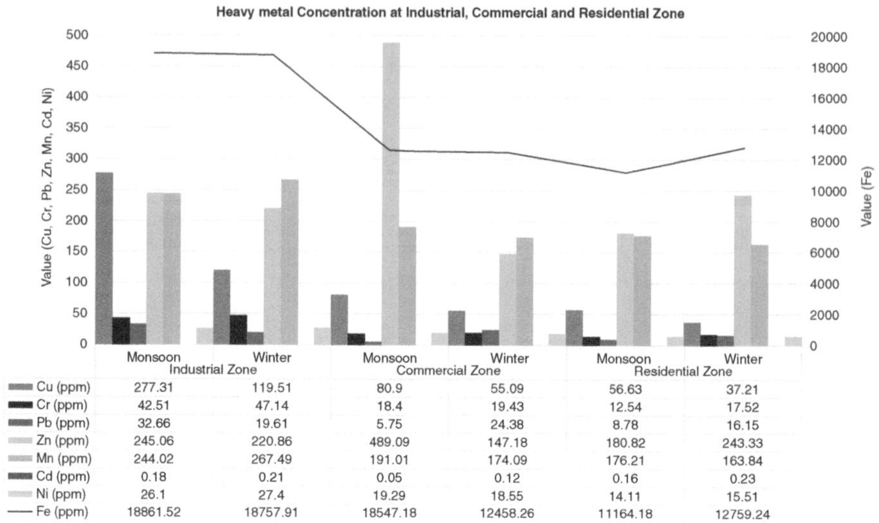

FIGURE 13.4 Heavy metal concentration in different land use and different seasons.

The levels of heavy metals that are present in road dust are depicted in Figure 13.4, which covers three different zones and two different seasons. All calculated heavy metals exceeded the background value when contrasted with the Ramna site's measurements. The industrial zone had greater Fe, Zn, Pb, and Cu levels during rainy seasons. Fe, Cu, Zn, Mn, and Ni were greater in the commercial zone during the monsoon season, whereas Cu, Mn, and Cd were more abundant in the residential zone.

Dhaka's street dust samples were compared to global data and many literature-based estimates for heavy metal concentration (Table 13.1). This study shows that road dust elemental makeup differs substantially per country. This investigation found higher Fe, Cu, and Ni concentrations than global statistics. Global data showed more Cr, Pb, Mn, and Cd than the study. Possible causes include (a) variations in countries' geochemical formation and (b) the impact of anthropogenic activities adjacent to highways. Moreover, it has been determined that the majority of elements in road dust samples exhibit concentrations that are situated within the median of the worldwide range (Table 13.1).

Environmental Pollution Indices Assessment

The impact of heavy metal pollution in road dust on the environment and ecology was evaluated using a number of indices, including the EF, Igeo, PLI, CF, and PERI.

Geo-Accumulation Index (Igeo) Assessment

Heavy metal Igeo values in roadside dust from 3 zones of Dhaka City — each with 2 unique seasons (winter and monsoon)—are shown in Figure 13.5. For all three zones of

TABLE 13.1 Heavy metal levels in Dhaka's road dust compared to background values and global distribution

LOCATION NAME AND RECOMMENDED BODY	*HEAVY METAL CONCENTRATIONS (PPM)*								*REFERENCES*
	CD	*CR*	*PB*	*FE*	*NI*	*ZN*	*CU*	*MN*	
Ramna Park, Dhaka (Background Value)	0.13	7.28	5.52	9,310.89	6.36	139.54	6.61	162.56	This Study
Dhaka, Bangladesh (Dust)	11.6	144.3	18.9	–	37.1	239.1	49.8	261.5	Rahman et al. (2019)
Urumqi, China (Dust)	1.17	54.2	53.5	–	43.2	294.4	94.5	927	
Japan	0.41	41.3	20.4	–	28.5	63.8	36.9	–	Hughes,2002
Islamabad, Pakistan	5	125	104	–	23	166	52	433	Faiz et al. (2012)
India	3.12	54	536	–	42	159	44	–	Chatterjee and Banarjee (1999)

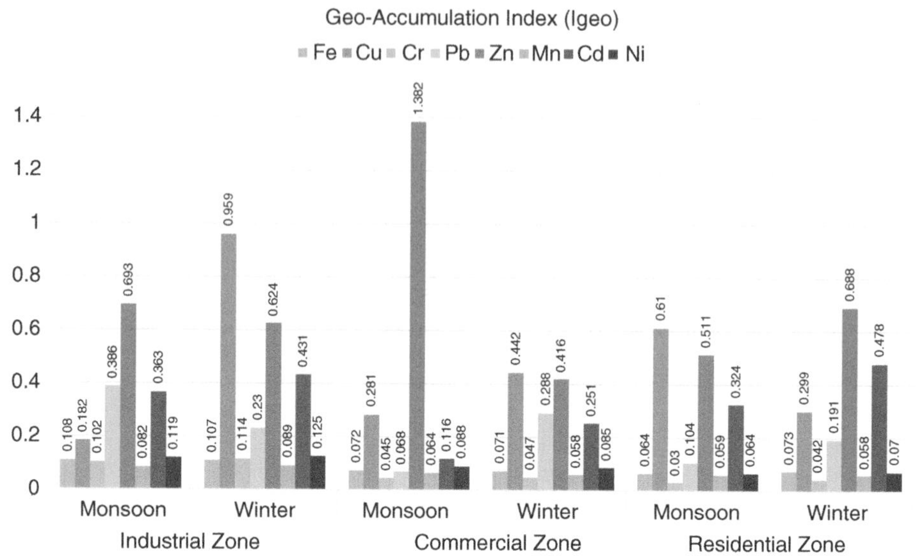

FIGURE 13.5 Geo-Accumulation Index based on land usage and season.

heavy metal, the Igeo quality is found to be unpolluted. The industrial zone has elevated concentrations of zinc (Zn), iron (Fe), and lead (Pb) compared to other metals in the period of rainy season. In the monsoon time, Fe, Pb, Zn, Mn, and Ni were much greater in the commercial zone, while Mn and Ni were higher in the residential zone.

The lack of proper procedures for the management and disposal of waste—which includes old tires, paints, oils, metal filings, greases, and unburned fuels—correlates with the socioeconomic activities in commercial areas. The contamination with Cu, Zn, and Cd was substantial and came from multiple sources. Vehicles tire particles, lacking exhaust systems, electroplating, coatings, brake pads, and roadside auto shops were among these causes.

Enrichment Factor (EF) Assessment

Figure 13.6 illustrates heavy metal enrichment factors (EF) utilizing the upper crustal composition of the Earth as a baseline and iron (Fe) as a reference material. Cu and Zn were identified in the moderate to severe group in all zones, while Pb, Cd, and Ni were detected in the moderate to minor group. Higher levels of Cu, Pb, and Zn were found during the monsoon season. In the winter, Cr, Pb, and Cd were more significant in the commercial zone, whereas Cr, Zn, and Cd were more prevalent in the residential zone. In conclusion, the greater Cu and Zn enrichment levels should be considered. Temperature inversion and other climatic phenomena may be too responsible for these seasonal fluctuations.

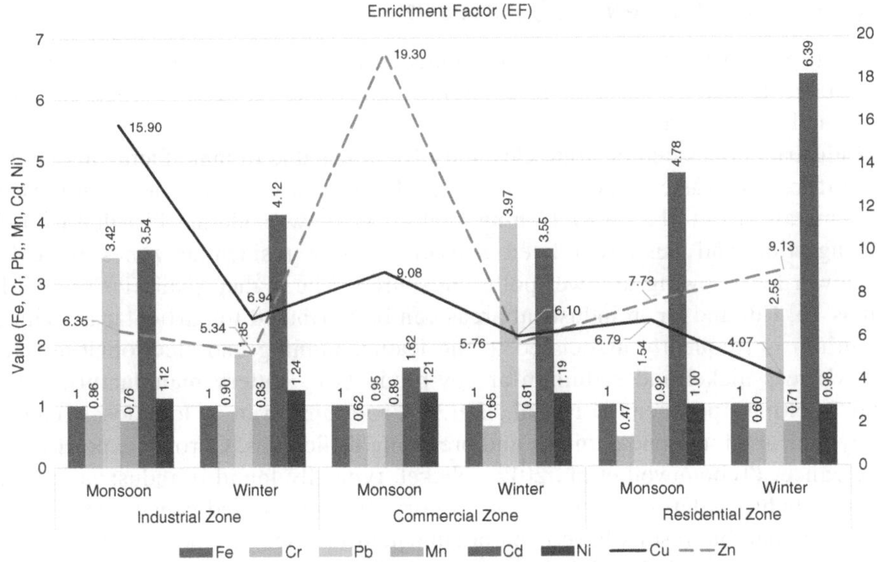

FIGURE 13.6 Land use and season-based enrichment factor assessment.

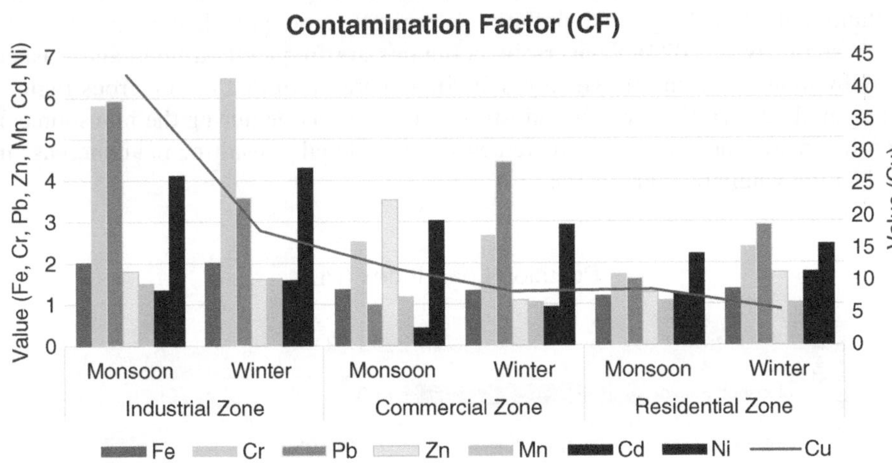

FIGURE 13.7 Heavy metal pollution in Dhaka City's roadway dust.

Contamination Factor (CF) Assessment

Cu was found in extremely high concentrations in all zones, as well as two seasons, according to the CF categorization. Other heavy metal genres were determined to be in the moderate to considerate category. The industrial zone had higher levels of Cr, Zn, Mn, Cd, and Ni during the winter, the commercial zone had higher levels of Cr, Pb, and Cd, and the residential zone had higher levels of Fe, Cr, Pb, Zn, and Ni (Figure 13.7).

Pollution Load Index (PLI) Assessment

Figure 13.8 depicts the pollutant load index (PLI) for three zones and seasons. Anthropogenic activity and seasonal environmental factors vary PLI in urban road dust across land use and seasons.

Industrial zones demonstrate elevated PLI values due to manufacturing, vehicular traffic, and waste disposal emissions, which release substantial quantities of heavy metals (Li et al., 2008). Commercial areas show moderate PLI due to traffic congestion and business-related activities, while residential zones typically have lower levels owing to fewer pollution sources The higher quantities of metals such as Ni, Cd, and Cr in industrial areas can be attributed to particular activities. Chromium is frequently associated to the leather tanning and electroplating sectors, whereas nickel and cadmium are by-products of battery manufacturing and metal smelting operations (Li et al., 2013). Cadmium, even at low doses, causes kidney malfunction, bone damage, and breathing difficulties. Chronic exposure can cause cancer (Tchounwou et al., 2012). Nickel, typically found in industrial pollutants, can induce allergic reactions, respiratory difficulties, and cancer (Das et al., 2008). Comparable results have been documented in research undertaken in urban environments, where industrial zones routinely demonstrate elevated contamination levels compared to residential and commercial areas (Li et al., 2001). As a result of dry conditions, limited pollutant dispersion, and atmospheric inversions that confine contaminants near the ground, pollution is typically more prevalent during the winter season (Yavuz, 2024). Conversely, pollutants are frequently diluted and washed away by monsoon rainfall, which results in a decrease in PLI in numerous regions (Yang et al., 2021). However, in industrial zones, discharge during the monsoon can mobilize heavy metals from untreated waste, occasionally resulting in an increase in the PLI (Onwughara et al., 2011).

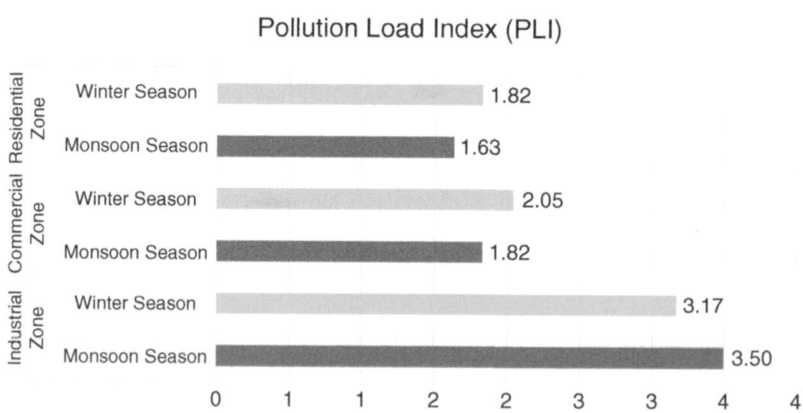

FIGURE 13.8 Land use and season-based pollution load index.

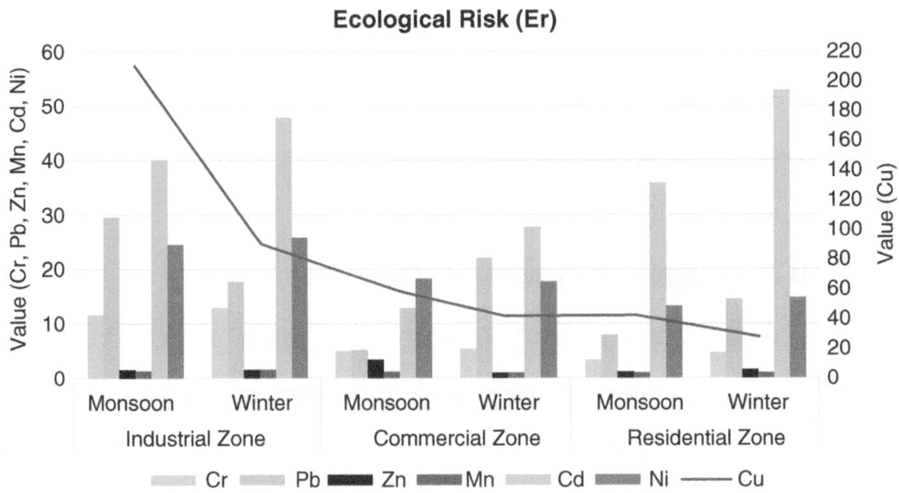

FIGURE 13.9 Ecological risk assessment for different land uses and seasons.

Ecological Risk (ER) Assessment

Figure 13.9 illustrates the ecological risks associated with dust from urban roadsides in Dhaka, as per Hakanson's research. Except for Cu, the results show that mean Er values were low for almost all of heavy metals. It was higher to considerable in the industrial zone, moderate in the commercial zone, and moderate to low in the residential zone.

Potential Ecological Risk Index/Risk Index (PERI/RI) Assessment

An ecological risk index (RI), as shown in Figure 13.10, was made to figure out how much pollution different heavy metals in urban street dust were causing. The industrial zone posed the greatest ecological risk during the monsoon season. Commercial and Residential zones were below the range in the summer but were higher in the winter.

CONCLUSIONS

The bioaccumulation potential and availability of heavy metals in road dust make them a frequent source of concern in metropolitan areas. Heavy metals that are potentially harmful were measured in a set of methodically collected samples from three separate zones, covering two different seasons. This study calculated road dust heavy metal concentrations and environmental impacts using a variety of mathematical methodologies. Compared to the findings from a background site (in this case, Ramna Park), all

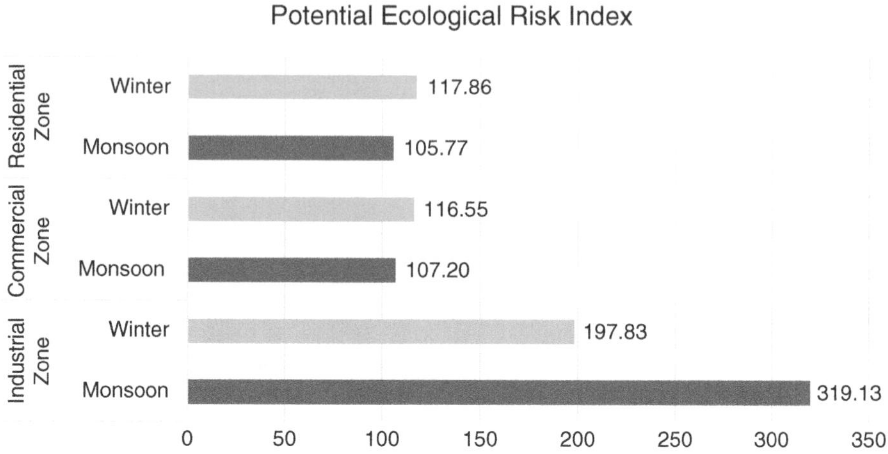

FIGURE 13.10 Land use and season-based potential ecological risk index.

estimated levels of heavy metals were greater than the background value. The industrial zone had increased Fe, Cu, Pb, and Zn levels during monsoons. Fe, Cu, Zn, Mn, and Ni were more in the commercial zone, whereas Cu, Mn, and Cd were higher in the residential zone throughout the season. Fe, Cu, and Ni levels were higher than global published data. The global data had more Cr, Pb, Mn, and Cd than the research. Igeo was clean for all heavy metals in 3 zones. In monsoon season, the industrial zone had greater Fe, Pb, and Zn levels than other metals. During monsoon, Fe, Pb, Zn, Mn, and Ni were higher in the commercial zone, while Mn and Ni were higher in the residential zone. Sources of high concentrations of Cu, Zn, and Cd contamination included brake drums or pads, tire's tiny fragments, road colorants, zinc plating, automobiles garage, and emissions from cars that don't use exhaust. In conclusion, the greater Cu and Zn enrichment levels should be considered. Cu was identified in extremely high concentrations in all zones and two seasons, according to the Contamination Factor classification. Industrial zone has the highest pollution load index during the rainy season due to heavy metal concentrations that cannot disperse in wet conditions. The residential zone had the lowest pollution load index value. Except for Cu, the analysis shows that the average values of ecological risk for most trace metals were low. In the Industrial Zone, it was substantially greater than in other areas. The industrial zone had the highest ecological risk during the monsoon season. Findings from our research can help local authorities take the necessary measures to reduce the current dust pollution problems. Future research should concentrate on $PM_{2.5}$ and PM10, two finer particle size fractions, and take into account the potential effects on human health. More study is required on site-specific exposure characteristics and transit factors to precisely determine the health risk and reduce the substantial ambiguity currently associated with risk estimations. The findings of this study will also assist researchers, planners, politicians, and public health officials in developing and implementing appropriate policies and activities to protect the environment and people's health.

REFERENCES

Abrahams, P.W. (2002). Soils: Their implications to human health. *Science of the Total Environment* 291: 1–32.

Adriano, D.C. (2001). *Trace Elements in Terrestrial Environments: Biogeochemistry, Bioavailability and Risks of Metals*. Berlin: Springer.

Ahmed, F. and Ishiga, H. (2006). Trace metal concentrations in street dust of Dhaka City, Bangladesh. *Atmospheric Environment* 40: 3835–3844.

Atiemo, M. S., Ofosu, G. F., Kuranchie-Mensah, H., Osei Tutu, A., Palm, N. D. M. L., & Blankson, S. A. (2011). Contamination assessment of heavy metals in road dust from selected roads in Accra, Ghana. *Research Journal of Environmental and Earth Sciences* 3(5): 473–480.

BRTA (2016). Bangladesh road transport authority. Air pollution in Bangladesh. BRTA Associated With Greater Dhaka Metropolitan Area Integrated Transport Study, Working Paper No. 23.

Chatterjee A. and Banerjee, R.N. (1999). Determination of lead and other metals in a residential area of determining heavy metal contamination of road dust in Delhi, India 233 greater Calcutta. *Science of the Total Environment.* 227: 175–185. doi:10.1016/S0048-9697(99)00026-1.

Chowdhury, Q.I., Rahman, G., Majumder, M.K., and Rana, B. (2000). Smoke and haze in Dhaka's concrete jungle. Air pollution in Asian cities. *Forum of Environmental Journalists of Bangladesh (FEJB)*, Dhaka, pp. 1–14.

Das, K.K., Das, S.N., and Dhundasi, S.A. (2008). Nickel, its adverse health effects & oxidative stress. *Indian Journal of Medical Research* 128(4): 412–425.

Diganta, M.T.M., Sharmi, T.T., Saifullah, A.S.M., Uddin, M.J., and Sajib, A.M. (2020). Appraisal of heavy metal contamination in road dust and human health risk in a municipality of Bangladesh. *Environmental Engineering & Management Journal* 19(12): 2165–2177.

Ergin, M. et al. (1991). Heavy metal concentrations in surface sediments from the two coastal inlets of the northeastern sea of Marmara. *Chemical Geology* 91: 269–285.

Faiz, Y., Siddique, N., and Tufail, M. (2012). Pollution level and health risk assessment of road dust from an expressway. *Journal of Environmental Science and Health, Part A: Toxic/Hazardous Substances and Environmental Engineering*, 47(6): 818–829. https://doi.org/10.1080/10934529.2012.664994

Ferreira-Baptista, L. and Miguel, E.D. (2005). Geochemistry and risk assessment of street dust in Luanda, Angola: A tropical urban environment. *Atmospheric Environment* 39: 4501–4512.

Gajbhiye, T. et al. (2016). Airborne foliar transfer of PM bound heavy metals in cassia siamea: A less common route of heavy metal accumulation. *Science of the Total Environment* 573: 123–130.

Hakanson, L. (1980). An ecological risk index for aquatic pollution control: A sedimentological approach. *Water Research* 14: 975–1001.

Iqbal, A., Afroze, S., and Rahman, M.M. (2019). Probabilistic health risk assessment of vehicular emissions as an urban health indicator in Dhaka City. *Sustainability* 11: 6427.

Jorfi, S. et al. (2017). Pollution load index for heavy metals in mian-ab plain soil, Khuzestan, Iran. *Data in Brief* 15: 584–590.

Kabir, M.H., Rashid, M. H., Wang, Q., Wang, W., Lu, S., and Yonemochi, S. (2021). Determination of heavy metal contamination and pollution indices of roadside dust in Dhaka City, Bangladesh. *Processes* 9(10): 1732.

Li, F., Fan, Z., Xiao, P., Oh, K., Ma, X., and Hou, W. (2008). Contamination, chemical speciation and vertical distribution of heavy metals in soils of an old and large industrial zone in Northeast China. *Environmental Geology* 57(8): 1815–1823. https://doi.org/10.1007/s00254-008-1469-8.

Li, X. Poon, C.S., and Liu, P.S. (2001). Heavy metal contamination of urban soils and street dusts in Hong Kong. *Applied Geochemistry* 16(11–12): 1361–1368.

Li, X., Poon, C., and Liu, P. (2013). Heavy metal contamination of urban soils and street dusts in Hong Kong. *Applied Geochemistry* 18(7): 1143–1158.

Mathur, R., Balaram, V., Satyanarayanan, M., and Sawant, S.S. (2016). Assessment of heavy metal contamination of road dusts from industrial areas of Hyderabad, India. *Environmental Monitoring and Assessment* 188: 514.

Men, C., Liu, R., Xu, F., Wang, Q., Guo, L., and Shen, Z. (2018). Pollution characteristics, risk assessment, and source apportionment of heavy metals in road dust in Beijing, China. *Science of the Total Environment* 612: 138–147.

Müller, G. (1969). Index of geo-accumulation in sediments of the Rhine River. *Geojournal* 2(3): 108–118.

Onwughara, N.I., Umeobika, U.C., Obianuko, P.N., and Iloamaeke, I.M. (2011). Emphasis on effects of storm runoff in mobilizing the heavy metals from leachate on waste deposit to contaminate Nigerian waters: Improved water quality standards. *International Journal of Environmental Science and Development* 2011: 55–63. https://doi.org/10.7763/ijesd.2011.v2.97

Rahman, M.S. et al. (2019). Assessing risk to human health for heavy metal contamination through street dust in the Southeast Asian megacity: Dhaka, Bangladesh. *Science of the Total Environment* 660: 1610–1622.

Rajaram, B.S., Suryawanshi, P.V., Bhanarkar, A.D., and Chalapati Rao, C.V. (2014). Heavy metals contamination in road dust in Delhi City, India. *Environmental Earth Sciences* 72: 3929–3938.

Selonen, V., Varjonen, R., and Korpimäki, E. (2014). Immediate or lagged responses of a red squirrel population to pulsed resources. *Oecologia* 177: 401–411.

Sobhanardakani, S. et al. (2016). Assessment of heavy metal contamination in surface sediment of the darreh-morad Beyg river. *Iranian Journal of Health Sciences* 4(3): 22–34.

Suryawanshi, P.V., Rajaram, B.S., Bhanarkar, A.D., and Chalapati Rao, C.V. (2016). Determining heavy metal contamination of road dust in Delhi, India. *Atmósfera* 29: 221–234.

Tchounwou, P.B., Yedjou, C.G., Patlolla, A.K., and Sutton, D.J. (2012). Heavy metal toxicity and the environment. *Experientia Supplementum* 101: 133–164.

Tomlinson, D.L. et al. (1980). Problems in the assessment of heavy metal levels in estuaries and the formation of a pollution index. *Helgoland Marine Research* 33: 566–575.

Trujillo-González, J.M., Torres-Mora, M.A., Keesstra, S., Brevik, E.C., and Jiménez-Ballesta, R. (2016). Heavy metal accumulation related to population density in road dust samples taken from urban sites under different land uses. *Science of the Total Environment* 553: 636–642.

U.S. Environmental Protection Agency (US EPA). (1996). Soil Screening Guidance: Technical Background Document. EPA/540/R-95/128.

Wijaya, A.R., Ouchi, A.K., Tanaka, K., Shinjo, R., and Ohde, S. (2012). Metal contents and Pb isotopes in road-side dust and sediment of Japan. *Journal of Geochemical Exploration* 118: 68–76.

World Health Organization (WHO). (2007). *Health Risks of Heavy Metals from Long-Range Transboundary Air Pollution*. Copenhagen: WHO Regional Office for Europe.

Yang, L., Li, J., Zhou, K., Feng, P., and Dong, L. (2021). The effects of surface pollution on urban river water quality under rainfall events in Wuqing district, Tianjin, China. *Journal of Cleaner Production* 293: 126136. https://doi.org/10.1016/j.jclepro.2021.126136

Yavuz, V. (2024). Unveiling the impact of temperature inversions on air quality: a comprehensive analysis of polluted and severe polluted days in Istanbul. *Acta Geophysica* 73: 969–986. https://doi.org/10.1007/s11600-024-01417-0

14 Revitalizing Traditional Wisdom

A Critical Analysis of Indigenous Water Conservation Practices and Their Contemporary Relevance

Dipankar Jena and Prakash Chandra Mishra

INTRODUCTION

Water is a vital natural resource that supports life systems, environmental health, ecological balance, and economic development. Access to clean water is essential for public health, agriculture, and biodiversity protection. For millennia, communities have relied on traditional water sources like wells, rivers, and waterfalls for drinking and domestic purposes. Indigenous Knowledge (IK) in water conservation, such as

DOI: 10.1201/9781003613183-14

irrigation management and watershed protection, has been passed down through generations (Mahlangu and Garutsa, 2014; Kugara et al., 2022). Water scarcity is increasing gradually due to various reasons, starting from conventional causes to modern-age hitches of climate changes, and water pollution leads to contamination and unequal freshwater distribution, especially in arid areas, posing a challenge for universal access to clean water. Conservation and careful use are essential for ensuring adequate supply (Jury and Vaux, 2007; Rogers and Leal, 2010; Moshtagh and Mohsenpour, 2019). Conservation of water traditionally offers valuable lessons, but current challenges require scientifically proven, integrated, and effective solutions. Tackling the water crisis demands a multifaceted approach that balances ecosystem preservation, equitable resource allocation, and global water-use efficiency, ensuring sustainable management for the future (Sivakumar, 2011). For decades, rising populations, climate change, and ecological deterioration have made water management one of the major environmental issues. While modern technology and innovative approaches to water conservation are frequently explored, IK is being acclaimed as valuable among various sustainable water management practices. Indigenous societies worldwide have devised sophisticated, context-specific strategies for preserving water bodies and managing water quality (Edgar et al., 2022; Kunjuzwa et al., 2021). Modern-day water management approaches preferentially emphasize quantitative measurements and engineering solutions. Indigenous knowledge systems are holistic, combining practical techniques with spiritual and social syncretism, which has grown over the ages (Janke, 2018).

Indigenous water conservation practices are deeply rooted in the connection between communities and their environments, reflecting local ecological knowledge and regional climate conditions. These practices are passed down through oral traditions like stories, songs, and rituals and adapt to changing environmental factors (Gadgil and Berkes, 1993). Additionally, community-driven approaches focus on collective water management, prioritizing communal responsibility rather than individual ownership or privatization. This holistic method supports sustainability and resilience within Indigenous communities. IK (IK) and Traditional Knowledge (TK) encompass the skills, philosophies, and understandings of Indigenous societies. Passed down through generations, IK is practical and localized. TK enhances local livelihoods, and collaboration with Western science can benefit Indigenous communities and researchers, fostering mutual learning and growth. It helps scientists to better understand landscape management and restoration, while Indigenous people gain insights into climate change and new technologies. TK has become increasingly significant in water resource management (Klubnikin et al., 2000; Rosenberg et al., 2008).

The TK and Indigenous communities encounter several significant challenges. To tackle pressing environmental issues, it's crucial to balance IK and contemporary scientific systems. When combining modern science with traditional wisdom, it's vital to uphold Indigenous communities' rights, interests, and values. For instance, Indigenous communities might focus more on preserving the broader resources of forests rather than individual animal or plant species. Combining local and traditional approaches with scientific strategies can produce the greatest benefits (Finn and Jackson, 2011; Altieri, 2002).

This study attempted to reveal how Indigenous or traditional practices were effective in water conservation and their relevance in modern-day water crises. It examines water management systems in different parts of the world where traditional practices are still prevalent and align with modern science. The study stresses ancient wisdom, modern conservation strategies, and their amalgamation for preserving freshwater ecosystems and combating water pollution.

INDIGENOUS KNOWLEDGE AND CONSERVATION OF NATURAL RESOURCES

IK encompasses the collective wisdom, practices, and worldviews developed over centuries of human interaction with nature. It has been crucial in sustainably managing and conserving natural resources. Rooted in deep connections to land, water, and ecosystems, IK blends practical insights with cultural, spiritual, and ethical values, gaining renewed attention for its potential to enhance modern conservation efforts amid growing environmental challenges (Gadgil and Berkes, 1993; Ngara and Mangizvo, 2013)

Core Principles of Indigenous Knowledge

IK systems view the natural world as interconnected, emphasizing the relationships between humans, animals, plants, and natural phenomena (McGregor, 2004). This holistic perspective stresses balance and respect in human–environment interactions, in contrast to the fragmented view of Western science. Indigenous communities have developed sustainable practices focused on long-term ecosystem health, such as rotational farming and selective harvesting. For many Indigenous people, sustainability extends beyond ecology to include cultural and spiritual aspects. Adaptive management strategies, a key element of traditional ecological knowledge (TEK), help communities cope with environmental changes (Nelson and Shilling, 2018). Protecting Indigenous land rights is essential for biodiversity conservation and Indigenous governance, which involves collaborative decision-making effectively managing natural resources while addressing both ecological and cultural needs. Thus, integrating IK into current research could bring wider advantages to environmental conservation and sustainable development efforts (IEN, 2020; SDG, 2017; World Bank, 2018).

Applications of Indigenous Knowledge in Conservation

Indigenous people have long contributed to biodiversity conservation through fire control, fishing quotas, and rotational hunting (Gadgil and Berkes,1993; Harvey, 2009). Aboriginal communities in Australia use controlled burns to regenerate plants and support fire-adapted species. Their agricultural systems, including intercropping and agroforestry, focus on biodiversity, water conservation, and soil health, offering greater climate resilience than industrial farming (UNDP, 2014). IK has also enhanced water management, with tribes like the *Hopi* creating intricate irrigation systems and Indian groups using rainwater collection tanks in arid areas. Sacred natural sites, such as Nepal's sacred groves, also protect biodiversity by conserving plant species and endangered animal habitats. These practices highlight the deep connection between indigenous cultures and environmental stewardship (Kothari and Jha, 2011).

Scientific Validation of Indigenous Knowledge

IK has proven to be a valuable resource for biodiversity conservation. Practices such as selective logging and maintaining "clam shell gardens" have effectively sustained marine species in the Pacific Northwest. Similarly, traditional forest management methods, like harvesting non-timber products and selective logging, are more successful than commercial logging in preserving biodiversity in the Amazon (Brightman and Lewis, 2017) (Figure 14.1).

However, urbanization, cultural integration, and globalization threaten the preservation of IK, as younger generations often lack the skills passed down through the ages, risking the loss of valuable conservation techniques (CBD, 2004; WRI, 2005). Moreover, IK is increasingly commercialized without proper recognition or compensation. A key challenge is the lack of legal recognition of Indigenous land rights, making it difficult for Indigenous people to protect their territories from external threats like mining and infrastructure development. Despite these obstacles, IK offers sustainable solutions for conservation. It must be protected, respected, and integrated with scientific approaches to be fully utilized in modern conservation efforts. Collaboration between Indigenous communities, scientists, and policymakers is essential to address contemporary environmental challenges (Welch et al., 2013; GEF, 2020).

FIGURE 14.1 Sources of drinking water among indigenous communities of Odisha, India: (a) Hand pump, (b) pond/open source, (c) well, (d) supply water by Panchayat, and (e) tankered water.

TRADITIONAL WATER CONSERVATION PRACTICES

Water scarcity is now beyond geopolitical borders and a challenge that almost all countries are facing. The underlying causes are common across regions, irrespective of socioeconomic differences, but the question is how to control and mitigate the situation quickly. Traditional water conservation skills are gaining recognition for their role in this context. Despite modern technologies, traditional practices remain vital due to their low cost, sustainability, and cultural significance in addressing challenges like pollution and water scarcity.

Traditional Water Conservation Methods: A Global Prospect

Rainwater Harvesting (RWH) Scenario in India and the Middle East

Rainwater harvesting (RWH) is an ancient and popular water conservation method, especially in regions with seasonal rainfall and water scarcity, such as India, the Middle East, and parts of Africa. It involves collecting rainwater from rooftops, catchment areas, or surfaces and storing it in ponds, wells, or cisterns. RWH systems require specialized design based on climate and location. Traditional systems like Talabs, Johads, Guls, and Kuls have been used in India for centuries. Johads, small earth dams across seasonal streams, help collect rainwater and recharge groundwater. These systems are vital in extremely arid regions to safeguard water obtainability during summer (Kumar, 2018). Similarly, qanats, which are underground channels that carry water from aquifers to surface reservoirs, have long been used in the Middle East to support agriculture in arid regions (Marzo, 2019).

Terracing and Water Management in Agriculture in Asia and Latin America

Terracing is an ancient agricultural practice commonly used in mountainous regions with limited flat land. It preserves soil and water by reducing runoff, enhancing infiltration, and preventing erosion. In places like the Andes and the Himalayas, ancient societies built vast terrace systems, such as those by the Incas, to manage water and optimize agricultural use. These terraces enabled crop cultivation even in areas with minimal rainfall. In the Himalayas, farmers developed gravity-fed irrigation systems that direct mountain stream water to fields, requiring minimal infrastructure. This technique demonstrates how traditional methods have effectively addressed difficult terrains and water scarcity (Silverman, 2020).

Indigenous Water Management Systems in Australia and North America

Globally, Indigenous groups have created highly advanced water management systems that are intricately woven into their traditional customs. Aboriginal people have managed water resources in Australia's arid interior using TK. A thorough awareness of hydrological systems of the land is shown by methods like drilling *mulga* wells, which are hand-dug wells that draw from subterranean aquifers (Jackson, 2021). Native American tribes in North America, such as the Hopi, Zuni, and Navajo, have long used rainwater catchment systems and floodwater farming methods. In order to facilitate agricultural production in arid regions, the Hopi, for example, invented the technique of "check dams", which are tiny stone constructions designed to redirect and store water (Butler, 2017).

Water-Saving Architecture in North Africa and the Mediterranean

Architecture in places like the Mediterranean and North Africa has greatly aided water conservation. One example of how cultural practices have impacted water storage and conservation is the long-standing use of wind catchers, silos, and cisterns in nations like Tunisia, Morocco, and Egypt. While wind catchers-ventilation systems integrated into homes helped cool buildings and lessen evaporation from water sources, cisterns were frequently used to store rainwater for domestic use. Fountain houses, or traditional public fountains, were positioned thoughtfully throughout the city of Fez, Morocco, to supply fresh water and facilitate water distribution and storage. In addition to maximizing water use, these systems were created to promote community social interaction (Ababsa, 2013) (Figure 14.2).

Cultural and Religious Practices of Water Conservations

Religious and cultural beliefs are closely linked to various water conservation practices. For instance, water sources like rivers, lakes, and wells are frequently regarded as sacred and jointly managed by Sub-Saharan African communities. Water conservation is a shared responsibility among communities in many parts of Africa, where customary laws govern water use. In Southeast Asia, religious convictions also frequently impact water conservation measures. For example, the traditional irrigation system known as the Balinese Subak system in Indonesia is a cooperative water management model rooted in Hinduism. The *Subak* system is used to manage rice field irrigation by channeling water from mountain springs through a series of canals and gates, ensuring that water is distributed equitably among the community (Goris, 2020).

FIGURE 14.2 Various Indigenous approaches are used for water conservation with present days' scientific relevance.

Scope of Integrating Traditional Practices with Modern Water Management

Traditional water conservation techniques have been effective for centuries, but urbanization, modernization, and climate change now threaten their survival. Rapid industrialization and carelessness toward traditional skills lead to the loss of many of these methods (Cohen, 2020). However, there is growing interest in blending traditional practices with modern water management approaches. For example, in India, initiatives are promoting the revival of traditional water harvesting techniques like Johads (Srinivasan, 2021) to address groundwater depletion. Similarly, global efforts are focusing on reviving traditional farming practices like crop rotation and terracing to acclimate to the extreme climate. TK-based modern technology approaches, such as using GIS and remote sensing to map and monitor traditional systems, offer great potential. Modern engineering can also enhance traditional practices like irrigation and rainwater harvesting, improving efficiency without compromising cultural value (Lee et al., 2022). These time-tested water conservation methods offer valuable insights into resilience and sustainability. These age-old techniques remain relevant despite technological advancements, particularly in water-scarce regions. Combining traditional and modern methods can provide cost-effective, culturally appropriate solutions to contemporary water challenges, making the revival and adaptation of these practices crucial for ensuring water security in the future (Rastogi et al., 2024).

CONTEMPORARY RELEVANCE OF INDIGENOUS WATER CONSERVATION PRACTICES

Indigenous communities have developed sustainable water conservation practices over centuries, grounded in deep knowledge of local ecosystems. These techniques, such as rainwater harvesting, traditional irrigation systems, sacred water sites, and agroecological practices, offer valuable solutions to today's global water crisis. With 2.3 billion people living in water-scarce regions, sustainable water management is critical (UN, 2020). Indigenous methods, like the Zuni people's rainwater harvesting and the Philippines' rice terraces, focus on minimizing water waste and preserving ecosystems (Schnieder and Hook, 2004). However, modern water management policies often overlook or undervalue these practices. Despite this, the continued relevance of Indigenous water conservation methods needs exploration to create more holistic, adaptable water management solutions. These time-tested practices, such as crop rotation and contour farming, highlight the importance of integrating TK into modern water conservation strategies for long-term sustainability (Muecke, 2004; Absolon, 2011).

Contemporary Relevance of Indigenous Practices

Indigenous water conservation methods provide useful solutions to contemporary water management problems, making them more than just historical oddities. Integrating TK with modern science has benefited water and food security, especially during unpredictable extreme climate conditions.

Enhancing Climate Resilience

Indigenous practices emphasize resilience to environmental variability like the Inca civilization's terraces, canals, and storage systems designed to manage water scarcity and excess. These traditional water resilience strategies are highly relevant today, as climate change increases extreme weather events. Modern societies can enhance climate resilience by re-evaluating and adopting these time-tested systems (Denevan, 2001).

Ecological Sustainability

Ecological balance and ecosystems' long-term health are often prioritized in Indigenous water management techniques. Traditional techniques like agroforestry and rainwater harvesting maximize water use while preserving biodiversity and soil integrity (Udawatta et al., 2019). This is especially crucial given the continuous loss of biodiversity and environmental degradation. Desertification can be lessened, for example, by the traditional "gully plug" method employed by some Indigenous African communities, which channels water into natural depressions to stop erosion (Petry and Grimes, 2009).

Integrated Water Management

Indigenous systems often use integrated water management techniques that connect people, land, and water. The Aztecs, for example, utilized the Chinampa system, creating productive islands in shallow lakes, irrigated by canals. These systems were self-sustaining, with wetlands filtering water. This approach is vital for modern water management, viewing water as an interconnected environmental system (Pohl, 1994).

Cultural and Community Engagement

Indigenous water conservation practices involve local communities, fostering shared responsibility for sustainable water use. These practices emphasize respect for nature and sacred sites. Engaging communities in water management is essential for ensuring long-term sustainability and promoting lasting behavioral change in conservation efforts (Berkes, 2008).

Ground-Level Challenges and Opportunities

Indigenous water conservation practices are vital for addressing modern water management challenges, but they face obstacles like undervaluation by mainstream bodies, lack of recognition, and legal barriers. Unrecognized land rights limit Indigenous communities' ability to manage water resources traditionally. Land tenure issues, especially in developing countries, hinder sustainable practices. Climate change, urbanization, and ecosystem degradation complicate traditional methods. However, these challenges present opportunities for collaboration between Indigenous communities and scientists to adopt these practices. Integrating IK into modern systems can create more resilient, sustainable, and culturally relevant solutions, ensuring water security in a changing climate through community-based management (Turner, 2005; Nixon, 2016).

CHALLENGES IN ADOPTING THE AGE-OLD PRACTICES OF WATER CONSERVATION

As global water scarcity, population growth, and climate change increase, conserving water is vital for sustainability. Ancient water conservation methods, often rooted in IK, have historically addressed water shortages. However, modern challenges, such as institutional barriers, technological advancements, socioeconomic factors, and TK loss, hinder their adoption today. Exploring solutions to these obstacles is crucial for effective water resource management.

Socioeconomic Inequalities

Socioeconomic factors hinder the adoption of traditional water conservation methods like terracing, check dams, and rainwater harvesting. These techniques require more labor and financial resources than modern methods, and rural, impoverished communities prioritize immediate survival over long-term environmental management. Traditional methods are localized, focusing on individual communities, making it difficult to scale them for larger populations. The shift from community-based to privatized, market-driven systems complicates integration efforts. Indigenous practices based on communal ownership often conflict with privatized water rights, further limiting their application in contemporary society (Liu et al., 2016; McGranahan et al., 2018).

Modern Technological Progress

Recent technological advancements in irrigation and water management have replaced traditional methods with mechanized irrigation, drip systems, and advanced water treatment. While these modern systems are more efficient, they are expensive and may not always align with local environmental needs or sustainable practices (Shah, 2014). Traditional water conservation methods, like wells and ponds, are being overshadowed by large-scale infrastructure projects, such as dams, which increase water storage but often cause environmental harm. The shift toward centralized systems reduces efforts to preserve community-based, decentralized water practices, potentially rendering them obsolete or harmful in the long run (Pereira, 2006).

Institutional Regulations and Public Policies

Government regulations and institutional support are crucial for integrating traditional water-saving techniques into modern systems. However, many regions lack the political will or appropriate legislation to do so. Modern water management often favors top-down approaches over community-based solutions, and current policies don't address challenges like urbanization and climate change. Outdated regulations may overlook Indigenous water rights and TK. This regulatory gap limits the adoption of traditional conservation practices. Furthermore, contemporary water management organizations often lack the cultural awareness needed to incorporate IK, undervaluing local customs and reducing their relevance in modern decision-making (Gadgil and Berkes, 1993; Berkes et al., 2000).

Damage to Traditional Knowledge Systems

"Cultural erosion" refers to the loss of TK, particularly ancient water conservation techniques, which hinders their revival. These practices are increasingly overlooked, especially by younger generations in rural and Indigenous communities, who view them as outdated due to modern education and technology. Migration from rural to urban areas further disrupts the transmission of this knowledge, weakening community bonds. The lack of formal documentation, especially in societies with oral traditions, exacerbates the risk of complete knowledge loss. Additionally, colonial governments often suppressed Indigenous water practices, considering them inferior to Western methods, resulting in the loss of these vital techniques (Hughes, 2006; Hodgson, 2003).

Climate Change and Environmental Stresses

Climate change threatens both traditional and modern water conservation methods. Unpredictable weather patterns, environmental degradation, and urbanization can undermine techniques like terracing and small reservoirs. Adapting conservation strategies to these challenges is highly necessary to address the rapid loss of small to medium water bodies in the face of natural calamities and anthropogenic irresponsibility.

Socio-cultural Resistance

Socio-cultural aspects are also detrimental in adopting and successfully implementing traditional water conservation techniques. Modern society often prioritizes convenience and technology over labor-intensive methods. In urban areas, water is seen as unlimited, diminishing conservation efforts. Resistance arises from perceptions of inefficiency, gender roles in water collection, and a preference for unsustainable technologies (Molden, 2007). Despite the success of traditional conservation practices globally, challenges such as socioeconomic issues, institutional barriers, and climate change persist. Addressing these requires respecting TK, integrating community-driven solutions into governance, and supporting technologies that combine the strengths of both modern and traditional methods for sustainable water management.

CONCLUSION

TK is found helpful in water conservation, offering sustainable practices rooted in cultural, spiritual, and ecological principles. Techniques like rainwater harvesting, check dams, and sacred groves reflect a deep understanding of hydrological cycles, promoting efficient, equitable, and sustainable water management. These practices are vital for addressing water scarcity and climate change. However, challenges in the present day like colonization, industrialization, and urbanization-based development have led to the erosion of these traditional methods. There is a growing demand for the need to assimilate IK with scientific advances for more holistic water management.

This collaboration can enhance water conservation, ensuring sustainability and equitable community distribution. Bridging traditional and scientific approaches can lead to innovative, inclusive, and resilient water management systems. Future research and policies must prioritize recognizing and revitalizing Indigenous water management practices to ensure their ongoing contribution to global water conservation efforts.

REFERENCES

Ababsa, M. (2013). *Urban Water Management in the Maghreb: From Tradition to Modernity*. Routledge, London.

Absolon, K. (2011). Indigenous water management and spirituality. *Journal of Indigenous Research*, 6(2), 98–107.

Altieri, M. A. (2002). Agroecology: The science of natural resource management for poor farmers in marginal environments. *Agriculture, Ecosystems & Environment*, 93(1–3), 1–24.

Berkes, F. (2008). *Sacred Ecology: Traditional Ecological Knowledge and Resource Management*. Taylor & Francis, New York.

Berkes, F., Colding, J., & Folke, C. (2000). Rediscovery of traditional ecological knowledge as adaptive management. *Ecological Applications*, 10(5), 1251–1262.

Brightman, M., & Lewis, J. (2017). *Introduction: The Anthropology of Sustainability: Beyond Development and Progress* (pp. 1–34). Palgrave Macmillan, New York.

Butler, B. (2017). *Traditional Water Conservation Methods in Native American Communities*. University of Arizona Press, Tucson, Arizona.

Cohen, M. (2020). Traditional water management in the age of modernity: Lessons from the past. *Water International*, 45(3), 296–312.

Denevan, W. M. (2001). *The Native Population of the Americas in 1492*. University of Wisconsin Press, Madison, Wisconsin.

Edgar, B. A., Eva, M. P., & Lee, K. S. (2022). Disappearance of African indigenous knowledge of water conservation and management in Limpopo province of South Africa: An IKS perspective. *African Journal of Development Studies*, 12(2), 248–261.

Finn, M., & Jackson, S. (2011). Protecting indigenous values in water management: A challenge to conventional environmental flow assessments. *Ecosystems*, 14(8), 1232–1248.

Gadgil, M., & Berkes, F. (1993). Indigenous knowledge for biodiversity conservation. *Ambio*, 22(2–3), 151–156.

Global Environmental Facility (GEF). (2020). *Indigenous Knowledge for the Conservation of Biodiversity: Learning from the Field*. Global Environmental Facility (GEF), Washington, DC.

Goris, J. (2020). Water and culture: The subak irrigation system of Bali. *Asian Journal of Cultural Heritage*, 18(4), 158–174.

Harvey, B. (2009). Indigenous knowledges, sustainable development and the environment: implications for research, education and capacity building. In Langdon, J. (ed), Indigenous Knowledges, Development and Education (pp. 57–71). Brill, New York.

Hodgson, D. (2003). Indigenous knowledge and biodiversity conservation: The role of local knowledge in the conservation of biological diversity. *International Journal of Environmental Studies*, 60(3), 241–257.

Hughes, J. (2006). The erosion of traditional knowledge in rural communities. *Rural Sociology*, 71(4), 542–559.

Indigenous Environmental Network (IEN). (2020). *Protecting Indigenous Lands and Resources: Indigenous Knowledge and Environmental Stewardship*. Indigenous Environmental Network (IEN), Bemidji, MN.

Jackson, B. (2021). Indigenous water conservation practices in Australia. *Australian Journal of Environmental Management*, 28(1), 45–59.

Janke, T. (2018). *Indigenous Knowledge: Issues for Protection and Management*. Commonwealth of Australia, Australia.

Jury, W. A., & Vaux Jr, H. J. (2007). The emerging global water crisis: Managing scarcity and conflict between water users. *Advances in Agronomy*, 95, 1–76.

Klubnikin, K., Annett, C., Cherkasova, M., Shishin, M., & Fotieva, I. (2000). The sacred and the scientific: Traditional ecological knowledge in Siberian river conservation. *Ecological Applications*, 10(5), 1296–1306.

Kothari, A., & Jha, S. (2011). Indigenous knowledge systems and their role in biodiversity conservation. *International Journal of Environmental Studies*, 68(5), 753–768.

Kugara, S. L., Kugedera, A. T. T., Sakadzo, N., Chivhenge, E., & Museva, T. (2022). The role of indigenous knowledge systems (IKS) in climate change. In Tshifhumulo, R., & Makhanikhe, T. (eds.), *Handbook of Research on Protecting and Managing Global Indigenous Knowledge Systems* (pp. 1–23). IGI Global, Hershey, PA.

Kumar, P. (2018). *Water Harvesting and Conservation in Rural India*. Springer, Cham.

Kunjuzwa, D., Scholtz, B., & Fashoro, I. (2021). An indigenous knowledge system for promoting awareness of water resource issues. In *2021 IST-Africa Conference (IST-Africa)* (pp. 1–9). IEEE, New York.

Lee, D., Yang, S., & Kim, T. (2022). Modern technology and traditional water management: A global perspective. *Journal of Water Resources*, 67(5), 1042–1055.

Liu, J., et al. (2016). Sustainability and socio-economic challenges in traditional agricultural practices. *Nature Sustainability*, 2(5), 273–279.

Mahlangu, M., & Garutsa, T. C. (2014). Application of indigenous knowledge systems in water conservation and management: The case of Khambashe, Eastern Cape South Africa. *Academic Journal of Interdisciplinary Studies*, 3(4), 151–160.

Marzo, J. (2019). The qanat system: A millennia-old approach to water management in arid zones. *Water History*, 11(2), 127–141.

McGranahan, G., et al. (2018). Urban water management: Challenges and opportunities. *The Lancet Planetary Health*, 2(10), e394–e401.

McGregor, D. (2004). Coming full circle: Indigenous knowledge, environment, and our future. *American Indian Quarterly*, 28(3/4), 385–410.

Molden, D. (2007). *Water for Food, Water for Life: A Comprehensive Assessment of Water Management in Agriculture*. Routledge, London.

Moshtagh, M., & Mohsenpour, M. (2019). Community viewpoints about water crisis, conservation and recycling: A case study in Tehran. *Environment, Development and Sustainability*, 21(6), 2721–2731.

Muecke, S. (2004). Water conservation and Aboriginal knowledge. *Australian Aboriginal Studies*, 1, 27–42.

Nelson, M. K., & Shilling, D. (eds.). (2018). *Traditional Ecological Knowledge: Learning from Indigenous Practices for Environmental Sustainability*. Cambridge University Press, Cambridge.

Ngara, R., & Mangizvo, R. V. (2013). Indigenous knowledge systems and the conservation of natural resources in the Shangwe community in Gokwe District, Zimbabwe. *International Journal of Asian Social Science*, 3(1), 20–28.

Nixon, R. (2016). Indigenous land rights and water conservation policies. *Environmental Policy Review*, 22(4), 61–75.

Pereira, L. S. (2006). Water management in agriculture: Traditional knowledge and modern practices. *Agricultural Water Management*, 80(1–3), 1–10.

Petry, L., & Grimes, M. (2009). Gully plugging techniques: Indigenous water management systems in Africa. *Water Management Journal*, 21(3), 235–243.

Pohl, M. (1994). The chinampa system: Water management and agricultural techniques in the Valley of Mexico. *Latin American Studies Review*, 28(3), 56–68.

Rastogi, M., Kolur, S. M., Burud, A., Sadineni, T., Sekhar, M., Kumar, R., & Rajput, A. (2024). Advancing water conservation techniques in agriculture for sustainable resource management: A review. *Journal of Geography, Environment and Earth Science International*, 28(3), 41–53.

Rogers, P., & Leal, S. (2010). *Running Out of Water: The Looming Crisis and Solutions to Conserve Our Most Precious Resource*. St. Martin's Press, New York.

Rosenberg, D. E., Howitt, R. E., & Lund, J. R. (2008). Water management with water conservation, infrastructure expansions, and source variability in Jordan. *Water Resources Research*, 44(11), 6519.

Schnieder, E., & Hook, W. (2004). Zuni water management: A model for desert agriculture. *Desert Agriculture Review*, 10(2), 45–52.

Secretariat of the Convention on Biological Diversity (CBD). (2004). *The Role of Indigenous Knowledge in Biodiversity Conservation and the CBD: A Guide for Indigenous Communities*. CBD, London.

Shah, T. (2014). Water conservation and modern irrigation technologies. *Water Resources Management*, 28(12), 4553–4568.

Silverman, H. (2020). Agricultural terracing and water conservation in the Andean highlands. *Latin American Antiquity*, 31(2), 275–290.

Sivakumar, B. (2011). Water crisis: From conflict to cooperation—An overview. *Hydrological Sciences Journal*, 56(4), 531–552.

Srinivasan, K. (2021). Reviving traditional water systems in India: A policy approach. *Water Policy Journal*, 33(3), 412–426.

SDG. (2017). *Sustainable Development Goals (SDGs) and Indigenous Knowledge: A Study from the UN's Permanent Forum on Indigenous Issues*. SDG, Thailand.

The World Bank. (2018). *Indigenous Peoples and Conservation: Policies, Practices, and Challenges*. World Bank Publications, Washington, DC.

Turner, N. J. (2005). The role of traditional knowledge in modern conservation practices. *Conservation Biology*, 19(6), 1792–1797.

Udawatta, R. P., Rankoth, L. M., & Jose, S. (2019). Agroforestry and biodiversity. *Sustainability*, 11(10), 2879.

UNDP (United Nations Development Programme). (2014). *The Role of Indigenous Knowledge in Disaster Risk Reduction*. UNDP, New York.

Welch, J. R., Brondízio, E. S., Hetrick, S. S., & Coimbra Jr, C. E. (2013). Indigenous burning as conservation practice: Neotropical savanna recovery amid agribusiness deforestation in Central Brazil. *PLoS One*, 8(12), e81226.

World Resources Institute (WRI). (2005). *Finding Common Ground: Indigenous Knowledge and Biodiversity Conservation*. WRI, Washington, DC.

15 Community-Based Approaches to Conserve Water in the Indian Himalayan Region

A Review of Traditional to Modern Strategies and Outcomes

Vidhu Gupta, Akshay Saini, Lalita Bisht, and Rakesh Kumar Maikhuri

INTRODUCTION

Indigenous communities of the Indian Himalayan region (IHR) play a crucial role in conserving and managing various natural resources of the mountain ecosystem with their traditional knowledge and practices as they develop these practices over the years

DOI: 10.1201/9781003613183-15

by understanding the unique challenges of the Himalayan ecosystem. Water resources and their availability are essential for the sustainable socioeconomic development of mountain regions.

IHR is a critical ecological zone, stretching over 2,500 km across 12 states and is home to about 39.6 million inhabitants (Singh, 2024), highlighting its sparse and rural nature. The major water resources of IHR range from glaciers, rivers, lakes, groundwater, and springs are crucial for drinking, domestic irrigation, hydropower, and industrial purposes. However, this diversity also brings challenges, particularly in water accessibility and management. The IHR contributes significantly to India's freshwater resources, with annual precipitation averaging around 4,000 km^3 and generating about 500 km^3 of freshwater annually. As per the inventory, there are 9,575 glaciers in the Indian part of the Himalayas (Joshi et al., 2016). It hosts three major river basins, including the Indus, Ganges, and Brahmaputra, with 117 rivers and approximately three million springs that are vital for domestic and agricultural use. Around 8,634 million m^3 of water flows annually through the Himalayan rivers (Negi, 2002). Remote Sensing Center in 2011 shows that in IHR 2028, glacial lakes and water bodies are present. According to the NITI Aayog report 2018, approximately three million springs were found in IHR. Despite this abundance in water resources, communities residing in various altitudinal levels of the IHR often face acute water shortages as water resources face challenges, especially from climate change and global warming. The erratic rainfall pattern, deforestation, land use change for infrastructural development, agricultural expansion, soil erosion, and unregulated urbanization put huge pressure on different water resources.

According to various research, climate change and global warming significantly increase the temperatures of the Himalayan region, retreating the Himalayan glaciers ranging from a few meters to 61 m/year (Buytaert and De Bievre, 2012; Xu et al., 2009; Negi et al., 2018). Kulkarni and Kayakarte reported that 13% of the glaciated area had been lost during the last 4–5 decades. The cumulative loss in glacial ice during the last four decades has been estimated at 19 ± 7 m. A decrease in inflows of glacier-fed rivers results in severe water shortage as well as potential food and energy insecurity (hydropower generation) and an increase in GLOFs (Chaturvedi et al., 2021), which eventually disturbs the water-food-energy nexus. For many Himalayan communities, springs are the sole water source for drinking and other domestic purposes. According to the NITI Aayog Report (2018), nearly 50% of the springs in IHR have already dried up or have become seasonal, and tens of thousands of villages are currently facing acute water shortages.

For centuries, local communities residing at different altitudinal levels of IHR have established different innovative and sustainable techniques for managing glacier meltwater, springs, rainwater harvesting, and groundwater recharge to ensure water security for local communities throughout the year. In higher altitudes, communities mainly rely on snow and glacier-related meltwater; however, mid- and lower-altitude communities focus on springs and rivers as they are the primary water source for rural households. However, with changing environmental conditions and the rise of modern technologies, these traditional methods face challenges regarding effectiveness, scalability, and relevance. Addressing these challenges requires an integrated approach combining traditional knowledge and modern water management practices. Historically central to sustainable resource use, Indigenous knowledge systems are increasingly at risk of erosion. Revitalizing and integrating these practices with modern technologies can help address water conservation challenges. Community participation is crucial for developing and implementing effective strategies for water allocation, conservation, and management (Vaidya, 2015; Sen and Kansal, 2019).

By examining traditional and modern approaches, this review aims to provide actionable insights into sustainable water management, addressing the unique challenges of the IHR and contributing to the broader discourse on water conservation in fragile mountain ecosystems. The following are some traditional to modern water conservation practices in different regions of IHR to address water-related challenges and secure a sustainable future for this critical region

TRADITIONAL COMMUNITY-INDUCED METHODS OF WATER CONSERVATION IN THE INDIAN HIMALAYAN REGION (IHR)

1. **Zing**: The Zing system is a traditional water conservation practice in Ladakh's cold desert region. Zings are typically small tanks in which melted glacier water collects through a network of guiding channels known as "Yuras". These reservoirs, operated by a particular person called "Churpun", ensure that water is equally distributed among community members. Modern mechanical systems offer better scalability, but integrating Zings with modern irrigation can provide low-cost, community-centered solutions (Figure 15.1).

FIGURE 15.1 Zing water harvesting structure in Leh, India. (*Photo Source*: Huda et al., 2019.)

FIGURE 15.2 (a) Kuhl system in Himachal Pradesh. (*Image Source*: *Photo Source*: CPREEC-EIACP.) (b) Guhls made by villagers in Jagatpur, Dehradun for paddy field irrigation. (*Source*: Dehradun, March 27, 2025.)

2. **Tapkos and Yuras**: Tapko is the main drainage line or stream that collects the flowing water from mountains and rivers, and Yuras are the smaller streams or channels that originate from Tapko to feed various areas (Boyd, 2007; Dawa et al., 2000). Yuras are made using local materials such as mud and stones, with villagers involved in constructing and maintaining these canals, thus making it a true community endeavor. The sites of these canals are selected in such a way that they help optimize water use and minimize soil erosion for agricultural practices in this region, which is characterized by very little precipitation.
3. **Kuhl, Kool, and Guhl System**: Traditional age-old irrigation systems transport water from streams or rivers through small hand-dug channels to agricultural fields. These structures, made from local materials, are efficient and sustainable and are managed collectively by village communities. They are known as Kools in the Udhampur District of J&K (Slathia et al., 2016), Kuhl in Himachal Pradesh, and Guhls in the Uttarakhand region (Sharma et al., 2023). Some Guhl systems also provide water for hydropower, water mills (Gharats), potter's wheels, and non-drinking household uses. Although labor-intensive, their revitalization with modern techniques can enhance efficiency (Figure 15.2).
4. **Baodis (Baori, Baoli) and Nawns (Naons, Nahuns)**: Baodi is a traditional water storage system in rural Himachal Pradesh, designed for collecting and storing water. It is rectangular or square, with a gentle slope leading to the center where water collects for natural percolation (Lal and Verma, 2008). The structure is made of stone and has a ladder-like structure, an outlet, and a roof to prevent water contamination (Figure 15.3). The main recharge source is groundwater, which naturally percolates into the system, making it a reliable drinking source. Baodis are smaller than other water storage systems, like lawns, and are located near residences, allowing easy access for villagers (Sharma and Kanwar, 2009). They are made by skilled masons who place stones in a proper sequence for continuous water percolation, reflecting the traditional knowledge and skill of the local community.

FIGURE 15.3 Baudi. (*Source*: (a) peoplescienceinstitute.org; (b) Lal and Verma, 2008.).

FIGURE 15.4 Nawn. (*Source*: peoplescineceinstitute.org.)

On the other hand, **Nawns** are traditional water storage structures in Himachal Pradesh, designed for sustainable water management (Figure 15.4). Larger than **Baudis**, they serve diverse needs, including providing water for

FIGURE 15.5 Zabo. (*Source*: Surjit Singh/Mongabay).

drinking, irrigation, and other domestic needs (Sharma and Kanwar, 2009). Typically located near groundwater sources or springs, Nawns are constructed using stones available locally to form a durable and efficient storage structure. These shallow wells or depressions, either natural or artificial, collect and store groundwater, often lined with stones to prevent contamination and soil erosion. Nawns recharge via rainwater and groundwater, with protective designs including roofs, walls, and sluices to maintain water purity. Separate channels prevent contamination during washing and bathing. Their maintenance, guided by village elders, is a communal effort that ensures sustainability and strengthens social bonds.

5. **Zabo**: The Zabo system, also known as the Ruza system, is a traditional water conservation method and integrated farming practiced by the Chakhesang tribe of Nagaland (Figure 15.5). The term Zabo means "impounding water" in the local dialect. The Zabo system integrates multiple components, including forested catchment areas, water collection ponds, cattle barns, and agricultural fields on lower slopes (Singh et al., 2012). Rainwater from the protected forest catchments flows downhill through terraced landscapes, accumulating in intermediary ponds. Along the way, the channels pass through cattle yards, incorporating nutrient-rich animal waste into the water. This enriched water ultimately reaches the paddy fields, where it is used not only for irrigation but also for raising fish and cultivating medicinal plants (Figure 15.6).

   Zabo's focus on forested catchments and organic nutrient cycling aligns with modern sustainable agricultural practices. The integration of organic farming techniques with modern irrigation could enhance water efficiency and soil health, making Zabo highly adaptable to contemporary farming methods.

6. **Apatani Wetland System (Arunachal Pradesh)**: The Apatani community of Arunachal Pradesh exemplifies integrating ecological knowledge and community-based water management (Figure 15.7). Their wet-rice cultivation system is based on water-sharing. It is a unique system of paddy cultivation and fish farming in wetlands developed by the Apatani tribe and

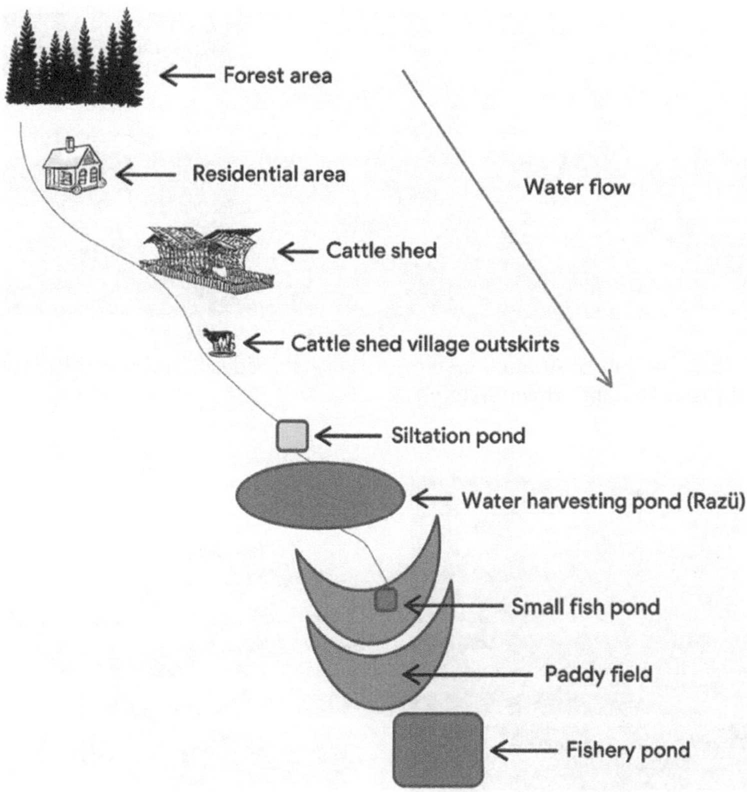

FIGURE 15.6 Land management in Zabo system. (*Source*: Amenla and Shuya, 2021.)

a headman oversees it. Water is tapped from small streams and springs by erecting temporary mud walls that act as barriers, diverting water flow to terraces and valley fields (Dolo, 2009). Bamboo plantations are also incorporated, conserving soil moisture and preventing erosion, while the community's reverence for sacred landscapes aids in water conservation and biodiversity preservation.

7. **Bamboo Drip Irrigation (Meghalaya)**: It is a centuries-old traditional practice used in northeast India, especially in Meghalaya's *Jaintia* and *Khasi* Hills (Figure 15.8). In this traditional method, spring water is directed to agricultural fields via bamboo pipes of various sizes (Dhiman and Gupta, 2011). Operating entirely by gravity, the system delivers water drop by drop directly to the plant roots, minimizing wastage and enhancing water absorption. The bamboo channels, which can span several meters to a few kilometers, are designed with varying diameters to regulate water flow across different sections, ensuring uniform distribution (Maurya and Singh, 2021). This method is beneficial in low-rainfall areas and allows farmers to control the water flows, ensuring that crops receive sufficient moisture during dry periods.

FIGURE 15.7 (a and b) Apatani system, Arunachal Pradesh. (*Source*: http://cpreecenvis.nic.in/Database/ApataniFarmingSystem.)

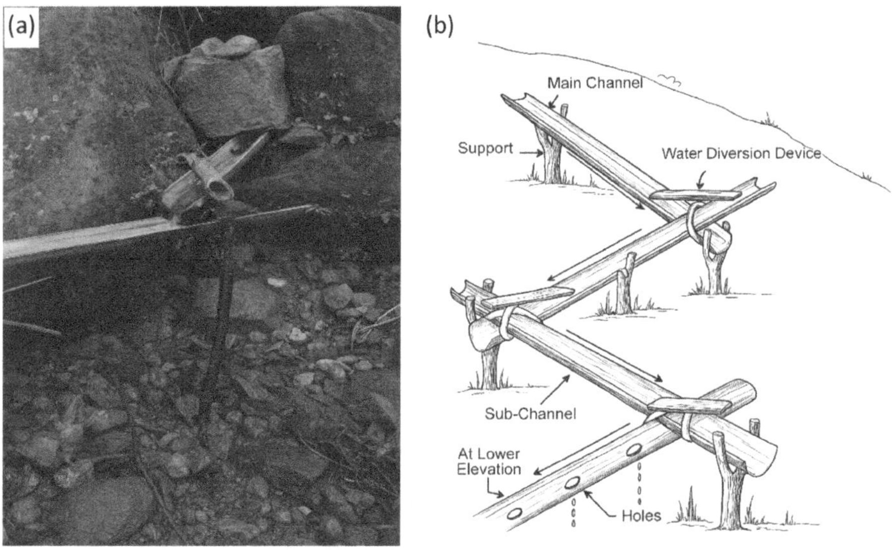

FIGURE 15.8 Bamboo drip irrigation. (*Source*: IHCUC Theme-V, 2022.)

8. **Dongs (ponds) Assam**: In Assam, the Bodo community has been practising a decade-old traditional method for decades to channel river water into their village communities (Figure 15.9). They construct mud canals to divert the river water, which is then stored in ponds locally known as *Pukhuri*. Often water is drawn directly from these mud channels for irrigation purpose. As highland streams flow down to the plains, they are guided into these canals, with their movement regulated by small dams built form tree branches, stones, and rocks. The main canal branches out into several smaller ones, known as *shakhas* and *prakhashas*, forming an extensive network that ensures water reaches every part of the community (Barman and Jha, 2020).

FIGURE 15.9 Dong. (*Source*: https://vishnuias.com/dong-traditional-water-management-system-of-bodo-people-in-water-scarcity-hit-assam/.)

FIGURE 15.10 Naula. (*Source*: IHCUC Theme-V, 2022.)

9. **Naula (Shallow Well)**: Naulas are traditional, shallow, stepped wells, especially common in the Kumaon region of Uttarakhand (Figure 15.10). The "Naula" usually collects water from subterranean seepages or springs and is vital for community water needs. Typically built in the shape of an inverted trapezoid, Naulas range in size, with larger and more ornate wells constructed by wealthy households or local rulers. Built within natural depressions and

FIGURE 15.11 Dharas. (*Photo Source*: IHCUC Theme-V, 2022.)

reinforced with stone masonry, their construction showcases local resources and traditional craftsmanship passed down through generations (Prasad and Sharma, 2019; Sharma et al., 2023). Many ancient functioning Naulas in Uttarakhand like the Badrinathji-ka-naula (7th century) in Chamoli district and Suryakot in Almora still exist.

10. **Dhara (Spout Spring)**: In Uttarakhand, *Dharas* (also referred as *mungurus*) serve as crucial freshwater sources (Figure 15.11). These springs typically emerge where the sloping ground surface intersects with the groundwater table. In most cases, the sources of these springs are unconfined aquifers, primarily formed within fractured rock masses. The volume and movement of water are governed by the aquifer's total yield and gravitational force (Sharma et al., 2023). Water is often extracted through man-made outlets, which may be simple pipes or sculpted forms resembling animals such as elephants, cows, and oxen. (Prasad and Sharma, 2019). Prominent examples include Parda dhara, Sipahi dhara, and the perennial dhara located near a Shiva temple in Gopeshwar. However, *dharas* are sensitive to environmental changes, as seen in the drying of Amni-ka-dhara in Tehri Garhwal following the 1991 Uttarkashi earthquake.
11. **Chaal and Khal System (Uttarakhand)**: The Chaal and Khal system is a traditional method of conserving rain water and recharging groundwater in Uttarakhand (Figure 15.12). **Chaal** refers to natural or manually dug depressions in the ground that capture and store rainwater. At the same time, **Khal** are small ponds or reservoirs, typically built-in valleys or flat areas, to gather surface runoff and rainwater (Negi and Kandpal, 2003). These structures are strategically located on natural slopes or depressions to maximize water collection. By holding and retaining rainwater, they help recharge groundwater aquifers. The water collected in these systems serves as a vital resource for drinking, irrigation, and livestock, supporting the daily needs of local communities.

FIGURE 15.12 Chaal and Khal. (*Source*: https://connectingheritage.com/article&news/articles/uttarakhand; Khal – picture taken in Bachher Van Panchayat, Dasholi Block, Chamoli.)

MODERN APPROACHES FOR WATER CONSERVATION IN THE INDIAN HIMALAYAN REGION

1. **Artificial Glaciers**: Artificial glaciers represent an innovative method for water harvesting, developed to store runoff as ice during the winter months on the slopes of mountains (Figure 15.13). These artificial glaciers are strategically constructed closer to the village and at a lower altitude, enabling them to begin melting much earlier typically around April or May, thereby providing supplementary irrigation water ahead of critical agriculture season. Construction typically takes place in November and December, where channels are built to divert runoff water to the shaded sides of the mountains, where lower temperature facilitates freezing. Along the slopes, a series of small retaining walls resembling miniature dams are constructed to further slow the flow of water and encourage stepwise freezing, ultimately forming an "artificial glacier". As temperature rises in April, the artificial glacier begins to melt, supplying much-needed water to the nearby fields precisely when crops such as barley (locally known as *Thachus*) require their initial irrigation. One successful example is the artificial glacier built in the village of Phugtal in Ladakh, which helps to irrigate barley fields. The glacier melts in late spring, providing essential water for crops during the dry months. Studies have shown that artificial glaciers can increase irrigation efficiency by 50%–60% in water-scarce areas. Implementing artificial glaciers is challenging due to the need for skilled labor, high initial construction costs, and the dependency on specific geographic features. The method may not be feasible in areas without the required climatic and topographical conditions.

FIGURE 15.13 Artificial glacier.

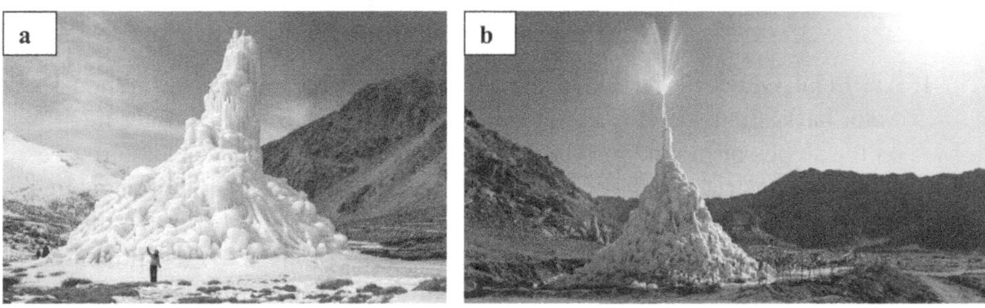

FIGURE 15.14 An ice stupa grows tall in Igoo, Ladakh. (*Source*: Rosa Oyarzún, 2022.)

2. **Ice Stupa**: Ice stupas are engineered ice structures designed to store winter runoff in the form of conical ice formation (Figure 15.14). This glacier grafting technique captures water that would otherwise be lost during winter months and preserves it as cone-shaped mounds. As temperature rises in the summer and water becomes scarce, the ice melts, providing a crucial source of irrigation for crops. The design of ice stupas is based on the principles of gravity and water pressure: water is channeled from a high-altitude stream through a pipeline elevated above ground level. The kinetic energy of the flowing water, driven by gravity and pressure, propels it through a sprinkler system at the construction site. Upon exposure to the cold air, the water freezes almost instantly, gradually forming a layered, conical ice structure. The first ice stupa was successfully constructed in Ladakh by Sonam Wangchuk in 2013.

It provides water for about 300 households and 2,000 acres of farmland. The ice stupa can store up to 1 million liters of water, which becomes available during the summer droughts (Wangchuk et al., 2015).

3. **Micro-Irrigation System**: Micro-irrigation, also called trickle or drip irrigation, is an efficient method of watering plants that conserves both water and fertilizer. It delivers water directly to the plant roots, either on the soil surface or directly into the root zone, using a network of valves, pipes, tubing, and emitters. Gravity pressure is typically used to maintain the water flow, and advanced emitters are designed to minimize water loss while ensuring consistent pressure, even on slopes. This innovative approach can reduce water usage by up to 80%, all while maintaining the ideal moisture levels needed for healthy plant growth and maximum crop yield.
4. **Rooftop Rainwater Harvesting**: Rooftop rainwater harvesting is a simple and effective way to collect rainwater from rooftops. The captured water is either stored in tanks for future use or directed into the ground to replenish groundwater reserves. In Shimla, rooftop rainwater harvesting systems have been implemented in several government buildings, with water storage capacity reaching up to 500,000 L per building. This has greatly reduced reliance on external water sources. This method is a sustainable solution to conserve water and reduce dependence on external water sources. Inadequate infrastructure and public awareness are significant barriers to large-scale implementation. The effectiveness of this method is also influenced by seasonal rainfall patterns.
5. **Other Water Conservation Methods**
    a. **Ponds and Reservoir**: Rainwater harvesting ponds and reservoirs can be constructed in hilly areas to collect runoff and recharge groundwater. In hilly areas with hills on three sides, embankment types of ponds are constructed for the collection of runoff by forming an embankment on the open side.
    b. **Farm Ponds**: A farm pond is a simple, dug-out water reservoir built on or near farmland. It captures and stores rainwater, serving as a valuable resource for various purposes. Farmers use it to irrigate crops, recharge groundwater, support aquaculture, and provide drinking water for livestock. This practical solution helps to ensure a steady water supply, especially during dry periods, promoting sustainable farming practices.
    c. **Check Dams**: These are small, artificial temporary dams constructed across streams or rivers to slow down water flow and increase groundwater recharge. It counteracts erosion by reducing water flow velocity, causing sediment traps. Materials used in the construction of check dams include rocks, sandbags, fiber logs, cement concrete, and other materials. The main function is to lower the velocity of water; infiltration and sedimentation are added benefits.
    d. **Percolation Tanks:** Percolation pits, also known as recharge pits, are man-made structures designed to collect and store rainwater runoff. Their main goal is to help rainwater seep slowly into the soil, replenishing

groundwater levels. In addition to boosting water reserves, these pits also play a role in conserving soil and maintaining ecological balance, making them a simple yet effective tool for sustainable water management.

e. **Borewell and Dug Well Recharge**: This involves directing rainwater into existing borewells or dug wells to recharge the groundwater through filtration systems.
f. **Contouring and Terracing:** These techniques in agriculture are designed to slow water runoff, enhance soil infiltration, reduce soil erosion, and boost groundwater recharge. One such method, contour-based storage, leverages the land's natural contours to create tiered reservoirs or ponds, particularly effective in hilly regions. These small, terraced storage systems help minimize soil erosion while maximizing water retention and promoting sustainable water and land management.
g. **Poly Tanks**: Rainwater harvesting is considered one of the most effective strategies for enhancing irrigation practices in the Himalayan region. The primary challenge in rain-fed agriculture lies in the storage of rainwater and ensuring its timely availability to crops. Poly tanks offers a cost-effective solution for this purpose (Kumar and Madhukar, 2020). These tanks can be constructed either temporarily or for a long-term use by lining pit or trench of varying sizes with polyethylene sheets. Water stored in poly tanks can then be distributed to valuable crops using a gravity-based supply system, ensuring efficient and sustainable irrigation.
h. **Reforestation and Watershed Management**: Implementing reforestation projects and effective watershed management practices can improve water retention and reduce soil erosion. Planting trees and restoring vegetation help stabilize the soil and enhance the natural water cycle, contributing to better water availability and quality.

CONCLUSION

Water is a fundamental human right that is crucial for life, health, and societal well-being. It is intrinsically linked to hygiene, sanitation, and overall quality of life. In the IHR, water resources are increasingly under threat due to impacts of climate change, population growth, and shifting socioeconomic dynamics. Recent studies highlight an urgent need to safeguard these critical freshwater sources, positioning water conservation as a central priority for policymakers and resource managers. Sustainable water management strategies must focus simultaneously on conservation and the equitable distribution of water, particularly recognizing the pivotal role women play in household water management. Policies must prioritize water conservation through integrated climate action plans, offering financial and institutional support for technologies such as micro-irrigation systems, rainwater harvesting, and artificial glaciers. These innovations should be contextually embedded within broader climate resilience frameworks to ensure long-term water security across vulnerable Himalayan communities. Importantly, policy framework must also address the gendered dimensions of water management. Women, as primary stewards of domestic water use, should be empowered through targeted training programs and their active inclusion in decision-making processes at both community and regional levels. Addressing the specific challenges faced by women can significantly enhance the effectiveness of water conservation initiatives. Community engagement remains fundamental to the success of water conservation strategies. Local communities must be actively involved in design, implementation, and maintenance of water management systems to ensure that interventions are culturally relevant, socially acceptable, and sustainable. Promoting participatory governance fosters a sense of ownership, which increases the long-term viability of conservation projects. Furthermore, investment in water infrastructure—such as a sustainable irrigation networks, decentralized water storage systems, and efficient wastewater treatment facilities—is crucial, especially in remote and ecologically sensitive areas of the IHR.

A holistic and integrated approach to water management is imperative—the one that synergizes modern scientific advancements with traditional ecological knowledge. Strengthening public awareness and re-engaging communities as active agents, rather than passive beneficiaries, is key to building resilient water management systems. Strong political will, robust information systems, and interdisciplinary research collaborations will be instrumental in achieving these objectives.

FUTURE DIRECTIONS

Future research efforts must intensify focus on the impacts of climate change on the IHR cryosphere, particularly glacial retreat, shifting precipitation patterns, and their cascading effects on agriculture and drinking water availability. The integration of satellite-based remote sensing with localized hydrological studies can provide real-time, high-resolution data essential for informed decision-making in water management. Additionally, more research should explore the potential of traditional water management practices—such as terracing, check dams, and community-managed irrigation systems—with modern technological interventions. Understanding how these hybrid models can enhance water conservation while preserving indigenous knowledge systems will be critical to fostering sustainable practices.

Ultimately, addressing the escalating water crisis in the IHR requires a comprehensive multi-scalar approach that combines governance, community participation, technological innovation, and a deep respect for traditional wisdom. Water must be recognized and managed as a shared, collective resource, with women and marginalized groups positioned at the center of conservation efforts. By uniting scientific research, traditional practices, and strong political commitment, it is possible to secure sustainable water future for generations to come.

REFERENCES

Barman, A., & Jha, R. (2020). Traditional stormwater management system in the watershed of wetland, a study, and overview of "Dong" systems in wetland fringe villages of Assam. *Hydro 2020 International 25th International Conference on Hydraulics, Water Resources and Coastal Engineering*, Odisha India.

Boyd, N. D. (2007). *Local Approaches for a Global Water Revolution: A Case Study of Communal Water Management in Ladakh, India.* Saint Mary's University, Halifax, NS.

Buytaert, W., & De Bievre, B. (2012). Water for cities: The impact of climate change and demographic growth in the tropical Andes. *Water Resources Research*, *48*(8), 11755.

Chaturvedi, A. K., Surendran, U., & Madhava Chandran, K. (2021). Exploring growth, physiological status, yield and water use efficiency of vegetables grown under wick method of irrigation. *Plant Physiology Reports*, *26*(1), 64–73.

Dawa, S., Dana, D., & Namgyal, P. (2000). Water harvesting technologies and management system in a micro-watershed in Ladakh, India. *Waters of Life: Perspectives of Water Harvesting in the Hindu Kush-Himalayas*, *2*, 235–253.

Dhiman, S. S., & Gupta, R. C. (2011). Traditional water harvesting structures and sustainable water management in India: A socio-hydrological review. *International Letters of Natural Sciences*, *37*, 33–42.

Dolo, M. (2009). Traditional irrigation system: A case of apatani tribe in Arunachal Himalaya, North East India. *Mountain Forum Bulletin*, *9*(1), 9–11.

Huda, M. B., Kumar, R., & Rather, N. A. (2019). Water resources scenario of Indian Himalayan region. *Applied Agricultural Practices for Mitigating Climate Change*, *2*, 257–274.

Joshi, R. K., Satyal, P., & Setzer, W. N. (2016). Himalayan aromatic medicinal plants: A review of their ethnopharmacology, volatile phytochemistry, and biological activities. *Medicines*, *3*(1), 6.

Kumar, A., & Madhukar, A. (2020). Management of traditional water systems and their conservation in North Eastern Region through local traditional wisdom. *International Journal of Environmental Science and Technology*, *17*(7), 2823–2832.

Lal, C., & Verma, L. R. (2008). Indigenous technological knowledge of soil and water management from Himachal Himalaya. *Indian Journal of Traditional Knowledge*, *7*(4), 618–624.

Maurya, R., & Singh, P. (2021). Bamboo drip irrigation: A potential tool for increasing crop productivity in Meghalaya. *The Agriculture Magazine*, *1*(1), 1–3.

Negi, G. C. (2002). Hydrological research in the Indian Himalayan mountains: Soil and water conservation. *Current Science*, *83*, 974–980.

Negi, G. C. S., & Kandpal, K. D. (2003). Traditional methods of water management in Central Himalayan agriculture. *Indian Journal of Traditional Knowledge*, *2*(3), 256–264

Negi, G. C. S., Samal, P. K., Kuniyal, J. C., Kothyari, B. P., Sharma, R. K., & Dhyani, P. P. (2018). Knowledge and practices of water harvesting for climate-resilient agriculture in the Indian Himalayan Region. *International Journal of Water Resources Development*, *34*(6), 933–952.

NITI Aayog. (2018). *Inventory and Revival of Springs in the Himalayas for Water Security.* National Institution for Transforming India, Government of India, New Delhi.

Oyarzún, R. (2022). *Ice Stupa in Igoo, Ladakh.* Kleinman Center for Energy Policy, Philadelphia, PA.

Prasad, C., & Sharma, R. C. (2019). Water resources and their traditional management in Kedarnath valley of Garhwal Himalaya, India. *International* Journal of Hydrogen *Energy*, *3*, 194–203.

Sen, S. M., & Kansal, A. (2019). Achieving water security in rural Indian Himalayas: A participatory account of challenges and potential solutions. *Journal of Environmental Management, 245*, 398–408.

Sharma, M., Sharma, N., & Sharma, A. (2023). Restoration and preservation of traditional water resources: A study of Uttarakhand. *West Science Social and Humanities Studies, 1*(5), 255–259.

Sharma, P., & Kanwar, N. (2009). Indigenous water conservation systems—A rich tradition of rural Himachal Pradesh. *Indian Journal of Traditional Knowledge, 8*(4), 510–513.

Singh, M. (2024). A review on traditional ecological knowledge of indigenous communities of Northeast India. In Borthakur, A., & Singh, P. (eds), *Learning 'from and 'with the Locals: Traditional Knowledge Systems for Environmental Sustainability in the Himalayas* (pp. 259–292). Springer, Cham.

Singh, R. A., & Gupta, R. C. (2002). Traditional land and water management systems of North-East hill region. *Indian Journal of Traditional Knowledge, 1*(1), 32–39.

Singh, R. K., Singh, V., Rajkhowa, C., & Deka, B. C. (2012). Zabo: A traditional way of integrated farming. In Deka, B. C., Patra, M. K., Thirugnanavel, A., Chatterjee, D., Borah Tasvina, R., & Ngachan S. V. (eds), *Resilient Shifting Cultivation: Challenges and Opportunities*, (pp.114–117). ICAR Research Complex for NEH Region, Nagaland Centre, Medziphema, Nagaland.

Slathia, P. S., Sharma, B. C., Paul, N., Kumar, R., & Gupta, S. K. (2016). Traditional water conveyance modes, their management and use in Udhampur district of Jammu and Kashmir. *Indian Journal of Traditional Knowledge, 15*(3), 503–508.

Vaidya, R. A. (2015). Governance and management of local water storage in the Hindu Kush Himalayas. *International Journal of Water Resources Development, 31*(2), 253–268.

Wangchuk, S., Rai, S. C., & Sharma, B. C. (2015). Traditional water management systems in the Sikkim Himalaya: A case study from the Rani Khola watershed. *Journal of Mountain Science, 12*(3), 567–579

Xu, J., Grumbine, R. E., Shrestha, A., Eriksson, M., Yang, X., Wang, Y. U. N., & Wilkes, A. (2009). The melting Himalayas: Cascading effects of climate change on water, biodiversity, and livelihoods. *Conservation Biology, 23*(3), 520–530.

Environmental Education and Youth Empowerment

A Pathway to Sustainable Conservation

16

Jibin Jose P. and Amutha Dhanaraj

INTRODUCTION

Environmental conservation has become a global concern in light of escalating ecological challenges. The stability of ecosystems that support life on the Earth is in jeopardy because of the intensification of climate change, biodiversity loss, and environmental degradation caused by rapid industrialisation, deforestation, and unsustainable resource extraction. The rise in global temperature, natural catastrophes, and deteriorating soil and water quality highlight the need for urgent conservation measures (Intergovernmental Panel on Climate Change, 2021). Since ecosystems deliver essential services such as pollination, water purification, and carbon sequestration, which are vital for food and water security, it is crucial to conserve biodiversity. Moreover, conservation fosters economic stability through green employment and sustainable industries while mitigating health risks associated with pollution and various diseases (World Wildlife Fund, 2022). Global undertakings

DOI: 10.1201/9781003613183-16

such as the Paris Agreement (UN Framework Convention on Climate Change, 2015) and the UN Sustainable Development Goals (UN Department of Economic and Social Affairs, 2015) highlight the collective need to preserve the environment. Humanity can ensure a sustainable future by adopting conservation, guaranteeing that ecosystems and natural resources will endure for future generations (Hulme, 2016).

Young people are essential to sustainability initiatives because they infuse environmental preservation with vitality, inventiveness, and a feeling of urgency. Youth comprise more than half of the world's population and are uniquely positioned to bring about revolutionary change inside and outside their communities. Their dedication to tackling urgent environmental issues, including pollution, deforestation, and climate change, is demonstrated by their active participation in sustainability efforts, ranging from grassroots movements to international advocacy campaigns (Skinner et al., 2012). Young people with access to education and technology use cutting-edge solutions, such as digital platforms and green technologies, to encourage sustainable habits and environmental awareness. Giving them leadership opportunities and environmental education (EE) ensures their voices are heard, resulting in a generation of environmentally conscious leaders (Tabanao et al., 2025).

Environmental education empowers youth to address environmental issues. Ecological literacy helps youth to comprehend environmental challenges like resource depletion, biodiversity loss, and climate change. Youth can champion sustainable practices and make educated choices by fostering critical thinking, problem-solving, and decision-making. Environmental education emphasises experiential and interactive methods that include young people directly in conservation efforts, going beyond theoretical education. Initiatives like garbage management, citizen science, and tree planting help people gain practical skills and cultivate a sense of environmental responsibility. Early environmental education empowers young people to lead sustainable initiatives and policy advocacy, fostering eco-awareness and preparing them to promote sustainable change.

This chapter examines the function of EE in enabling youth to serve as agents of sustainable conservation. The investigation explores how environmental education empowers youth by facilitating participation, skill development, and knowledge-sharing to promote ecological stewardship. It emphasises youth-led conservation efforts, experiential learning, inclusive environmental education, and culturally sensitive approaches in fostering agency, promoting sustainability, and ensuring equitable access to resources. It suggests that investing in EE serves as a bridge to a sustainable and resilient future, underscoring the transformational potential of youth engagement.

FOUNDATIONS OF ENVIRONMENTAL EDUCATION

The dynamic process of environmental education equips people to make wise decisions and act responsibly while promoting awareness of the environment. According to the Tbilisi Declaration (Gillett, 1977), EE empowers people to solve environmental issues, advance sustainable practices, and improve ecological literacy. The objectives of environmental education include raising awareness of environmental challenges, cultivating attitudes that promote sustainable living, and improving critical thinking and problem-solving skills. Environmental education promotes sustainable development by incorporating environmental factors into everyday activities and decision-making procedures, aiming to create an informed, environmentally conscious society to address global concerns like resource depletion and climate change.

The historical development of environmental education shows the necessity to address ecological challenges through educational initiatives. Early programmes that focused on environmental studies and conservation education originally emerged in the late 19th and early 20th century to promote an appreciation of natural resources. However, the contemporary idea of EE gained traction in the 1960s and 1970s as environmental deterioration and pollution worries increased. Since then, important international frameworks have influenced the evolution of environmental education. The Stockholm Conference on the Human Environment in 1972 emphasised the significance of education in environmental awareness (United Nations, 1972). Its wide focus lacked methods for actionable learning, even while it emphasised the importance of raising knowledge about environmental challenges. The Tbilisi Declaration (Gillett, 1977) improved this by outlining EE's goals – awareness, knowledge, skills, and participation – while encouraging a multidisciplinary approach (Gillett, 1977). Despite its importance, EE was implemented unevenly and was frequently overlooked in the national curriculum.

The United Nations' Decade of Education for Sustainable Development, 2005–2014 emphasised the importance of environmental education in advancing sustainability and global ecological responsibility with a focus on inclusion, local relevance, and critical thinking (UNESCO, 2005). However, current issues such as the digital divide, cultural opposition, and limited finance have highlighted inadequacies in these frameworks. Further, developing concerns like climate change, technological improvements, and socio-economic inequities necessitate adaptive, technology-driven, and equity-focused EE policies. These frameworks must prioritise global accessibility, localised content, and active youth involvement to handle environmental concerns and guarantee sustainability effectively. Environmental education fosters the values, attitudes, abilities, and knowledge necessary for sustainable living, linking it to sustainable development. It empowers people to foster sustainable behaviours by integrating critical thinking, problem-solving, and participatory methods of learning. It helps to fulfil the Sustainable Development Goals (SDGs) of the UN, especially SDG 4 (Quality Education), SDG 13 (Climate Action), and SDG 15 (Life on Land), by providing students with ecological literacy and sustainable habits (UN Department of Economic and Social Affairs, 2015). Ultimately, EE promotes a sustainable culture that motivates both individual and group efforts to build a resilient future.

YOUTH EMPOWERMENT IN CONSERVATION

Youth empowerment is defined as the process of giving necessary information, abilities, and self-assurance to the youth to take charge of their lives and make significant contributions to their communities (Jennings et al., 2006). Empowerment in conservation refers to developing agency, or the capacity to act autonomously and make well-informed decisions while confronting environmental issues (Arnold et al., 2009). Empowered youth stop being passive spectators and start actively suggesting solutions. An essential component of youth empowerment is capacity building. It emphasises the development of ecological knowledge, critical thinking, and the ability to use cutting-edge technology to support conservation efforts (Rap and Bodas, 2024).

Youth empowerment requires young leaders to motivate and inspire their peers, interact with legislators, and spearhead grassroots projects. It produces a generation of proactive environmental stewards dedicated to sustainable conservation by promoting agency and capacity (Chawla and Cushing, 2007). The convergence of EE and youth empowerment promotes ecological stewardship. While youth empowerment guarantees that young people actively utilise this information to effect change, EE offers the fundamental knowledge and skills required to comprehend environmental concerns (Aboukhousa, 2024). EE helps empowered youth cultivate a feeling of accountability and ownership over their surroundings to participate in conservation initiatives. The intersection of EE and youth empowerment is critical for developing a new generation of devoted ecological stewards committed to maintaining long-term environmental sustainability.

EE's ROLE IN YOUTH EMPOWERMENT

Awareness and knowledge are the cornerstones of environmental education essential to youth empowerment. EE promotes awareness of pollution, deforestation, climate change, and biodiversity loss by promoting ecological literacy. This makes youth take responsible action and make well-informed decisions in sustainability initiatives. EE helps them understand the relationship between human activity and environmental health, emphasising the impact of individual and collective actions on ecosystems. They understand important concepts like the carbon cycle, the significance of biodiversity, and the need for sustainable resource management. This encourages youth to critically analyse the environmental implications of daily activities like waste production and energy consumption (Ruiz-Mallén et al., 2022). Awareness of environmental challenges creates a feeling of urgency that inspires youth for conservation initiatives. Environmental Education, through online platforms, community workshops, or school curricula, ensures that youth are knowledgeable and equipped to handle environmental issues in the future (Figure 16.1).

Skill development empowers youth for proactive sustainability. EE equips youth with vital skills like problem-solving, critical thinking, and practical approaches like energy conservation, sustainable agriculture, and waste management. Critical thinking equips them to analyse environmental challenges, evaluate solutions, and comprehend causes and effects. Project-based learning, fieldwork, and experiential activities improve problem-solving abilities, promoting inventive solutions to local ecological concerns. EE also fosters advocacy and leadership skills, empowering youth to

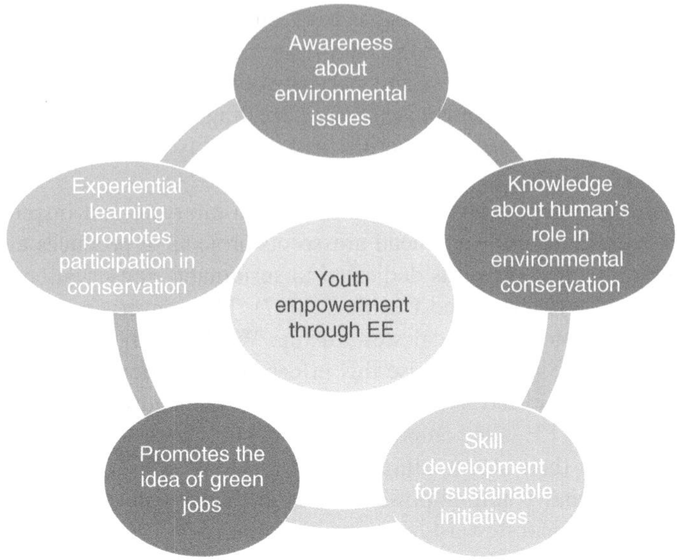

FIGURE 16.1 Youth empowerment.

create community-based initiatives, raise awareness, and push for legislative changes. Hands-on exercises develop practical skills in resource management, recycling, and sustainable practices, increasing confidence and effectiveness (Karim et al., 2022). Furthermore, environmental education promotes collaborative skills through group sustainability projects and trains youth to manage initiatives that bring about the revolutionary changes required for a sustainable future.

The interdisciplinary nature of EE strengthens its potential to tackle global environmental issues successfully. EE advances the idea of green jobs and sustainable economic behaviours by training youth for occupations in sustainable agriculture, renewable energy, and environmental preservation. EE tackles cultural resistance in sociology by incorporating regional customs and beliefs into instructional materials. Honouring cultural settings in conservation efforts promotes more acceptance and involvement in sustainability activities. Artificial Intelligence (AI) makes EE more interactive and accessible by personalising learning experiences, analysing environmental data, and simulating ecological scenarios despite obstacles like the digital divide and ethical constraints.

Participatory techniques in environmental education are crucial to engage youth in experiential learning for environmental protection. Participation in field excursions, community cleanups, tree plantings, and citizen science projects as part of experiential learning assists students in bridging the gap between theoretical understanding and practical actions. The participatory approach of community-based learning motivates youth to help solve local problems like trash management, water conservation, or biodiversity preservation, and it cultivates a feeling of community action (Restrepo-Mieth et al., 2023). Advocacy programmes and youth-led initiatives empower young people to lead environmental movements, raise public awareness, and influence policy, fostering a new generation of sustainable leaders. Environmental education is revolutionised by technology and digital platforms, which increase the effect and reach of programmes aimed at empowering young people. Utilising innovative, interactive, and scalable pedagogical techniques facilitated by technology integration in EE, youth can confront ecological concerns in engaging and accessible manners. Young people improve their ecological literacy with the use of digital technologies like interactive websites, online courses, and mobile applications (Yadav et al., 2022). Regardless of their geographical location, youth explore environmental concepts and global issues through virtual field trips and online educational modules. These tools provide up-to-date information on biodiversity, pollution, and climate change to participate in sustainability conversations and make wise decisions (Xia, 2024).

Digital networks such as Instagram, Twitter, and TikTok allow young people to share conservation activities, increase awareness, and motivate global action. Crowdsourcing, virtual reality, and gaming enable immersive learning experiences that encourage engagement in citizen science and sustainability programmes. These tools enable a connected and informed generation to drive the global environmental movement. Despite its revolutionary potential, technology in environmental education has constraints, such as the digital divide, which prevents kids in underprivileged areas from accessing gadgets or internet connectivity. Furthermore, dependence on digital platforms might reinforce prejudices since the material may represent cultural or regional attitudes that exclude different voices. Accessibility hurdles for marginalised populations and an overemphasis on technology over hands-on learning can diminish inclusivity and impede greater participation in environmental education and sustainability activities.

YOUTH-LED CONSERVATION: CASE STUDIES

The Fridays for Future campaign and Greta Thunberg are prime examples of how environmental education can empower young people (Salvatore and Wolbring, 2021). Thunberg is a Swedish environmental activist who worked to address the problem of climate change and inspired hundreds of thousands of students worldwide. The campaign, started by Thunberg's 2018 school strike, inspired millions of youths worldwide to call for immediate action on climate change. Thunberg used grassroots organising and social media to establish a global forum for youth-led advocacy. Inspiring worldwide awareness and action on urgent environmental challenges, this example highlights how EE empowers youth to advocate for sustainability (Britannica, 2020). The Roots and Shoots programme, launched by the Jane Goodall Institute in 1991, empowers young people to address global environmental, social, and animal issues. It promotes empathy, knowledge, leadership, critical thinking, advocacy, and problem-solving skills, fostering proactive, knowledgeable, and environmental stewards (Jane Goodall Institute, 1991). The Eco-Schools project, initiated by the Foundation for Environmental Education (FEE) in 1994, is a global initiative promoting environmental education through school-based action in over 70 nations. It promotes student leadership in sustainability initiatives including energy saving, trash reduction, and biodiversity restoration, incorporating environmental values into their classrooms and local communities. Eco-schools enable students to tackle real-world issues and foster a lifetime of environmental responsibility by fusing education with hands-on experience (Foundation for Environmental Education, 2024).

Initiatives for community-led conservation in places like Africa, Latin America, and India demonstrate how EE can empower youth and promote local participation. These initiatives combine contemporary conservation techniques with ancient ecological knowledge, frequently highlighting the role of youth as change agents. Programmes like the Chipko Movement in India, which engaged young people in spreading knowledge about sustainable forestry, demonstrated the need for grassroots education in forest protection (Petruzzello, 2020). Nepali Youth for Climate Action focuses on advocacy, capacity building, and grassroots activities like reforestation, waste management, and disaster preparedness, emphasising the importance of youth in promoting sustainability in climate-vulnerable areas (Nepalese Youth for Climate Action, 2021). In Bangladesh, Earth Guardians is a youth-led network that focuses on mangrove restoration, river cleanliness efforts, and educational activities to help alleviate the effects of climate change and environmental deterioration (Earth Guardians, 2020). African Wildlife Foundation engages youth in wildlife conservation in Africa through seminars and community initiatives to empower young people to protect biodiversity by fusing ecological education with real-world activities like habitat restoration and anti-poaching campaigns (African Wildlife Foundation, 2024).

The Green Generation Initiative, established by Elizabeth Wathuti in Kenya, emphasises the importance of youth in ecological restoration and climate resilience by including them in environmental awareness campaigns and tree-planting events (Green Generation Initiative, 2024). The youth-led Rise Up Movement in Uganda, led by Vanessa Nakate, draws attention to how vulnerable areas are disproportionately affected by climate change and gives young people the tools they need to advocate for sustainable policies and renewable energy sources through action and education (Nakate, 2023). In South Africa, the Eco Warriors Movement engages youth in community education on Sustainability, recycling campaigns, and urban cleanup initiatives. The movement encourages environmental care and waste management while instilling a feeling of accountability for the ecological well-being of their local communities (United Nations, 2019). The goal of Amazon Watch in Latin America is to empower indigenous communities – including young people – to save the Amazon rainforest by fusing EE with advocacy training (Amazon Watch, 1996). EE has proven to be a powerful tool in empowering youth and promoting sustainability. Case studies like the Fridays for Future campaign, Roots & Shoots, Eco-schools, and the Chipko Movement in India demonstrate how EE can transform theoretical objectives – like ecological consciousness, critical thinking, and proactive involvement – into practical conservation results. These initiatives empower youth to be active change agents, bridging the gap between knowledge and action for a sustainable future.

BARRIERS TO EFFECTIVE ENVIRONMENTAL EDUCATION

There are several challenges in implementing environmental education for youth empowerment despite its importance in encouraging sustainability and ecological stewardship. These challenges can be broadly categorised as structural, economic, cultural, and educational barriers. One of the most significant barriers to the implementation of EE projects is the lack of funding. Many educational institutions, particularly those in low-income countries, are unable to include environmental concerns in their curricula due to financial constraints. There is also a scarcity of competent instructors with the expertise to run successful EE programmes. Many educational institutions sometimes view environmental education as an elective rather than a necessary part of the curriculum. As a result, environmental concerns receive fragmented or cursory coverage that is frequently divorced from the larger objectives of growth and sustainability (Žalėnienė and Pereira, 2021). Successful EE initiatives utilize funding, collaborations, and government assistance to overcome obstacles. They include EE in mandatory courses and educate teachers to increase proficiency. Community-driven strategies and robust national policies, such as those found in Eco-schools, guarantee sustainability and inclusion. Without robust national policies to promote EE, local initiatives have trouble finding long-term sustainability (Figure 16.2).

Cultural opposition to EE exists in some communities when it goes against customs or regional lifestyles. Environmental goals such as conservation or sustainable resource use could conflict with local customs and economic interests in rural communities that rely on unsustainable practices for their livelihoods. Effective initiatives handle cultural

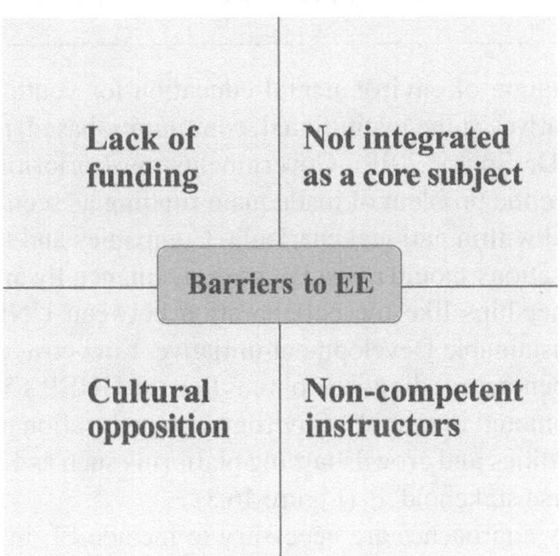

FIGURE 16.2 Barriers to EE.

objections by incorporating local customs and ecological knowledge into EE, increasing community buy-in. Socioeconomic disparities impede access to digital environmental education in underserved communities, where young people from low-income households frequently lack the requisite technology, exacerbating environmental knowledge and engagement gaps (Pihkala, 2020). Subsidies, cooperation with non-governmental organisations, and inclusive policies all help to reduce socioeconomic obstacles. Community-led conservation and grassroots efforts promote accessibility and relevance by filling knowledge gaps in underprivileged areas.

Occasional exclusion of youth-driven initiatives from formal educational frameworks is another challenge. Many youth programmes emphasize passive learning more than encouraging leadership, critical thinking, and active involvement. Consequently, young people do not feel motivated or equipped to take on environmental concerns (Marques and Xavier, 2020). Top-down approaches to EE, in which adults dictate what children should learn, might limit youth involvement and their sense of agency in enacting real change. A multimodal strategy, including youth-focused programmes, curriculum reform, technology access, and more funding, is required to promote EE and empower young people to lead sustainability efforts. Effective strategies overcome these challenges by incorporating youth-led activities into formal curricula and encouraging active learning. Approaches such as project-based learning and experiential activities promote leadership and critical thinking. Roots and Shoots, for example, empower adolescents by prioritising agency and cooperation, through inclusive, youth-led approaches.

EFFECTIVE EE: STRATEGIES AND RECOMMENDATIONS

Effective implementation of environmental education for youth empowerment necessitates strategic initiatives at the institutional, community-based, technology, and policy levels (Richter and De Sousa, 2019). Governments must prioritise EE in national and local budgets to solve the problem of inadequate funding as seen in Finland, where EE funding is embedded within national curricula. Companies and non-governmental and international organisations should assist EE projects financially and technically through public–private partnerships like the collaboration between UNESCO and Toyota on the Education for Sustainable Development initiative. Low-cost, eco-friendly materials sourced locally or open-access digital resources like the UNEP's Youth for Environment toolkit should be promoted in schools. Environmental education projects can also benefit from grant possibilities and crowdsourcing platforms such as EarthRangers to gather resources from diverse stakeholders (Figure 16.3).

Interdisciplinary approaches are necessary to include EE in the basic curriculum. For example, teaching renewable energy in economics or climate science in physics fills in knowledge gaps. Programmes for teacher training, like India's Green Teacher

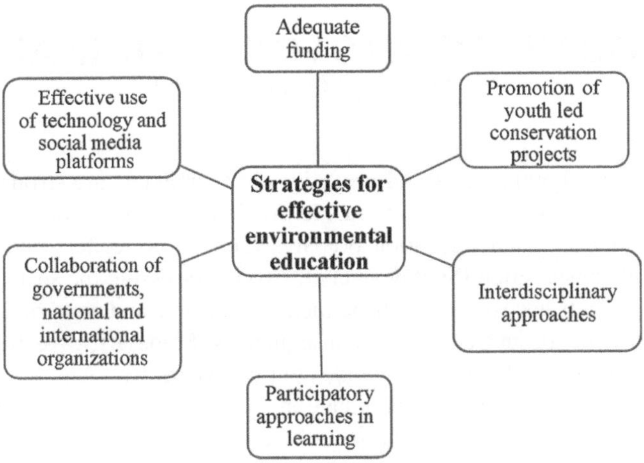

FIGURE 16.3 Strategies for effective environmental education.

Programme, prepare teachers to integrate EE into their classroom lessons smoothly. Clear EE policies, like Bhutan's Gross National Happiness curriculum, should be established by governments to provide equal access in all schools. EE initiatives need to be culturally aware and tailored to local circumstances to overcome cultural opposition (Permanasari et al., 2021). Co-creating instructional materials with local communities, as in Kenya's wildlife conservation curricula, ensures that EE is in line with regional demands, traditions, and financial realities. Community-based learning projects like waste management programmes or local conservation efforts like Sri Lanka's Mangrove Restoration Project can promote greater community engagement and practical skills. Expanding access to digital resources is crucial. Governments and non-governmental organisations can set up technology-equipped public access centres, as done in Brazil's Digital Inclusion Initiative to enable wider engagement in digital environmental education projects. In regions with poor internet access, offline materials like instructional films, podcasts, and mobile applications ensure inclusivity.

Environmental education programmes should encourage youth leadership in environmental projects, promoting ownership and accountability through youth-led initiatives like neighbourhood conservation and sustainability campaigns. Educators should allow youth to organise and work together on community-based solutions to initiate conversations, increase awareness, and impact local policy. Peer-to-peer education also enables young people to instruct one another in sustainable behaviours and bolsters empowerment. Social media can assist in the creation of transnational networks of environmentally conscientious youngsters, while online forums like iNaturalist and virtual seminars enable information sharing, collaborative projects, and awareness-raising efforts. In addition, crowdsourcing as demonstrated in Nepal's community-based forest monitoring can also be used to collect data, mobilise local populations, and advance environmental initiatives efficiently.

YOUTH: AGENTS FOR SUSTAINABLE CONSERVATION

Young people, as digital natives and proactive changemakers, are strong advocates for sustainable conservation, using EE to tackle problems like pollution, biodiversity loss, and climate change. Their education prepares them to be leaders and champions of environmental conservation (Shutaleva et al., 2021). Assessing the sustainability effect of youth-led projects necessitates well-defined frameworks and metrics that encapsulate environmental, social, and economic consequences. Some effective frameworks and indicators for assessing the impact include community involvement, biodiversity restoration, carbon savings and emissions reduction, and technology integration. Reductions in greenhouse gas emissions brought about by energy efficiency projects, the adoption of renewable energy sources, or tree planting can be measured through tools like carbon calculators. The quantity of trees planted, CO_2 sequestration rates, or kilowatt-hours of renewable energy produced can be considered examples of metrics. Indicators under the paradigm of biodiversity restoration include increases in native species populations, changes in species diversity, and habitat restoration areas. Participation rates in campaigns, seminars, and citizen science projects can be monitored to evaluate how efforts involve local communities. Increased environmental knowledge or behavioural changes can be assessed using surveys or feedback methods. Innovative methods could be demonstrated by the use of mobile apps for data collection or GIS mapping for habitat restoration. The quantity of data crowdsourced or the number of app users can be considered examples of metrics.

Greta Thunberg's climate activism and Fridays for Future are instances of youth-led movements that demonstrate the importance of youth in raising awareness about environmental concerns and holding government and business organisations accountable. These initiatives demonstrate that young people are dedicated to taking action and are conscious of the importance of conservation (Reis, 2020). They use social media and technology to cooperate on conservation initiatives, exchange ideas, and raise their voices to motivate others. Encouraging young people to be change agents guarantees that conservation initiatives are not only practical but long-lasting, opening the door to a sustainable future.

CONCLUSION

This chapter emphasises how environmental education enables youth to be important change agents for sustainable conservation. EE equips youth to tackle urgent global issues including pollution, biodiversity loss, and climate change by cultivating ecological awareness, critical thinking, and practical skills. Incorporating participatory methodologies, such as youth-led initiatives, community-driven projects, and experiential learning, further enhances their ability to engage as proactive environmental stewards. In contemporary environmental education, technology and digital platforms have become crucial instruments that allow youth to advocate, invent, and work together globally.

The chapter also recognises important barriers like lack of resources, gaps in the curriculum, cultural hurdles, and uneven access to technology. It is imperative to tackle these challenges through strategic actions like increased funding, curriculum reform, community involvement, and bridging the digital divide to guarantee the availability and efficacy of environmental education for everyone. With enthusiasm, ingenuity, and devotion, young people act as catalysts for long-term conservation. Their role in environmental movements and ability to motivate collective action are critical to improve long-term ecological sustainability. Inclusive and comprehensive environmental education will equip the present generation to construct a resilient and equitable future, guaranteeing a sustainable world.

REFERENCES

Aboukhousa, Heba. Promoting climate action through youth empowerment and environmental education: Insights from Mohamed Bin Zayed University for humanities. *Journal of Infrastructure, Policy and Development*, vol. 8, no. 11, 2024, p. 8636, https://doi.org/10.24294/jipd.v8i11.8636.

African Wildlife Foundation. *Empowering Leadership*. African Wildlife Foundation, Nairobi, Kenya, 2024, https://www.awf.org/leading-for-wildlife/empowering-leadership.

Amazon Watch. *Protecting the Amazon and Our Climate in Solidarity with Indigenous Peoples*. Amazon Watch, Oakland, CA, 1996, https://amazonwatch.org/work.

Arnold, Heather E., Fay G. Cohen, and Alan Warner. Youth and environmental action: Perspectives of young environmental leaders on their formative influences. *The Journal of Environmental Education*, vol. 40, no. 3, 2009, pp. 27–36, https://doi.org/10.3200/JOEE.40.3.27-36.

Britannica. Fridays for Future: International Activist Movement. Britannica, Chicago, IL, 2020, https://www.britannica.com/topic/Fridays-for-Future.

Chawla, Louise, and Debra Flanders Cushing. Education for strategic environmental behavior. *Environmental Education Research*, vol. 13, no. 4, 2007, pp. 437–52, https://doi.org/10.1080/13504620701581539.

Earth Guardians. *Who Is Earth Guardians?* Earth Guardians, Laurel, MD, 2020, https://www.earthguardians.org.

Foundation for Environmental Education. *Eco Schools*. Foundation for Environmental Education, Denmark, 2024, https://www.ecoschools.global.

Gillett, Margaret. The Tbilisi Declaration. *McGill Journal of Education/Revue Des Sciences de léducation de McGill*, vol. 12, no. 002, 1977, https://mje.mcgill.ca/article/view/7156.

Green Generation Initiative. *About Us*. Green Generation Initiative, Nairobi, Kenya, 2024, https://greengenerationinitiative.org/.

Hulme, Mike. 1.5°C and climate research after the paris agreement. *Nature Climate Change*, vol. 6, no. 3, 2016, pp. 222–24, https://doi.org/10.1038/nclimate2939.

Intergovernmental Panel on Climate Change. *Climate Change 2021: The Physical Science Basis*. Cambridge University Press, Cambridge, 2021, https://doi.org/10.1017/9781009157896.

Jane Goodall Institute. *Roots & Shoots*. Jane Goodall Institute, Washington, DC, 1991. https://janegoodall.org/our-work/our-approach/roots-shoots/.

Jennings, Louise B., Deborah M. Parra-Medina, Deanne K. Hilfinger-Messias, and Kerry McLoughlin. Toward a critical social theory of youth empowerment. *Journal of Community Practice*, vol. 14, no. 1–2, 2006, pp. 31–55, https://doi.org/10.1300/J125v14n01_03.

Karim, Norhafizah, Hidayatulfathi Othman, Zul-'Izzat Ikhwan Zaini, Yanti Rosli, Muhammad Ikram A. Wahab, Al Mumin Al Kanta, Syamimi Omar, and Mazrura Sahani. Climate change and environmental education: Stance from science teachers. *Sustainability*, vol. 14, no. 24, 2022, p. 142416618, https://doi.org/10.3390/su142416618.

Marques, Ronualdo, and Claudia Regina Xavier. The challenges and difficulties of teachers in the insertion and practice of environmental education in the school curriculum. *International Journal on Social and Education Sciences*, vol. 2, no. 1, 2020, pp. 49–56.

Nakate, Vanessa. *Your Biggest Climate Crisis Questions, Answered.* Time, Delhi, India, 2023, https://time.com/collection/earth-awards-2023/6272883/climate-questions-answered-by-vanessa-nakate/.

Nepalese Youth for Climate Action. *About Us*. NYCA, Lalitpur, Nepal, 2021, https://nyca.net.np/about-us.html.

Permanasari, Gayatri Hanna, et al. The implementation of environmental education to achieve sustainable development: literature review. In Tri Retnaningsih Soeprobowati et al., (Eds.), *E3S Web of Conferences,* vol. 317, EDP Sciences, France, 2021, https://doi.org/10.1051/e3sconf/202131701069.

Petruzzello, Melissa. Chipko movement. *Saving Earth: Encyclopedia Britannica*, 2020, https://explore.britannica.com/explore/savingearth/chipko-movement.

Pihkala, Panu. Eco-anxiety and environmental education. *Sustainability*, vol. 12, no. 23, 2020, p. 10149, https://doi.org/10.3390/su122310149.

Rap, Shelley, and Moran Bodas. Innovate for impact: Young adults education and empowerment for climate action. *Australian Journal of Environmental Education*, vol. 40, no. 4, 2024, pp. 784–94, https://doi.org/10.1017/aee.2024.46.

Reis, Pedro. Environmental citizenship and youth activism. In Andreas Ch. Hadjichambis et al., (Eds.), *Conceptualizing Environmental Citizenship for 21st Century Education*, Springer International Publishing, Cham, Switzerland, 2020, pp. 139–48, https://link.springer.com/10.1007/978-3-030-20249-1_9.

Restrepo-Mieth, Andrea, et al. Community-based participatory climate action. *Global Sustainability*, vol. 6, 2023, p. e14, https://doi.org/10.1017/sus.2023.12.

Richter, Barry Wilhelm, and Luiza De Sousa. The implementation of environmental education to promote sustainability: An overview of the processes and challenges. *International Journal of Sustainable Development & World Ecology*, vol. 26, no. 8, 2019, pp. 721–31, https://doi.org/10.1080/13504509.2019.1672220.

Ruiz-Mallén, Isabel, et al. Community climate resilience and environmental education: Opportunities and challenges for transformative learning. *Environmental Education Research*, vol. 28, no. 7, 2022, pp. 1088–107, https://doi.org/10.1080/13504622.2022.2070602.

Salvatore, Chiara, and Gregor Wolbring. Children and youth environmental action: The case of children and youth with disabilities. *Sustainability*, vol. 13, no. 17, 2021, https://doi.org/10.3390/su13179950.

Shutaleva, Anna, et al. Environmental behavior of youth and sustainable development. *Sustainability*, vol. 14, no. 1, 2021, p. 250, https://doi.org/10.3390/su14010250.

Skinner, Ellen A., et al. Intrinsic motivation and engagement as "active ingredients" in garden-based education: Examining models and measures derived from self-determination theory. *The Journal of Environmental Education*, vol. 43, no. 1, 2012, pp. 16–36, https://doi.org/10.1080/00958964.2011.596856.

Tabanao, Eden Grace V., et al. Youth engagement in environmental sustainability: Bridging action and awareness for SDG 11. *Ignatian International Journal for Multidisciplinary Research*, vol. 3, no. 2, 2025, pp. 335–345, https://doi.org/10.17613/GHYQ0-8V059.

UN Department of Economic and Social Affairs. *The 17 Goals: Sustainable Development*. United Nations, Stockholm, 2015, https://sdgs.un.org/goals.

UN Framework Convention on Climate Change. *The Paris Agreement*. United Nations Climate Change, Paris, France, 2015, https://unfccc.int/process-and-meetings/the-paris-agreement.

UNESCO. *UN Decade of Education for Sustainable Development, 2005–2014: The DESD at a Glance*. UNESCO, Paris, France, 2005, https://unesdoc.unesco.org/ark:/48223/pf0000141629.

United Nations. *From the Field: "Eco-Warriors" Fight Climate Change in South Africa*. United Nations, Stockholm, 2019, https://news.un.org/en/story/2019/08/1044101.

United Nations. *United Nations Conference on the Human Environment,* Stockholm, 1972, https://www.un.org/en/conferences/environment/stockholm1972.

World Wildlife Fund. *Living Planet Report 2022: Building a Nature-Positive Society*. WWF International, Gland, Switzerland, 2022.

Xia, Xiaomeng. Enhancing environmental education through digital tools. *International Journal of New Developments in Education*, vol. 6, no. 9, 2024, pp. 212–16, https://doi.org/10.25236/IJNDE.2024.060932.

Yadav, Shailesh Kumar, et al. Environmental education for sustainable development. In Manoj Kumar Jhariya, Ram Swaroop Meena, Arnab Banerjee, and Surya Nandan Meena (Eds.), *Natural Resources Conservation and Advances for Sustainability*, Elsevier, Amsterdam, Netherlands, 2022, pp. 415–31, https://linkinghub.elsevier.com/retrieve/pii/B9780128229767000107.

Žalėnienė, Inga, and Paulo Pereira. Higher education for sustainability: A global perspective. *Geography and Sustainability*, vol. 2, no. 2, 2021, pp. 99–106, https://doi.org/10.1016/j.geosus.2021.05.001.

17

Training for Water Resources Management

Bridging Knowledge Gaps for Sustainable Development

Laxmi Lahalya Boddeti

INTRODUCTION

The management of water resources has a serious impact on ecosystems, economies, and communities global wide, making it a vital component of sustainable development. Global issues such as water pollution, flooding, water scarcity, climate change, and rapid urbanisation are getting worse, especially in developing countries as compared to developed ones (Zhang, 2016). Umair Ahmed (2019) asserts that this rapid population increase exerts significant pressure on the environment, particularly on water resources, leading to substantial issues in ecology and hydrology. The significant pressure from population density and the growing demands for infrastructure and housing in rapidly developing urban areas lead to changes in land use and an increased demand for natural resources (Manju Mohan, 2020). The extent of water bodies, cultivable land, and open

DOI: 10.1201/9781003613183-17

spaces has significantly decreased and been overutilised, intensifying water-related issues such as polluted water sources, inadequate potable water supply, stormwater inundation, and waterborne illnesses in swiftly growing urban regions (Mijic, 2019; Harikumar, 2023). As a result, effectively managing water resources has become serious, and this can be managed by a comprehensive approach that encompasses proper training to enhance technical expertise at all levels, with efficient governance and sustainable methods.

In the last few decades, geo-spatial technology (GST) software has advanced all over the world, integrating remote sensing (RS), geographic information systems (GIS), and global navigation satellite systems (GNSS). The GST has been successful at mapping and keeping an eye on all natural resources, including management of water resources and important parts of the hydrological cycle (Chakraborti, 1999).

These training programmes should be devised for water resources management to effectively bridge the knowledge gaps between technical breakthroughs and their practical application in real-world scenarios. These programmes are essential for enabling a diverse array of stakeholders, including engineers, policymakers, urban planners, and community leaders, to manage the intricacies of water systems and to guarantee that solutions are both scientifically valid and contextually suitable for the region (Navaneeth, 2024).

The circumstances underscore the pressing necessity for adequate training in water resources management. This chapter addresses a significant knowledge gap by exploring the necessary technical and non-technical training tools required for professionals in developing countries like India, Korea, China (excluding Hong Kong), etc., and water-sensitive cities, particularly in larger metropolitan areas, to effectively manage water resources for sustainable development. This chapter examines numerous training programmes for each wet utility issue, emphasising software such as GIS, remote sensing technologies, and hydrological and hydraulic modelling tools while addressing a wide range of audiences. These technologies enable users to evaluate, monitor, and control water resources in real time, addressing issues such as flood risk management, water shortages, and drought mitigation methods.

This chapter also discusses the technical and collaborative abilities essential for linking scientific understanding with policy execution. The effort corresponds with the UN SDGs by promoting sustainable water utilisation, involving stakeholders, and deploying water-conserving technologies.

Digital tools and e-learning platforms can enhance the scope and efficacy of training, facilitating regional capacity-building. These initiatives enable individuals and organisations to establish robust water management systems that ensure fair access and sustainable water security for communities and future generations.

In the following chapters, our goal is to dig further into the complications of effective water resource planning and management, describe the required skill sets, and consider how computer-based programs and models might help practitioners with practical assistance and tools to improve methods of water management.

UNDERSTANDING THE CHALLENGES IN WATER RESOURCES MANAGEMENT

Data Gaps and Technological Barriers

Many areas worldwide are facing problems with lack of data collection to design the most efficient decisions related to water management. The issue arises from uneven monitoring of water quantity, quality, and usage, leading to gaps in proper planning and inefficiencies. In addition, the use of modelling techniques such as GIS, remote sensing, and water resource software technologies has become limited. This is mainly due to financial problems or the lack of infrastructure; as a result of these obstacles, the ability to predict and react to natural untruths, such as droughts and floods, has become challenging.

FRAGMENTATION OF GOVERNANCE AND POLICY

Various entities oversee distinct facets of water management, exhibiting minimal collaboration. Environmental regulators, urban water supply agencies, and irrigation authorities may function independently, resulting in inefficiencies and redundancies. The technical deficiency hinders forecasting and response to floods and droughts. The lack of integration obstructs the formulation of comprehensive policies that meet the needs of local consumers, industry, and agriculture.

ENVIRONMENTAL DEGRADATION AND ECOSYSTEM STRAIN

The excessive extraction of groundwater, pollution of rivers, and deterioration of wetlands are jeopardising ecosystems. Since 1900, over 64% of global wetlands have been eradicated, despite their critical roles in water filtration, aquifer replenishment, and flood mitigation. Dams and industrial effluents endanger river basins, imperilling aquatic life. Ecosystem degradation escalates, threatening the communities reliant on them.

SOCIOPOLITICAL CONFLICTS OVER WATER RESOURCES

Disputes regarding water resources among local, regional, and international stakeholders are common. Transboundary rivers like the Nile, Indus, and Mekong exemplify the possibility for conflict between nations regarding shared resources. Minor disputes between urban and rural consumers, or between upstream and downstream areas, exemplify the overarching water distribution challenge. Addressing these challenges requires a focus on technical solutions, diplomatic efforts, and the engagement of participatory governance.

RAPID URBANISATION AND RISING DEMAND

The continuous growth of urban areas is exerting considerable pressure on water infrastructure. The growing need for sanitation, safe water, and industrial applications in urban environments frequently surpasses the capacities of current systems. A high proportion of urban residents reside in slums, which in many cases have inadequate water and sanitation facilities, thereby escalating public health risks. Enhancing infrastructure and encouraging sustainable use are important for addressing urban water challenges.

CLIMATE CHANGE AMPLIFYING RISKS

The impacts of climate change increase the challenge in water resources. Temperature increases lead to higher evaporation rates and precipitation shifts, causing water shortages and floods in specific areas. At the same time, glaciers—those incredible natural water reservoirs—are melting at a slower pace, which means there's even less water flowing downstream. Strategic management techniques that anticipate and manage risks are crucial in fighting climate change.

The complex difficulties illustrate the need for integrated water resource management. Training programmes aimed at addressing these concerns can aid stakeholders in managing the intricacies of water systems. Understanding the roots and interconnections of these difficulties is the first step towards developing new and sustainable solutions (Figure 17.1).

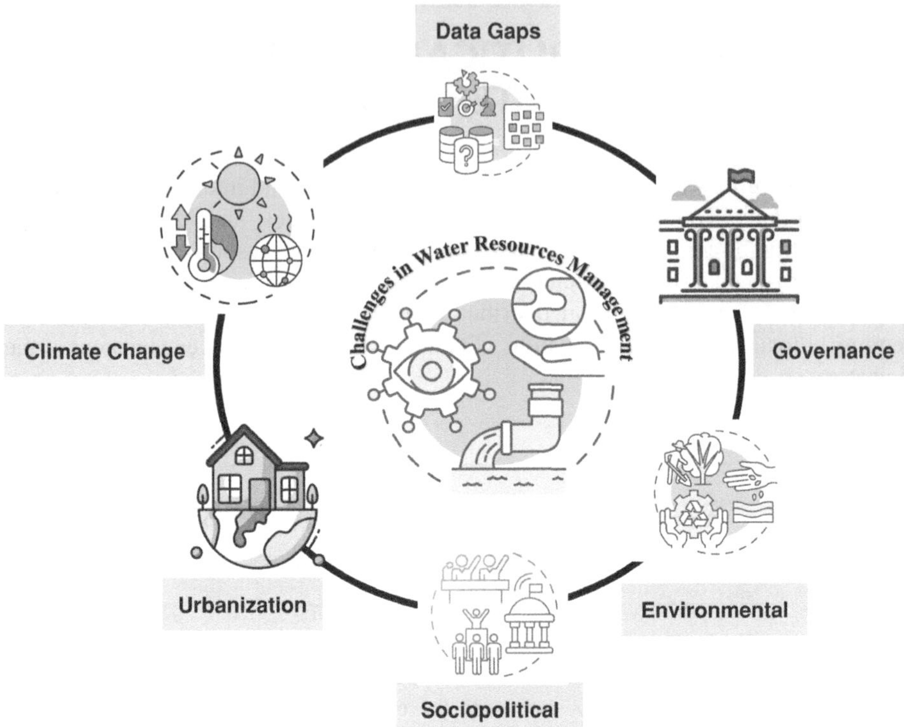

FIGURE 17.1 Interconnected challenges in water resources management: governance, data gaps, environmental, sociopolitical, urbanisation, and climate change.

BUILDING CAPACITY FOR EFFECTIVE WATER MANAGEMENT: KEY SKILLS AND KNOWLEDGE AREAS

Efficient management of water resources necessitates a synthesis of interdisciplinary expertise and technical proficiency to tackle water scarcity, pollution, and the impacts of climate change. Professionals should have widespread awareness about the strategies required for their effective management and adaptation along with an understanding of different characteristics of surface water, groundwater, and recycled water sources. This encompasses instruction in both conventional water management techniques and contemporary technological innovations that improve decision-making and operational efficacy (Figure 17.2).

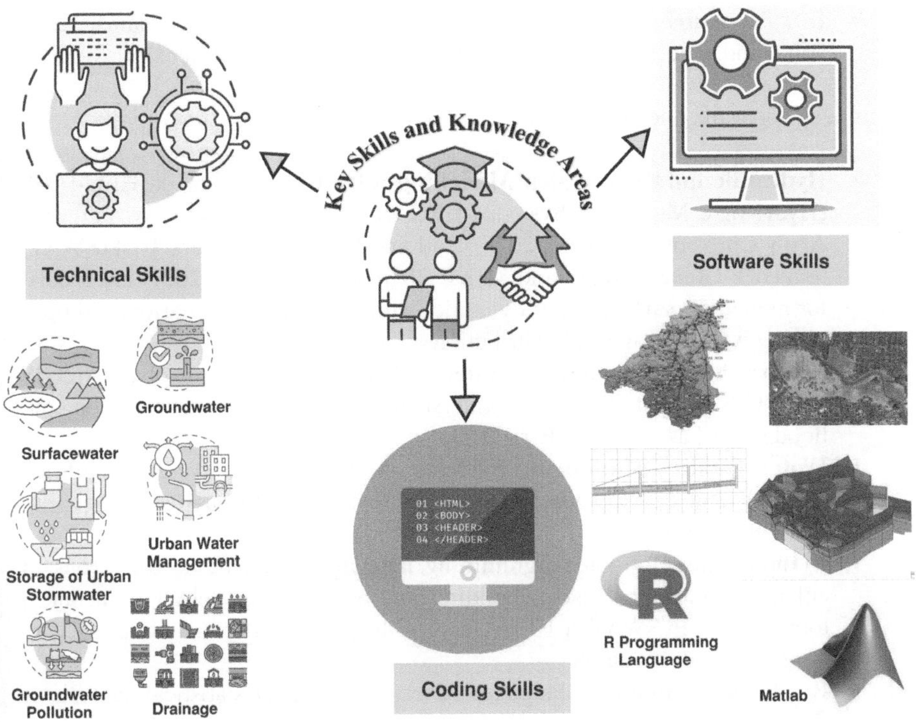

FIGURE 17.2 The figure illustrates key skills and knowledge areas for effective water management.

SURFACE WATER MANAGEMENT

Key skills and knowledge areas:

- **Water Cycle and Hydrological Processes**: Precipitation patterns, runoff dynamics, and flood controlling strategies are the basic requirements for understanding the water cycle and hydrological processes.
- **Water Quality Monitoring**: Knowledge of criteria determining water quality, i.e., pH, turbidity, dissolved oxygen, and contaminants.
- **Flood Risk Management**: A study of flood dynamics, river or coastal basin management, and flood management techniques enables the control of flood risk.
- **Ecosystem Management**: An additional ecosystem management aspect entails training on harmonizing water extraction processes with the conservation of aquatic ecosystems.

- **Infrastructure Design**: Expertise in developing designs for dams, reservoirs, and water distribution systems focused on surface water storage.

Technical, software, and coding skills required:

- **Hydraulic and Hydrological Modelling Software**: Software like HEC-HMS (Hydrologic Modelling System), HEC-RAS (River Analysis System) (US Army Corps of Engineers, 2024), Soil and Water Assessment Tool (SWAT) (Sánchez-Gómez, 2025), and MIKE 21/MIKE 3 (Haiyan, 2021) are crucial for modelling surface water dynamics, simulating flood scenarios, and optimising water storage and distribution.
- **Geographical Information System (GIS)**: GIS usually involves ArcGIS and QGIS for spatial analysis, watershed mapping and characterisation, and flood hazard assessment (Esri, 2021).
- **Water Quality Monitoring Systems**: Aqua prototype systems reinforce research and development in treatment operations through water quality monitoring systems.
- **Python**: Thus, Python programming language would be advantageous in enhancing data processing, building and analysing models. For instance, tools such as PySWAT or HEC-GeoRAS help make different model integrations and analyses easier.
- **R**: The R programming language is very effective for statistical analysis of hydrological data and evaluations of water quality.

GROUNDWATER MANAGEMENT

Key skills and knowledge areas:

- **Hydrogeology**: Examining groundwater flow, recharge, and aquifer properties..
- **Sustainable Extraction**: Controlling rates of extraction guarantees the long-term viability of sustainable groundwater supplies.
- **Groundwater Quality Monitoring**: Methods of assessing salt and pollutant levels in groundwater.
- **Hydrogeological Modelling**: Using models to replicate groundwater properties and processes over many geological formations.

Technical, software, and coding skills required:

- **Groundwater Modelling Software**: Proficient in groundwater modelling software, including MODFLOW, GMS, and Feflow, for simulating groundwater flow and solute transport.

- **Geospatial Tools**: ArcGIS and QGIS will be used for the mapping of aquifer recharge zones and groundwater levels. Groundwater data management and integration with other hydrological models benefit from advanced ArcGIS training.
- **Data Collection and Monitoring Software**: Training is vital for correct data collection via municipal water resource departments or websites as well as for following pertinent national laws and standards.
- **Python**: Python packages like pyMODFLOW and PySAL help with groundwater modelling and geospatial analysis. Python allows PyModflow to build, calibrate, and post-processes MODFLOW, MOD PATH, MT3DMS, and PEST groundwater flow and transport models. Through building and managing models, researchers and experts can use its easy-to-use interface to replicate groundwater flow and pollution movement.

  The open-source geospatial data science library PySAL has specialisation in vector data. It can find clusters, hot spots, and outliers, create spatial graphs, perform spatial regression, and statistically model spatially embedded networks.
- **MATLAB**: MATLAB meets requirements for more comprehensive advanced groundwater simulation modelling. It uses strong numerical algorithms, robust visualisation, and links with other modelling tools to become a powerful modelling tool for groundwater. The scripting allows task automation and customisation for improved efficiency and accuracy in simulation.

WASTEWATER AND RECYCLED WATER MANAGEMENT

Key skills and knowledge areas:

- **Wastewater Treatment Processes**: Examining several phases of wastewater treatment – primary, secondary, and tertiary – helps one to better understand them.
- **Water Reuse Technologies**: Understanding techniques for the treatment and reuse of wastewater for both non-potable and potable uses helps one to employ water reusing technologies.
- **Regulatory Compliance**: Knowledge of guidelines for treated wastewater and water quality criteria.
- **Public Education**: Expertise in educating communities and companies on the importance of water recycling.

Technical, software, and coding skills required:

- **Wastewater Treatment Simulation Software**: Skilled in employing tools like BioWin and WEST to model wastewater treatment processes and optimise treatment plant efficiency.
- **Water Quality Monitoring**: Hach WIMS (Water Information Management Solution) improves water quality monitoring by centralising data collection, reporting, and analysis to assure compliance and efficiency. This integration improves wastewater and reclaimed water monitoring and management, protecting human health and the environment.
- **Recycling Process Optimisation**: Aqualytics solutions use machine learning and data analytics to monitor, evaluate, and optimise water usage and recycling. Leak detection, inefficiency identification, and actionable insights for better water management improve efficiency and sustainability.
- **Python**: Python is an effective programming tool for automating data processing, doing simulations, and integrating different phases of wastewater treatment using libraries like NumPy and SciPy.
- **MATLAB**: This programming language can be utilised to simulate and analyse the efficacy of wastewater treatment systems, facilitating the optimisation of treatment procedures.

INTEGRATED WATER RESOURCES MANAGEMENT (IWRM)

Key skills and knowledge areas:

- **Water Allocation and Policy**: An assessment of those policies and initiatives geared towards equitable water distribution among various sectors will prove to be an important step for Integrated Water Resources Management (IWRM), providing assurance that the allocation and access to water resources are sustainable and just for all parties concerned
- **Stakeholder Engagement**: Expertise in incorporating local communities, industry, and governmental entities into the decision-making process.
- **Conflict Resolution**: Resolving disputes arising from competing demands for water supplies.
- **Monitoring and Evaluation**: Methodologies for evaluating water management systems and analysing their sustainability.

Technical, software, and coding skills required:

- **IWRM Decision Support Systems (DSS)**: Proficient in utilising technologies such as WEAP and WaterGAP to enhance decision-making in IWRM procedures.
- **Geospatial Analysis Software**: Expertise in utilising ArcGIS and QGIS for watershed management, cartography, and conflict assessment. These tools are immensely helpful in integrated water resources management, accounting for hydrological modelling, spatial analysis, and decision-making tools for strategic water resource management and illustration.
- **Stakeholder Collaboration Platforms**: Proficient in employing communication platforms like Microsoft Teams or Slack to promote dialogue and improve collaboration.
- **Data Visualisation**: Training in Power BI or Tableau to proficiently illustrate water data and improve decision-making processes.
- **Python**: The automation of data collection, analysis, and integration for IWRM platforms, like WEAP and WaterGAP, can be achieved using the Python programming language.
- **R**: R Programming Language allows complete analyses and visualisation of water data, facilitating the identification of trends. Proper data communication by way of maps, charts, and graphs supports evidence-based decision-making in water resource management.

Improving the management of water resources requires a blend of traditional knowledge and growing modern technical expertise. Engaging with various water resource disciplines, including surface water, groundwater, and recycled water, requires extensive expertise in multiple software and technical skills, guided by manuals and advisories that adhere to national standards in accordance with government policies (Figure 17.3).

In addition to the Manual and Advisory recommendations for groundwater, which are restricted to Table 17.1, this information can be obtained from the Central Ground Water Board (CGWB).

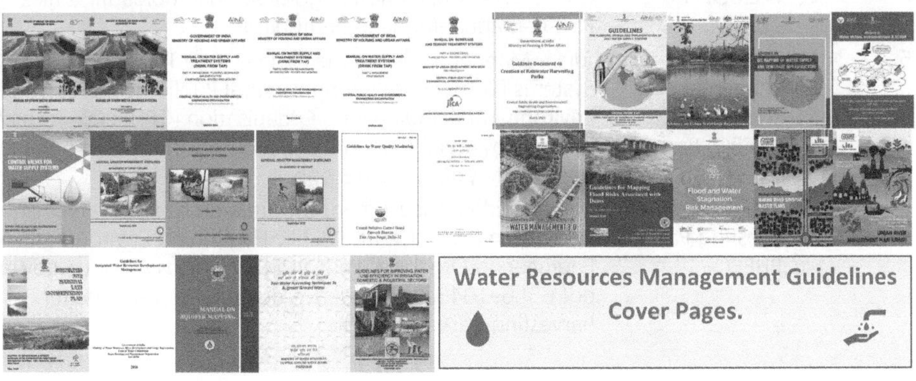

FIGURE 17.3 Water resources management guidelines cover pages.

TABLE 17.1 India's water resources management manuals and advisories

SURFACE WATER, GROUNDWATER, WASTEWATER AND SEWERAGE, WATER DISTRIBUTION AND SUPPLY, IRRIGATION AND AGRICULTURAL, DROUGHT, INTEGRATED WATER RESOURCES MANAGEMENT

SL.NO.	*MANUALS AND ADVISORIES*	*ORGANISATIONS AND LINKS*
1	Storm Water Drainage Systems (CPHEEO) Part-A	India – Manual on Storm Water Drainage Systems, Volume-1 Part A: Engineering Design , Central Public Health and Environment Engineering Organization (CPHEEO) – https://mohua.gov.in/publication/manual-on-storm-water-drainage-systems--2019.php
2	Storm Water Drainage Systems (CPHEEO) Part-B, C	India – Manual on Storm Water Drainage Systems, Volume-II Part B: Operation and Maintenance, Part C: Management, Central Public Health and Environment Engineering Organization (CPHEEO) – https://mohua.gov.in/publication/manual-on-storm-water-drainage-systems--2019.php
3	Water Supply and Treatment Systems (CPHEEO) Part-A	India – Manual on Water Supply and Treatment Systems (Drink from Tap) Part A: Engineering – Planning, Design and Implementation, Central Public Health and Environmental Engineering Organisation (CPHEEO) – https://mohua.gov.in/upload/uploadfiles/files/Part-A-Manual-Engineering-Planning-Design-and-Implementation(3).pdf
4	Water Supply and Treatment Systems (CPHEEO) Part-B	India – Manual on Water Supply and Treatment Systems (Drink from Tap) Part B: Operation and Maintenance , Central Public Health and Environmental Engineering Organisation (CPHEEO) – https://mohua.gov.in/upload/uploadfiles/files/Part-B-Manual-Operation-and-Maintenance(5).pdf
5	Water Supply and Treatment Systems (CPHEEO) Part-C	India – Manual on Water Supply and Treatment Systems (Drink from Tap) Part C: Management , Central Public Health and Environmental Engineering Organisation (CPHEEO) – https://mohua.gov.in/upload/uploadfiles/files/Part-C-Manual-Management(4).pdf
6	Wastewater Management Systems (CPHEEO) Part-A	India – Manual on Sewerage and Sewage Treatment Systems Part A: Engineering, Central Public Health and Environmental Engineering Organisation (CPHEEO) –https://mohua.gov.in/publication/manual-on-sewerage-and-sewage-treatment-systems--2013.php
7	Rainwater Harvesting Parks (CPHEEO)	India – Guidance Document on Creation of Rainwater Harvesting Parks, Central Public Health and Environmental Engineering Organisation (CPHEEO) – https://mohua.gov.in/pdf/6566e1048ab41guidance-document-on-rainwater-harvesting-parks-final.pdf

(*Continued*)

TABLE 17.1 (*Continued*) India's water resources management manuals and advisories

SURFACE WATER, GROUNDWATER, WASTEWATER AND SEWERAGE, WATER DISTRIBUTION AND SUPPLY, IRRIGATION AND AGRICULTURAL, DROUGHT, INTEGRATED WATER RESOURCES MANAGEMENT		
SL.NO.	*MANUALS AND ADVISORIES*	*ORGANISATIONS AND LINKS*
8	24×7 Water Supply Systems (CPHEEO)	India – Guidelines for Planning, Design and Implementation of 24×7 Water Supply Systems (Drink from Tap), Central Public Health and Environmental Engineering Organisation (CPHEEO) – https://mohua.gov.in/pdf/62023520928a0advisory-on-water%20supply-final-new.pdf
9	Urban Waterbody Rejuvenation (CPHEEO)	India – Advisory on Urban Waterbody Rejuvenation, Central Public Health and Environmental Engineering Organisation (CPHEEO) – https://mohua.gov.in/pdf/659b7be5e21b3 Advisory-on-Urban-Waterbody-Rejuvenation.pdf
10	GIS Mapping of Water Supply and Sewerage Infrastructure (CPHEEO)	India – Advisory on GIS Mapping of Water Supply and Sewerage Infrastructure, Central Public Health and Environmental Engineering Organisation (CPHEEO) – https://mohua.gov.in/pdf/5ed0d05ccd83cFinal-Advisory%20on%20GIS%20Mapping%20of% 20 Water%20Supply%20and%20Sewerage%20 Infrastructure.pdf
11	Water Meters, Instrumentation and SCADA (CPHEEO)	India – Advisory on Water Meters, Instrumentation and SCADA Central Public Health and Environmental Engineering Organisation (CPHEEO) – https://mohua.gov.in/pdf/5efc2893a6f0bAdvisory_on_Water_Meter_Instrumentation_and_SCADA_Final.pdf
12	Control Valves for Water Supply Systems (CPHEEO)	India – Advisory on Control Valves for Water Supply Systems Central Public Health and Environmental Engineering Organisation (CPHEEO) – Ministry of Housing and Urban Affairs government of india
13	Management of Urban Flooding (NDMA)	India – National Disaster Management guidelines – Management of Urban Flooding, National DisasterManagementAuthorityGovernmentofIndia(NDMA)– https://ndma.gov.in/sites/default/files/PDF/Guidelines/management_urban_flooding.pdf
14	Management of Floods_(NDMA)	India – National Disaster Management guidelines – Management of Floods, National Disaster Management Authority Government of India (NDMA) – https://ndma.gov.in/sites/default/files/PDF/Guidelines/flood.pdf

(*Continued*)

TABLE 17.1 (*Continued*) India's water resources management manuals and advisories

SURFACE WATER, GROUNDWATER, WASTEWATER AND SEWERAGE, WATER DISTRIBUTION AND SUPPLY, IRRIGATION AND AGRICULTURAL, DROUGHT, INTEGRATED WATER RESOURCES MANAGEMENT		
SL.NO.	*MANUALS AND ADVISORIES*	*ORGANISATIONS AND LINKS*
15	Management of Drought	India – National Disaster Management guidelines – Management of Drought, National Disaster Management Authority Government of India (NDMA) – https://ndma.gov.in/sites/default/files/PDF/Guidelines/droughtguidelines.pdf
16	Water Quality Monitoring	India – Guidelines for Water Quality Monitoring, Central Pollution Control Board (CPCB). Microsoft Word – Guidelines_for_WQ_Monitoring [1].doc
17	Drinking Water Specification.	Indian Standard Drinking Water- Specification – IS 10500:2012. IS 10500 (2012): Drinking water
18	Best Practices in Water Management 3.0	Compendium of Best Practices in Water Management 3.0. NITI Aayog. COMPENDIUM-OF-BEST-PRACTICES-IN-WATER-MANAGEMENT-3.0_Water-Resources-Vertical_2_8_23.pdf
19	Mapping Flood Risks Associated with Dams	India – Guidelines for Mapping Flood Risks Associated with Dams, Central Water Commission (CWC) Ministry of Water Resources, River Development and Ganga Rejuvenation Government of India. Guidelines-for-Mapping-Flood-Risks-Associated-with-Dams.pdf
20	Flood and Water Stagnation Risk Management	Training Manual on Flood and Water Stagnation Risk Management. Ministry of Housing and Urban Affairs Government of India. WM4 - Flood.pdf
21	Making River-Sensitive Master Plans	India – Strategic Guidelines for Making River-Sensitive Master Plans. National Institute of Urban Affairs. making-river-sensitive-master-plans.pdf
22	Managing Urban River Stretches in the Ganga River Basin	India – A Strategic Framework for Managing Urban River Stretches in the Ganga River Basin. National Institute of Urban Affairs 2263.pdf
23	Lake Conservation Plan	India – Guidelines for National Lake Conservation Plan, Ministry of Environment and Forests. AHEC Guidelines 2008.p65
24	Integrated Water Resources Development and Management	India – Guidelines for Integrated Water Resources Development and Management. Government of India Ministry of Water Resources, River Development and Ganga Rejuvenation Central Water Commission. Guidelines-for-Integrated-Water-Resources-Development-and-Management.pdf

(*Continued*)

TABLE 17.1 (*Continued*) India's water resources management manuals and advisories

SURFACE WATER, GROUNDWATER, WASTEWATER AND SEWERAGE, WATER DISTRIBUTION AND SUPPLY, IRRIGATION AND AGRICULTURAL, DROUGHT, INTEGRATED WATER RESOURCES MANAGEMENT		
SL.NO.	*MANUALS AND ADVISORIES*	*ORGANISATIONS AND LINKS*
25	Aquifer Mapping	Manual on Aquifer Mapping. Government of India Ministry of Water Resources Central Ground Water Board. 16861227981023889265file.pdf
26	Rainwater Harvesting Techniques to Augment Groundwater	Ministry of Water Resources Central Ground Water BOARD SKM_458e19080210341
27	Water Use Efficiency in Irrigation, Domestic and Industrial Sectors	India – Guidelines for Improving Water Use Efficiency in Irrigation, Domestic and Industrial Sectors. Central Water Commission (CWC), Ministry of Water Resources. Guidelines-for-Improving-Water-Use-Efficiency-in-Irrigation-Domestic-Industrial-Sectors-MoWR.pdf

The increasing importance of data in the sustainable management of water resources requires professionals to have a robust comprehension of software tools, modelling platforms, and programming languages such as Python and R, which enable effective data analysis and improve water management strategies. Empowering professionals with critical skills enables us to bridge knowledge gaps and improve the efficacy of global water resource management.

DEVELOPING A STRATEGIC FRAMEWORK FOR SUSTAINABLE WATER RESOURCE MANAGEMENT

The 1992 International Conference on Water and the Environment produced the Dublin Principles for Sustainable Water Management, which promoted new techniques for evaluating, developing, and managing freshwater resources (see Figure 17.4). Traditional water management strategies that just evaluate water availability, without considering social and ecosystem impacts, are no longer sufficient.

Therefore, an in-depth approach is crucial for the long-term stewardship of water resources. This plan seeks to combine scientific knowledge, sound policies, and active community participation to tackle challenges like water scarcity, environmental

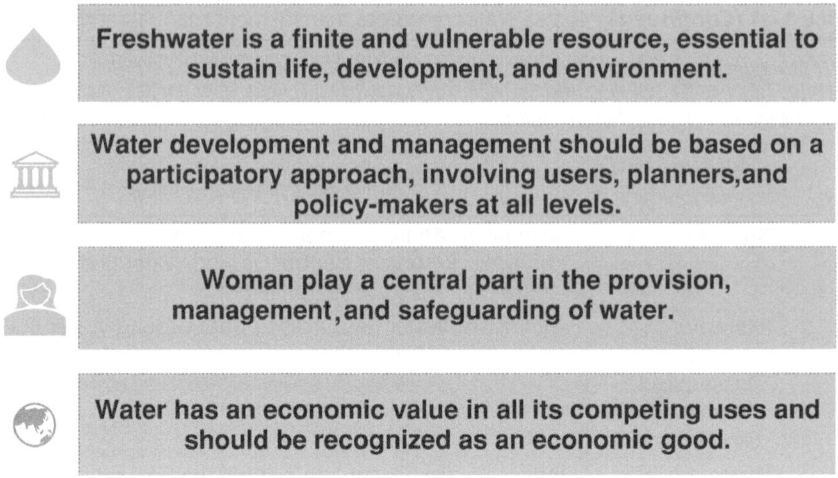

FIGURE 17.4 Dublin principles for sustainable water management, 1992.

degradation, and the rising demand for water resources. Sustainable management of water resources necessitates a perspective that transcends short-term fixes. It is essential to ensure the availability of high-quality water for the foreseeable future.

1. **Vision, Goals, and Policy Integration**:

   A strategic framework commences with a vision that aligns with the community's and environment's long-term objectives. This vision should prioritise:

   - Guaranteeing fair access to clean and safe water for all communities.
   - Developing innovative techniques and technologies for the sustainable use of water sources.
   - Reconciling water extraction with conservation initiatives to protect ecosystems.

   Such a policy framework will align with the national, regional, and international policies. The focus will be on compliance with international mandates, including the UN SDGs, towards achievement in water efficiency, conservation, and waste management.
2. **Stakeholder Involvement and Collaboration**:

   Water management requires teamwork among various players, such as industries, government authorities, local communities, and environmental representation groups. Through coordination, it can achieve a well-rounded approach that fits the needs and concerns of diverse stakeholders. This section focuses on:

   - Establishing calls for communication channels that permit dialogue and facilitate decision-making.
   - Organising workshops, consultations, and public meetings with a view to gathering opinions and gradually reaching an agreement.

- Establishing forums for information exchange among sectors to align policies with real-world conditions.

  Inclusive decision-making fosters greater accountability while ensuring that the interests of marginalised communities are prioritised.

3. **Integrating Ecosystem-Based Approaches**:

   Ecological health should be part of the decision-making in water management. Implementation of ecosystem-based approaches will further recognise that natural ecosystems, such as wetlands, forests, and rivers, have a significant role in the process of water filtration, storage, and flood control. The framework should focus on:
   - Protection and recovery of natural water ecosystems providing essential services.
   - Using nature-based solutions for water problems and enhancing biodiversity.
   - Integrating conservation and restoration initiatives into comprehensive water management plans.
4. **Adaptation to Climate Change and Resilience Building**:

   Considering the escalating hazards linked to climate change, formulating adaptive measures is essential for safeguarding the resilience of water systems. This section underscores the significance of:
   - Employing climate models to predict effects on water resources, including modified precipitation patterns and increased occurrences of droughts or floods.
   - Formulating adaptive solutions to address fluctuations in climate trends, including alterations to irrigation practices and diversification of water sources.
   - Constructing robust infrastructure capable of enduring the heightened occurrence of extreme weather phenomena.

   Continuous monitoring and updating the strategic framework in response to evolving climatic trends guarantees the operational integrity of water systems amid unpredictability.
5. **Financial Sustainability and Investment Models**:

   An effective water management framework necessitates sustainable financing. Planning financial models and obtaining adequate funds are crucial for sustaining infrastructure and executing water conservation initiatives. Key factors include:
   - Leveraging public–private partnerships to fund large-scale initiatives.
   - Investigating novel financing frameworks, such as green bonds and climate finance.
   - Establishing cost recovery procedures that guarantee the financial viability of utilities while maintaining cost-effective water access for all consumers.

   Financial measures must guarantee the sustainability of water services while avoiding excessive burdens on vulnerable populations.

6. **Data-Driven Decision-Making and Monitoring**:
   An effective water management strategy depends on precise, real-time data. The strategic foundation should establish the framework for:
   - Initiating data acquisition systems aimed at monitoring water quality, quantity, and usage across various locations.
   - Utilising geospatial instruments, remote sensing technologies, and hydrological models to improve decision-making processes.
   - Establishing performance indicators to evaluate the efficacy of water management strategies and policies.

   Monitoring on a regular basis allows changes in the water systems to be detected early; hence, management plans can be responded to and adjusted accordingly in time.
7. **Capacity Development and Knowledge Sharing**:
   Continuous training and education are essential for empowering those engaged in water management. To address current knowledge gaps, the framework must emphasise:
   - Capacity-building programmes to be undertaken for the enhancement of skills in hydrology, policy development, and water quality evaluation.
   - Collaborative efforts of different sectors together and sharing effective strategies and cutting-edge technologies.
   - Activities that encourage the participation of inhabitants in the conservation of water, making them join different sustainable practices regarding water use.

   By developing essential human capital, water management experts can address the intricate problems presented by evolving environmental conditions.

A strategic framework for sustainable water resource management is an integrated and dynamic approach, combining scientific knowledge, community engagement, and adaptive strategies. It also emphasises collaboration with stakeholders in the development of resilience, data-driven decision-making, and skill acquisition. These long-term solutions implement practical solutions to the water issues of today while ensuring access for future generations. Moreover, this methodology maintains efficiency over time and aligns with global development objectives towards a sustainable future in water resource management for all.

CASE STUDIES IN WATER RESOURCES MANAGEMENT

This section highlights key success stories and strategies in water management:

URBAN FLOOD MANAGEMENT IN INDIA: A CASE STUDY OF THE THREE VALLEYS SYSTEM AND SUDS APPLICATIONS

Bangalore, popularly known as the Silicon Valley of India, has indeed witnessed an avalanche of urban flooding caused by the twin factors of rapid urbanisation coupled with inefficient stormwater drainage systems. To deal with the growing flood problem, the city uses a three-valley system combined with Sustainable Drainage Systems (SuDS). An integrated urban flood model developed by the Indian Institute of Science (IISc) and Karnataka State Natural Disaster Monitoring Centre (KSNDMC) consists of real-time rainfall monitoring, rainfall forecast systems, and a hydrologic flood model. This approach has considerably mitigated the flooding impact in Bangalore (Mohan Mujumdar, 2021).

DROUGHT MITIGATION STRATEGIES IN ARID ECOSYSTEMS: TWO CASE STUDIES FROM RAJASTHAN AND GUJARAT, INDIA

Rajasthan and Gujarat, two of the most drought-prone states in India, have implemented a number of such strategies to mitigate the impact of drought. These strategies include rain harvesting, watershed management, and encouraging drought-resilient crops. The International Water Management Institute (IWMI) has conducted a detailed study on drought policies and their impact in Rajasthan, underscoring community participation and government aid in drought management (Rathore, 2005).

TRANSBOUNDARY COOPERATION: INSIGHTS GAINED FROM THE NILE BASIN INITIATIVE AND THE INDUS WATER TREATY

The Nile Basin Initiative (NBI) and the Indus Water Treaty are instances of successful transboundary water management agreements. The NBI, which began in 1999, aims to foster dialogue and cooperation among Nile Basin nations for the purpose of promoting sustainable water resource management. The Indus Water Treaty, signed in 1960 between India and Pakistan, has played an important role in managing water resources in the Indus River Basin. Cooperation and shared benefits in the management of transboundary water resources are given high emphasis by both agreements (Paisley, 2013).

COMMUNITY-BASED WATER MANAGEMENT: SUCCESS STORIES OF RURAL WATER HARVESTING SYSTEMS IN INDIA

Community-based water management initiatives have made significant strides in various parts of rural India. Such initiatives involve local communities through all steps of project planning: implementation and operation of the water harvesting systems like check dams, percolation tanks, and ponds. These systems help replenish groundwater, maintain moisture in the soil, and provide secure sources of water for agricultural and household needs. The role of the local communities engages in a way that guarantees the sustained use of the water, conserving its effectiveness in such practices (Gupta, 2011).

CHALLENGES AND BARRIERS IN TRAINING IMPLEMENTATION

- **Funding Constraints**: Inadequate financial resources for nationwide training initiatives.
- **Technological Barriers**: Restricted access to contemporary tools and technologies in developing areas.
- **Knowledge Gaps**: Integrating conventional approaches with contemporary technologies like SuDS for comprehensive sustainable water management.
- **Resistance to Change**: Mitigating institutional rigidity and promoting innovative methodologies.

ROLE OF PUBLIC–PRIVATE PARTNERSHIPS (PPPs) IN CAPACITY BUILDING

- **Collaborative Initiatives**: The partnership between commercial businesses and governmental agencies in the co-funding and co-design of training initiatives facilitates the accurate assessment of water resources management.

- **Corporate Social Responsibility (CSR)**: Corporate Social Responsibility (CSR) efforts foster sustainable water resource management by urging enterprises to use eco-friendly practices, invest in water conservation projects, and collaborate with communities to ensure responsible water utilisation.
- **International Organisations**: UNESCO, UNDP, and the World Bank augment worldwide water training initiatives by supplying resources, expertise, and financial support, assisting nations in executing efficient and sustainable water management policies.

FUTURE TRENDS IN WATER RESOURCES MANAGEMENT TRAINING

- **Digital Transformation**: In the future, virtual and augmented reality will improve water resource management by providing immersive training and real-time data visualisation.
- **AI-Driven Learning Platforms**: AI-driven learning systems enhance training for water resource management by providing real-time insights, automating data processing, and personalising the learning experience, resulting in more effective and efficient outcomes.
- **Global Networking and Knowledge Sharing**: Global connections and Knowledge-sharing platforms will enhance future water resource management by promoting collaboration, sharing best methods, and providing access to advanced research and technologies.
- **Sustainability Certifications**: Sustainability certifications ensure that professionals have the knowledge and techniques necessary to effectively manage water resources by standardising water management abilities and talents, leading to more consistent and long-lasting outcomes in the future.

CONCLUSION

In conclusion, appropriate training for decision-makers, professionals, and communities is important to effectively deal with water-related issues such as floods, water pollution, water pollution, and climate change. Proper skills and knowledge may hinder governance and policy decisions from attaining sustainable water management. Training courses covering hydrological and hydraulic modelling skills, disciplines wise all water-related software, GIS, and real-world case studies constructively address major knowledge gaps and translate technical proficiency into actionable solutions.

Training must therefore be tailored to the specific needs of developing countries, or for water-sensitive cities, to arm them with localised methodologies for comprehensive sustainable water management, if long-term success is to be achieved. International organisations like World Bank, UNDP, and UNESCO are important in providing financial assistance, conducting research, and increasing these initiatives by providing policy advice. Cooperation between public and private sectors enhances the impact by encouraging innovation and expanding access to various resources. Moreover, water sustainability certificates establish standardised competencies, ensuring that water managers possess skills that are recognised globally.

Ultimately, guaranteeing a robust and water-secure future relies on ongoing education, data-informed decision-making, and adaptable measures. Investing in comprehensive water management training fosters ecological preservation, economic stability, and long-term sustainability for future generations.

REFERENCES

Chakraborti, A. K. (1999). Achievements in use of satellite remote sensing in practical water resources management in India. *Selected Papers from Activities Held in 1999. Seminars of the United Nations Programme on Space Applications, Volume 11.* United Nations, New York, p. 85.

Esri. (2021). *ArcGIS Pro: Geospatial Analysis and Mapping*. Environmental Systems Research Institute (ESRI). Retrieved from https://www.esri.com/en-us/arcgis/about-arcgis/overview

Gupta, A. (2011). Drought disaster challenges and mitigation in India: Strategic appraisal. *Environmental Science, Agricultural and Food Sciences*, 100, 1795–1806.

Haiyan, Y. (2021). Modelling impacts of water diversion on water quality in an urban artificial lake. *Environmental Pollution*, 276, 116694.

Harikumar, M. A. (2023). Occurrence, distribution, and ecological risk of heavy metals and persistent organic pollutants (OCPs, PCBs, and PAHs) in surface sediments of the Ashtamudi wetland, south-west coast of India. *Regional Studies in Marine Science*, 64, 103044.

Manju Mohan, A. P. (2020). Urban sprawl during five decadal period over National Capital Region of India: Impact on urban heat island and thermal comfort. *Urban Climate*, 33, 100647.

Mijic, K. B. (2019). Integrating green and blue spaces into our cities. *Land Use Policy*, 124, 106455.

Mohan Mujumdar, P. (2021). Development of an urban flood model for Bengaluru city, Karnataka, India. *Current Science*, 120, 1441–1448.

Navaneeth, A. (2024). Unlocking sustainable groundwater governance in secondary cities: Lessons from the assessment of groundwater vulnerability in a Coastal City of India. *Environmental Science, Geography*, 57, 102116.

Paisley. R. K. (2013). Transboundary governance of the Nile River Basin: Past, present and future. *Environmental Development*, 7, 59–71.

Rathore, M. S. (2005). State Level Analysis of Drought Policies and Impacts in Rajasthan, India. Working or Discussion Paper. https://doi.org/10.22004/ag.econ.92401.

Sánchez-Gómez, A. A. (2025). Modelling water management using SWAT+: Application of reservoirs release tables and the new water allocation module in a highly managed River Basin. *Water Resources Management*, 39, 2357–2399.

Umair, A. (2019). Efficient water quality prediction using supervised machine learning. *Water*, 11(11), 2210.

US Army Corps of Engineers. (2024). *HEC-HMS Users Manual*. Retrieved from https://www.hec.usace.army.mil/confluence/hmsdocs/hmsum/latest/release-notes/v-4-13-0-release-notes

Zhang, X. Q. (2016). The trends, promises and challenges of urbanisation in the world. *Habitat International*, 54, 241–252.

Environmental Education for Quality Conservation

18

Surya Pratap Singh, Diwesh Kumar, and Sudarshana Banerjee

INTRODUCTION

Importance of Conservation in the 21st Century: In today's hastily budding world, technological advancements are transforming in every aspect of human life. While this progress is indispensable, it has come at a noteworthy cost to our environment. The consequences—air pollution, water pollution, pollution due to solid waste, global warming, climate change, and increased greenhouse gas emissions—are already being marked (Carter 2020).

If this trajectory continues unchecked, we may soon face a future where survival on this planet becomes increasingly difficult. To prevent such a scenario, we must take immediate action to improve and protect our environment (Griscom et al. 2017).

One of the most effective strategies is Environmental Quality Education (Rickinson 2001). By fostering awareness and equipping individuals with the knowledge and skills to address environmental challenges, we can inspire sustainable practices and drive positive change (Ramesh Kumar et al. 2024). This education will be pivotal in ensuring a healthier, more sustainable planet for future generations, as shown in Figure 18.1.

For ecosystem conservation, it is essential to conserve biodiversity (Chandra and Ruffalo 2024). With the huge depletion in climate, it has become crucial to maintain

DOI: 10.1201/9781003613183-18

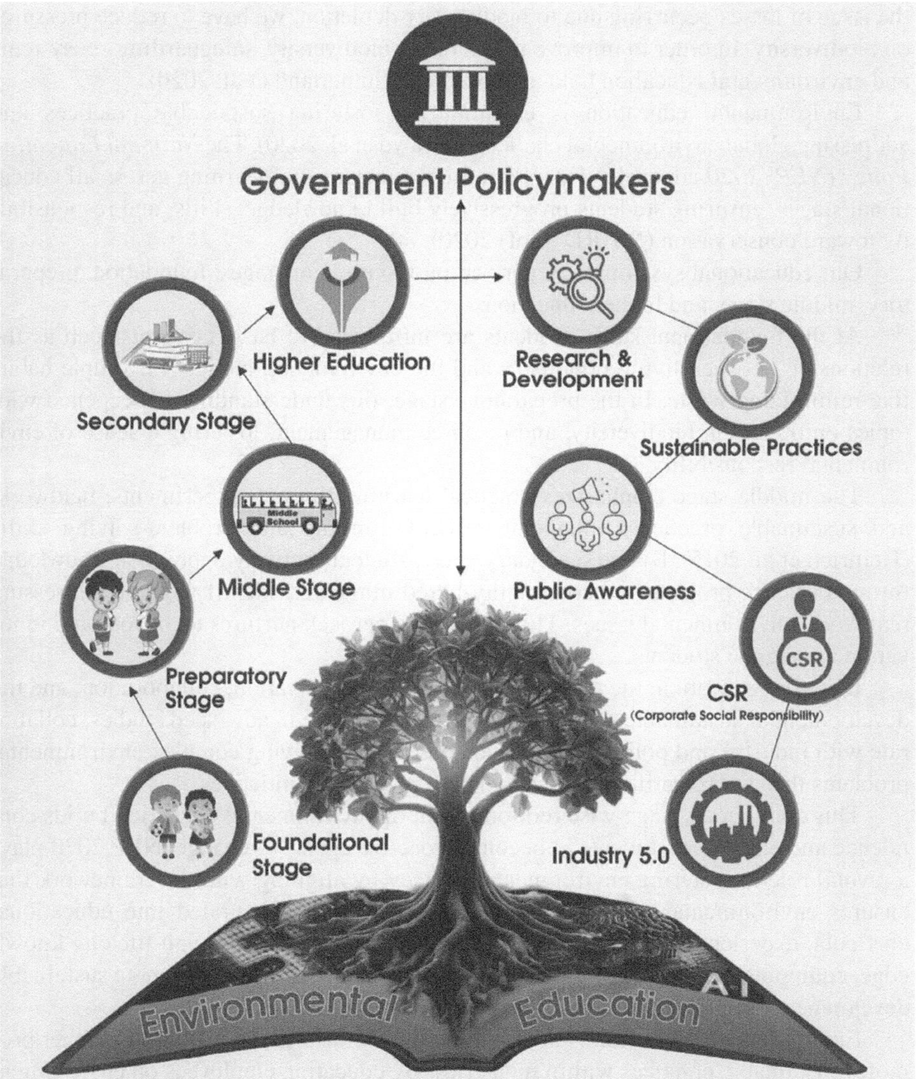

FIGURE 18.1 Importance of environmental education in the digital world.

the conservation of ecosystems in a planned manner (Patil 2024). Conservation of clean air and safe drinking water has become a prerequisite for human well-being. As all the natural resources are at their utmost lowest levels, having negative impacts on environment and human life, adoption of sustainable practices for conservation of air and water can ensure long-term viability (Schoene 2024).

From the pandemic response and reshaping conservation, the ability to influence human behavior embarks major focus (Srisathan and Naruetharadhol 2022). To address the issue of losses occurring due to biodiversity depletion, we have to reduce pressures on biodiversity. In order to improve the status of biodiversity, safeguarding ecosystems and environmental education hold a pivotal role (Sutherland et al. 2020).

Environmental education is essential for fostering sustainable practices and addressing global environmental challenges (Boyd et al. 2020). The *National Education Policy (NEP) 2020* advocates for integrating environmental learning across all educational stages, ensuring students progressively build knowledge, skills, and responsibility toward conservation (MHRD (GoI) 2020).

Our educational system is categorized into four major stages: foundation, preparatory, middle stage, and higher education.

At the foundational stage, students are introduced to basic concepts such as the relationship between living organisms and the environment, along with simple habits like minimizing waste. In the preparatory stage, this understanding is deepened with topics on pollution, biodiversity, and resource management, fostering a sense of environmental responsibility.

The middle stage emphasizes practical learning through experiments, fieldwork, and sustainable practices, enhancing critical thinking and problem-solving skills (Heinrich et al. 2015). By the secondary stage, students actively apply their knowledge through science projects and community-based initiatives, identifying and addressing real-world environmental issues. This hands-on approach nurtures teamwork and innovation among the students.

In higher education, the focus shifts to research, innovation, collaboration, and the development of sustainable technologies. Students engage in advanced studies, collaborate with industry and policymakers, and contribute to solving complex environmental problems through scientific research and entrepreneurial initiatives.

This continuous, stage-wise reinforcement of environmental education builds confidence and empowers students to become proactive agents of change. NEP 2020 plays a pivotal role in fostering environmental literacy by aligning with the framework that ensures environmental awareness through sustainability integrated into educational curricula, experiential learning, and local ecological contexts through lifelong knowledge, equipping future generations to lead conservation efforts and promote sustainable development (MHRD (GoI) 2020).

Environmental education is not bound only to educational institutions, rather promotes sustainable practices within industries. By educating employees on environmental challenges and solutions, industries can adopt strategies that trim down resource utilization, waste management, and energy consumption. These practices not only lead to significant cost savings but also minimize the ecological footprint, ensuring that businesses can thrive without depleting natural resources. Furthermore, sustainable practices contribute to improved efficiency, waste reduction, and a stronger and more resilient bottom line for industries (van de Sande 2024).

Environmental education also stimulates innovation in green technologies, encouraging industries to develop eco-friendly products, processes, and solutions. A well-educated workforce is more likely to identify opportunities for integrating environmentally sustainable technologies, such as renewable energy systems or green manufacturing methods. This fosters innovation, enabling companies to stay competitive in a rapidly evolving market that increasingly demands sustainability and eco-conscious alternatives (OECD 2018).

Integrating environmental education into Corporate Social Responsibility (CSR) strategies enables industries to demonstrate their commitment to sustainability. Industries that actively promote environmental stewardship through their CSR initiatives gain the trust of consumers, investors, and communities. This not only enhances brand reputation but also fosters positive relationships with stakeholders, positioning the company as a leader in responsible and ethical business practices (Gigauri and Khan 2024).

As industries evolve toward *Industry 5.0*, which emphasizes human-centric, sustainable, and resilient practices incorporating technological advancements, environmental education becomes even more crucial (Budhwar et al. 2023). It equips industries with the knowledge to combine cutting-edge technologies, like machine learning, Artificial Intelligence (AI), Internet of Things (IoT), and automation, with sustainability goals. This integration ensures that businesses not only remain efficient but also align their operations with global environmental standards, creating a balance between technological advancement and ecological responsibility (Berry and Rondinelli 1998).

Role of Education in Environmental Awareness

The rapid growth in urbanization and industrialization exacerbated impacts on the environment. Education is widely recognized as a key factor for economic and social development. For the pursuit of sustainable development, environmental conservation and education sensitizes about the several dimensions of the environment and its developments. The awareness and understanding of the environment is the rationale for sustainable development. Value education fosters and plays a crucial role for the awareness and adoption of strategies to implement different policies (Zhao et al. 2023).

Environmental education is a learning process that incorporates sensitization and awareness by lifelong learning and builds a comprehension about several challenges and strategies to overcome it. Research indicates that awareness and environmental education is effective, encouraging a lifelong commitment to environmental stewardship. The United Nations Educational, Scientific and Cultural Organization (UNESCO) states that environmental education functions crucial in imparting an inherent reverence for temperament amid society and in enhancing community environmental awareness (Chan 2024).

Several approaches like project-based learning, hands-on training, and awareness drives add on critical thinking and develop problem-solving skills. With environmental education, people's thoughtfulness of their day-to-day environment is improved, and they critically contribute to policies that foster social and economic betterment, especially those designed to encourage sustainable development and bring into reality the principle of making fair use of nature. Despite having several positive impacts on society, environmental education is confronted with challenges such as resource constraints, which easily led to the overall failure of the institution, incorporation into various series of curricula advancement and children's physical health issues, and evaluation consistently difficult but vital for success. Hoping to change this situation calls for more intensified policy support as well as hard physical work, technology, and innovation (Lang et al. 2012).

Education motivates people to start taking action in a sustainable direction. Education must adapt to and weather these tests if it is to be a powerful obligation for environmental stewardship and buoyancy. But what is necessary is a clear strategy on the part of all educating authorities, as well as making calculations about what reforms are or are not feasible.

Adopting methods of education that are inevitable and universal, a future where all can live well with each other might be accomplished (UNESCO 2024).

FUNDAMENTALS OF AIR AND WATER QUALITY CONSERVATION

Without clean air and water, life is impossible. Its basic significance to maintaining ecological balance and nurturing living creatures in our environment can never be underestimated. As resources are essential for human survival, both natural and man-made forms face major challenges. The four key challenges identified in implementing effective environmental education for quality conservation are lack of resources, lack of qualified educators, limited human capacity, and poor transformation of knowledge to practice. Addressing these challenges requires a holistic approach that integrates environmental consciousness, sustainability, and advanced problem-solving skills. Environmental education aims to develop a society that is aware of environmental issues and has the skills and commitment to work toward solutions. As the world faces unprecedented environmental degradation, 21st-century education must evolve to emphasize these critical aspects, preparing future generations to tackle these pressing issues effectively (Carter et al. 2015).

Overview of Air and Water Quality Challenges

The deterioration of air and water has become a critical problem caused by pollution, urbanization, and the changing climate. The combustion of fossil fuels, industry, and vehicles are responsible for high levels of carbon monoxide, nitrogen oxides, sulfur dioxide, and particulate matter, which negatively impacts air quality. Moreover, climate change further complicates existing air quality issues through more wildfires, more smog, and more disturbances to the atmospheric arrangement (Munsif et al. 2021). The water quality, in contrast, is affected by pesticides, fertilizers, and more through farm runoffs as well as industrial effluents and the serious issue of plastic pollution. Such pollutants degrade the biosphere while ameliorating the demand for clean and potable water, hence increasing health risks. All these issues are so intertwined that they require education systems that are geared toward measures and addressing cross-cutting themes and have a high degree of creativity to design solutions suitable for such distinctions (Dwivedi et al. 2021).

Key Threats to Air and Water Resources

Numerous key threats diminish the sustainability of air and water resources. Deforestation has worsened on a planetary scale because it diminishes the amount of carbon dioxide that can sequester, worsening the greenhouse effect. Water resources do not fare well either, as new research indicates that they are in danger due to excessive abstraction, agriculture and industrialization pollution, and habitat destruction (Saatsaz 2020). This does not help the case, since climate change damages water resources even more due to crustal expansion, causing increased chances of very hot or dry weather, potential flood danger, and changes to the balance of precipitation. These inter-related threats reveal the need to revamp the education systems of countries in such a way that learners possess scientific knowledge, advanced technologies, and the concepts of conservation in the end. Uniquely positioned people are the linchpin for addressing and finding effective approaches to environmental challenges (Ogonsola 2023).

Global and Local Impacts of Pollution

The effects of air and water pollution are multifaceted. This is demonstrated both at the global and local levels. Global pollution is accelerating climate change, causing ozone loss and disrupting the ecosystem resulting in loss of biodiversity. Rising global temperatures due to greenhouse gas emissions affect agriculture marine ecosystem and human health. Plastic pollution and ocean acidification also threaten aquatic life, fisheries, and food security. Locally, pollution has a direct impact on public health. It causes respiratory diseases and waterborne diseases and reduces life expectancy in areas with high levels of pollution. Urban centers battle smog, heat islands, and water shortages. Meanwhile, rural areas face contamination of drinking water sources from farm run-off of these challenges but also gain inspiration and skills to implement sustainable solutions.

The Demand for 21st-Century Education

Air and water pollution represent the danger the quality of our living environment is faced with and this, therefore, necessitates a total reform in all levels of institutional knowledge (Pezzey and Toman 2017). The new pattern of education of the 21st century is to give priority to environmental education so that every student becomes equipped with the right knowledge and skills to tackle life-threatening subjects such as these (Yadav et al. 2022). For instance, one can steer away from the mechanical paradigm and steer the way to the practical and process-based model in which thinking critically and grappling with real-world problems become the focal point. Besides this, tech skills like the use of digital means for the analysis of prodigious amounts of data and computer simulations are crucial for understanding multifaceted environmental issues and coming up with ingenious remedies. In addition to that, education needs to be about enhancing global responsiveness and cross-cultural competence as well, and it is the idea of activating students toward thinking to new horizons and working as a team on problems with worldwide locations and common origins (Tõnurist and Hanson 2020).

Moreover, sustainability concepts should be included in this curriculum of contemporary learning, and students can be helped to develop sustainable behaviors that will conserve natural resources and mitigate environmental impacts. Aside from that, the students should also develop skills of emotional intelligence and ethical responsibility that will enable them to be empathetic and accountable toward the issues (McPhail 2004). Through the inclusion of these factors, school systems will be able to form a group of leaders who have the skills and knowledge to address the issues of air and water quality conservation and carry out the development of sustainable and resilient communities.

The change of this educational paradigm is not only a mere response to the crises of the environment but also a foresight in the bright future to come. By encouraging innovation, collaboration, and sustainability, 21st-century education can become one of the main players in solving the challenges of air and water quality conservation, with the welfare of the present and future generations being the prerogative (Sanabria et al. 2024).

ENVIRONMENTAL EDUCATION: GOALS, SCOPE, AND IMPACT

Environmental education is not limited to the classroom, rather it is a collective effort for a balanced society (Waite 2019). It is a multidisciplinary field integrating disciplines such as Physics, Chemistry, Mathematics, Biology, Ecology, Earth science, Atmospheric science, Geology, and Geography. The United Nations Educational, Scientific and Cultural Organization (UNESCO) states the role of environmental education in safeguarding future global developments of societal quality of life through the incorporation of sustainable development goals by 2030 (Corlett 2020).

Objectives of Environmental Education

Environmental education holds a broad objective to impart knowledge about the principles required in the sense of nature. Through environmental education, the conservation of natural resources and its explanation is being provided, which holds good for the utilization of natural resources and the progress of mankind as a primary goal.

FIGURE 18.2 Key objectives of environmental education.

Maintaining a friendly and well-balanced relationship between man and nature in the realization of environmental education is important. The primary objectives of environmental education are illustrated in Figure 18.2.

Scope: Formal and Informal Education Platforms

This is concerned with our environment and the state of things around us, as well as how it works or the kind of interaction one gets from the environment. It is also important that we have the ability to learn how to care for the environment; for this reason, it is essential that we have the ability to care for the environment so that the current generation and the generations that follow can sustain the environment for the time being (Farley and Smith 2020). School curriculum, college education, university lectures, and even public office communications often engage the environment education topics, such as pollution effects, environmental problems, and the means to conserve the environment.

Formal Education

It is introduced at the foundational level and progresses to higher education level. Almost every country included environmental education as an objective or even incorporated them in the curriculum as a form of integration into existing modules. However, most countries used this approach as a way to engage in environmental education in an existing way, while the others adopted a different approach in doing so. The objectives of environmental education include the four major points: to raise awareness about the significance of science in daily life and its role in society (Kamil et al. 2020); to make people familiar with the importance of science as a basis for decision-making and other activities of their daily lives; to comprehend natural events that occurs; and to provide an outlook that highlights how different disciplines of science have been approached.

Informal Education

The mode of education is not related to the formal educational process, but rather is a combination of various approaches and projects (AI, machine learning, IoT, etc.) for value education and knowledge building. The informal activities about the environment were carried out in parallel with formal education through various informal practices like extracurricular activities and mass media. One can note the differences in what might be done by different people or organizations to solve their own problems or find solutions to their problems and what others are doing.

Informal education plays a pivotal role in fostering environmental awareness and sustainability. To maintain sustainable practices, there are several acts to conserve the environment among which Environment Protection Act, 1986 holds major significance (Nassani et al. 2022). Informal education has several constituents like Adult Education, Tribals and Forest Dwellers, Children Activities, Eco-development Camps, Training Executives, Research and Development Programmes, Foundation Courses for the probationers, Development of Educational Material and Teaching Aids, and World Environmental Day celebrations.

Measuring the Impact of Environmental Education on Conservation

Value-added environmental education enhances attitude, values, and knowledge, as well as builds skill that prepares individuals and communities to collaboratively undertake positive environmental action. Environmental education also facilitates acquaintances, involving actionable research findings and on-the-ground practices, creating synergistic spaces where stakeholders team up to address dynamic environmental issues eventually. Because of this pledge to relevance and iteration, environmental education can result in direct payback to the environment and deal with conservation issues concretely. Yet, the path to achieving those substantial impacts can be winding, with vigorous data documenting changes demanding to fabricate (Ardoin et al. 2020).

Key Metrics for Measuring Impact

The impact of environmental education on conservation can be presented through key metrics, as shown in Figure 18.3. Carbon footprint reduction highlights the decrease in greenhouse gas emissions achieved through sustainable practices. *Resource conservation* emphasizes reduced water and energy usage and lower waste generation

FIGURE 18.3 Key metrics for measuring impact.

(Singh et al. 2021). Biodiversity preservation showcases efforts to protect ecosystems, habitats, and species. Finally, the social and economic benefits reflect increased community engagement, job creation, and financial investments in sustainability initiatives (Kopnina 2020).

Examples of Measuring Impact

Sustainability can be modeled through an examination where one can choose to apply the theoretical approach. Employment of organic products to reduce the dependence on chemicals and low water costs are used in production while minimizing the use of chemicals, fair practices, and support of workers involved are promoted as well as environmental protection. In the same way, to measure the effectiveness, the brand can calculate the degrees (Celsius) of carbon emissions produced, or the total amount of water resources consumed, or the increase in workers' wages. A thorough assessment aids the companies in determining strategic decisions to adopt sustainability as the concept of entrepreneurship of sustainable growth (Chaturvedi et al. 2021).

STRATEGIES FOR INTEGRATING CONSERVATION INTO ENVIRONMENTAL EDUCATION

This requires introducing conservation actions for education on environmental matters, which promote a generation of responsible citizens who act on their responsibility to sustain earth's resources (Figure 18.4). There is a need to ensure that education policies are developed in a way that addresses formal education institutions in conjunction with other Community Organizations and technology, or other similar interactive platforms for improving learning prospects in different communities.

Curriculum Development for Schools and Universities

Implementation of conservation subjects in the curriculum of schools and universities can hold a path to achieve the goal. The primary aspect is to reach the level of learning

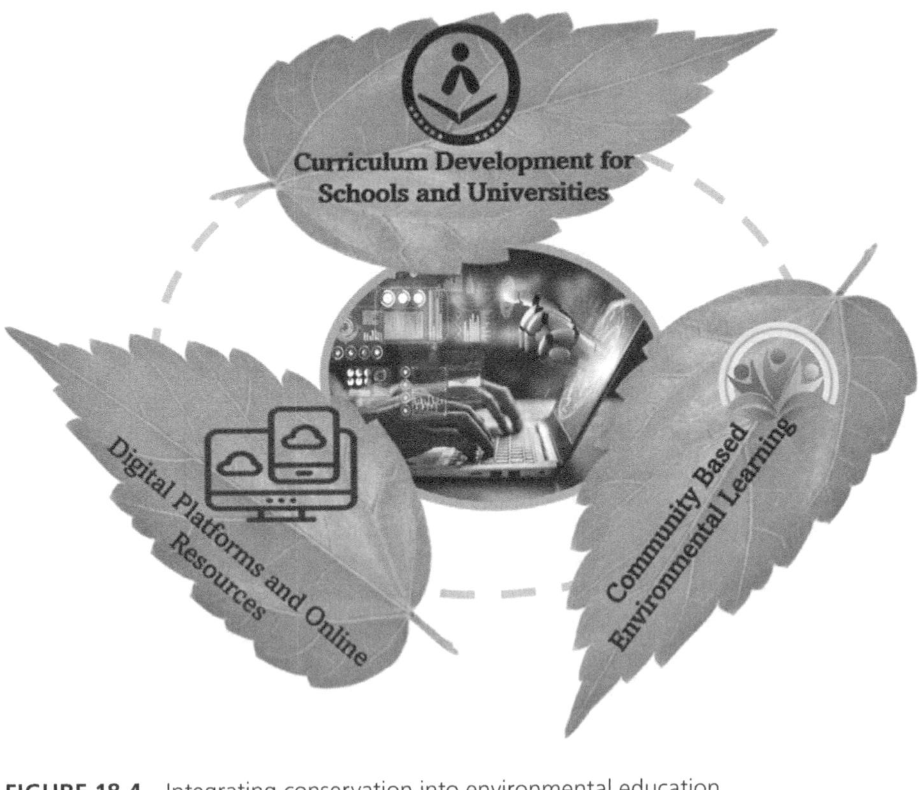

FIGURE 18.4 Integrating conservation into environmental education.

so that the receiver can have an elaborate and pertinent understanding, such as climate change, environmental policy, solar energy, and sustainability. Conservation education is not limited to classes alone rather expanded to prime goals of the society (Glavič 2020). Educational institutions have the capacity to design programs that are tailored to specific training areas such as conservation science, legal education and engineering and place graduates in positions like conservation, legal, and engineering. Similarly, collaborations with firms and government organizations can enhance the use of conservation education among tertiary users.

Community-Based Environmental Learning

Conservation can be done on all aspects of the education system even though education is not enough. Community centers, NGOs, and local group work are the most effective in conveying messages and facilitating people's actions. These actions taken by NGOs can increase consciousness of the environment (McCormick 1999), not only among the general public but also among individuals within the same society, fostering unity and a more conscious population. Participatory models of participation are shown through initiatives such as neighborhood beautification, community gardening, and other citizen projects that aim to gather environmental information. Apart from improving the sustainability of organizations, these programs are social programs that foster cooperation and management of the organization's environment.

Digital Platforms and Online Resources

In the current digital age, Leveraging technology is essential in expanding environmental education and reaching a global audience. Digital tools for environmental education include several online courses. Educational applications' virtual reality (VR) experience and interactive simulations make learning engaging and accessible for a wide range of learners (Asad et al. 2021). Online platforms such as Coursera, edX, and YouTube offer courses and videos on environmental topics. Helping individuals around the world gain access to high-quality educational content. VR can transport students to remote ecosystems. It provides an immersive experience that deepens our understanding of conservation. Challenges and Solutions' social media and online communities play an important role in spreading environmental awareness. Campaigns on platforms like Instagram, Twitter, LinkedIn, and Facebook can reach millions of people. Inspire collaborative action and share success stories of conservation efforts.

CHALLENGES IN IMPLEMENTING ENVIRONMENTAL EDUCATION FOR CONSERVATION

Implementation of environmental education for conservation is essential to addressing the growing ecological crisis facing the world today. However, many challenges present it, especially in terms of resources (Figure 18.5). Institutional framework and public

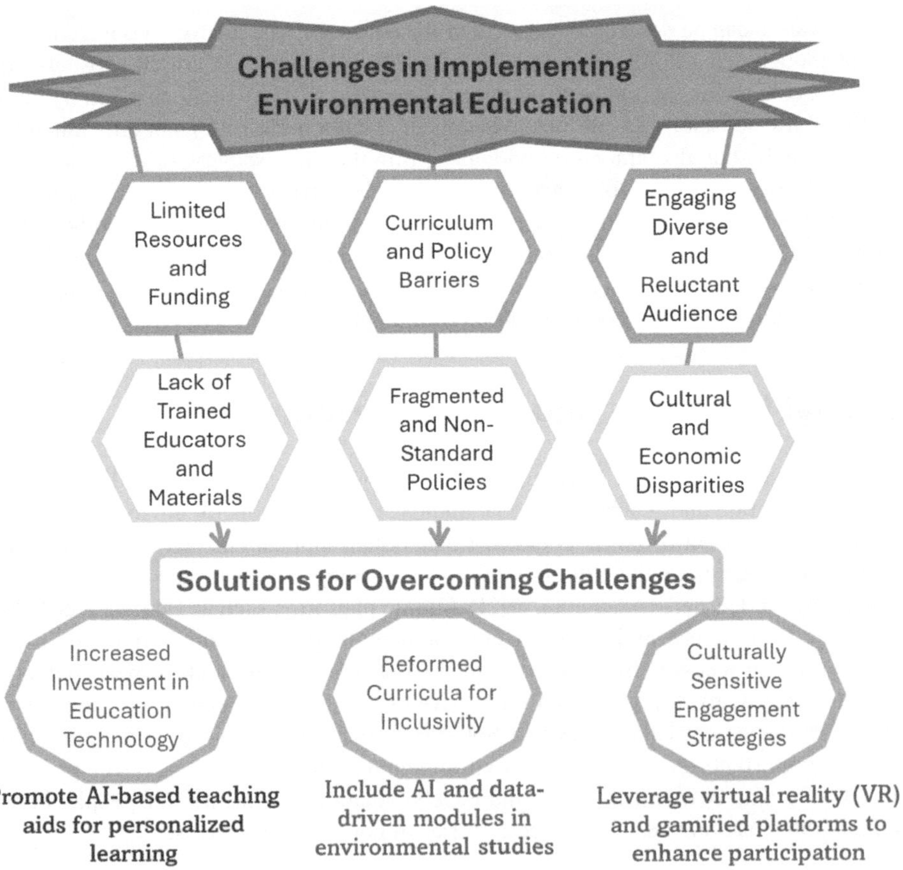

FIGURE 18.5 Flowchart of challenges and solutions in implementing environmental education for conservation.

participation all limit efficiency and accessibility. One of the most important barriers is a lack of resources and funding. Many schools especially in developing regions lack adequate teaching materials Trained teachers and the necessary infrastructure to provide effective environmental education. Governments and private institutions often focus on traditional academic subjects. This leaves environmental education underfunded and neglected. This leads to subpar programs that are unable to bring about long-term changes.

The inclusion of conservation in the school curricula meets several hindrances, including policy and curricular obstacles. Environmental education, for instance, does not get as much "attention" as core subjects where policies are either non-specific or disregarded. Basic materials are scarce, as the standard curriculum is primarily aimed at the content of core disciplines, thus having very limited spaces for interdisciplinary theory, where science, economics, and ethics are combined for portraying knowledge of conservation problems. Disparities in quality and access to environmental education between regions results from inconsistent policy enforcement and a lack of common frameworks for teaching.

One more obstacle that should be gouged is making various and reluctant people interested in it all. The differences in social welfare, culture, and immediate needs among certain populations usually override environmental issues (Ojala et al. 2021). Therefore, in some cases, we have to deal with fake news and prejudice from politicians who are the key drivers of climate change and environmental degradation. As a result, people are less involved and more against environmental protection. Cultural understanding, inclusiveness, and relatability are the major components for breaking the communication barriers and ensuring wide dissemination.

The urgent necessity for increased financial support, education reform in the direction of fundraising, and targeted approaches toward diverse audiences in the environmental domain are the solutions to these problems. Elimination of those barriers is vital in giving power to people and communities to take serious steps toward sustainability and being good environmental guardians in the world today.

RECOMMENDATIONS AND FUTURE DIRECTIONS

To overcome the global environmental challenges, a big investment in education that can take the form of a collective approach is now seriously required. The decision to give practical and effective ways the priority of course leverages the length and success of conservation education for the teachers and for environmental organizations.

Recommendations for Educators and Environmental Organizations

Making an interesting curriculum with environmental literacy at the core is definitely the first step to take. Some improvements worth mentioning are making the course two-related, introducing one of the political sciences, one of morality, and one of social studies courses, and, in general, knowing the conservation. Some of the fruitful experiential learning means, i.e., field visits, the student's active learning process by making their own natural crafts, and other effective technology tools to compel the students to expose them to the system of learning.

It is the local community's active involvement as another main criterion for the promotion of more healthy environmental reasons. Teachers should connect with the local civic groups that support students' participation in activities like tree-planting, clean-up, and conservation (Rustamova 2023). Appropriately adjusting tactics that address the environmental challenges in the nearby area makes it visible and thus the ownership of the community on the solution.

Policy Recommendations for Supporting Environmental Education

Policy changes may bring the biggest impact on educational systems worldwide. The definitive requirements in the educational arena are monetary allocations for the development of good teaching materials and for the support of teacher training in sustainability. Public–private partnerships (PPPs) can add a much-needed injection of both money and technology to the projects. They can also be arbitrators in disputes concerning resource management and the implementation of measures for effective and environmental management.

It is important that the government pass laws that will include conservation education on the programs as they are so important to the future of our planet. Lawmakers have to motivate companies to use eco-friendly practices and give resources to educational projects in the field. Through global cooperation among countries, the transfer of ideas, tools, and technologies about environmental education will be possible and thus wider audiences will be reached (Affolderbach 2022).

In conclusion, the future of conservation education depends on the adoption of innovative teaching, strong policy support, and the involvement of an active community. By generating environmental stewardship from scratch at the local level and expanding it through consortium iterations, mankind can give impetus to young generations for the conservation of our planet's sustainability.

CONCLUSION

Environmental awareness is a must because the air and water pollution decrease the quality of life, thus making it impossible for natural resources and humans to coexist. The present-day environmental degradation is mainly triggered by pollution and climate change. Environmental education is the biggest means to teach people about the causes of the problem and help them with new solutions. Lessons can be learnt from both the formal and informal education sectors such as schools and communities to change these behaviors through involvement in outreach and digital platforms. People and environmental organizations can form workable partnerships to enhance environmental education and silence the constraints such as limited resources, curriculum barriers, and audience engagement. This chapter publishes practical tips on how to become more effective in environmental education by educators, policymakers, and environmental organizations. The main goal of environmental literacy is to have a society where both environmental education programs are included in the curriculum and other learning avenues are the basic foundation for multifaceted solutions that simultaneously involve action, ecosystem adaptation, and human nature nurturing.

REFERENCES

Affolderbach, Julia. 2022. Translating green economy concepts into practice: Ideas pitches as learning tools for sustainability education. *Journal of Geography in Higher Education* 46(1): 43–60. doi:10.1080/03098265.2020.1849063.

Ardoin, Nicole M., Alison W. Bowers, and Estelle Gaillard. 2020. Environmental education outcomes for conservation: A systematic review. *Biological Conservation* 241: 108224. doi:10.1016/j.biocon.2019.108224.

Asad, Muhammad Mujtaba, Aisha Naz, Prathamesh Churi, and Mohammad Mehdi Tahanzadeh. 2021. Virtual reality as pedagogical tool to enhance experiential learning: A systematic literature review. *Education Research International* 2021: 1–17. doi:10.1155/2021/7061623.

Berry, Michael A., and Dennis A. Rondinelli. 1998. Proactive corporate environmental management: A new industrial revolution. *Academy of Management Perspectives* 12(2): 38–50. doi:10.5465/ame.1998.650515.

Boyd, Claude E., Louis R. D'Abramo, Brent D. Glencross, David C. Huyben, Lorenzo M. Juarez, George S. Lockwood, Aaron A. McNevin, et al. 2020. Achieving sustainable aquaculture: Historical and current perspectives and future needs and challenges. *Journal of the World Aquaculture Society* 51(3): 578–633. doi:10.1111/jwas.12714.

Budhwar, Pawan, Soumyadeb Chowdhury, Geoffrey Wood, Herman Aguinis, Greg J. Bamber, Jose R. Beltran, Paul Boselie, et al. 2023. Human resource management in the age of generative artificial intelligence: Perspectives and research directions on ChatGPT. *Human Resource Management Journal* 33(3): 606–659. doi:10.1111/1748-8583.12524.

Carter, Jeremy G., Gina Cavan, Angela Connelly, Simon Guy, John Handley, and Aleksandra Kazmierczak. 2015. Climate change and the city: Building capacity for urban adaptation. *Progress in Planning* 95: 1–66. doi:10.1016/j.progress.2013.08.001.

Carter, Peter. 2020. Environmental health risk assessment for global climate change and atmospheric greenhouse gas pollution. *Climate Action* 1: 413–423. doi:10.1007/978-3-319-95885-9_95.

Chan, Mina. 2024. *What Can We Learn From Nature: Values-Based Education for The Development of Holistic Education*. Trinity Western University, Canada. chrome-extension://efaidnbmnnnibpcajpcglclefindmkaj/https://www.twu.ca/sites/default/files/2024-06/Mina Chan-Final%2BCorrectedKTProject.pdf.

Chandra, Ankit, and Will Ruffalo. 2024. Understanding carbon credits in agricultural voluntary carbon markets: Ecosystem actors and their roles. *CSA News* 69(12): 3–7. doi:10.1002/csan.21428.

Chaturvedi, Preeti, Parul Shukla, Balendu Shekher Giri, Pankaj Chowdhary, Ram Chandra, Pratima Gupta, and Ashok Pandey. 2021. Prevalence and hazardous impact of pharmaceutical and personal care products and antibiotics in environment: A review on emerging contaminants. *Environmental Research* 194: 110664. doi:10.1016/j.envres.2020.110664.

Corlett, Richard T. 2020. Safeguarding our future by protecting biodiversity. *Plant Diversity* 42(4): 221–228. doi:10.1016/j.pld.2020.04.002.

Dwivedi, Yogesh K., Laurie Hughes, Elvira Ismagilova, Gert Aarts, Crispin Coombs, Tom Crick, Yanqing Duan, et al. 2021. Artificial intelligence (AI): Multidisciplinary perspectives on emerging challenges, opportunities, and agenda for research, practice and policy. *International Journal of Information Management* 57: 101994. doi:10.1016/j.ijinfomgt.2019.08.002.

Farley, Heather M., and Zachary A. Smith. 2020. *Sustainability:If It's Everything, Is It Nothing?* Second edition. Routledge, New York. doi:10.4324/9781351124928. .

Gigauri, Iza, and Ali Junaid Khan (eds.). 2024. Navigating corporate social responsibility through leadership and sustainable entrepreneurship. In *Advances in Marketing, Customer Relationship Management, and E-Services*. IGI Global, London, UK. doi:10.4018/979-8-3693-6685-1.

Glavič, Peter. 2020. Identifying key issues of education for sustainable development. *Sustainability* 12(16): 6500. doi:10.3390/su12166500.

Griscom, Bronson W., Justin Adams, Peter W. Ellis, Richard A. Houghton, Guy Lomax, Daniela A. Miteva, William H. Schlesinger, et al. 2017. Natural climate solutions. *Proceedings of the National Academy of Sciences* 114(44): 11645–11650. doi:10.1073/pnas.1710465114.

Heinrich, William F., Geoffrey B. Habron, Heather L. Johnson, and Lissy Goralnik. 2015. Critical thinking assessment across four sustainability-related experiential learning settings. *Journal of Experiential Education* 38(4): 373–393. doi:10.1177/1053825915592890.

Kamil, Puspita Annaba, Eka Putri, Syahrul Ridha, Sugeng Utaya, Sumarmi, and Dwiyono Utomo. 2020. Promoting environmental literacy through a green project: A case study at adiwiyata school in Banda Aceh City. *IOP Conference Series: Earth and Environmental Science* 485(1): 012035. doi:10.1088/1755-1315/485/1/012035.

Kopnina, Helen. 2020. Education for the future? Critical evaluation of education for sustainable development goals. *The Journal of Environmental Education* 51(4): 280–291. doi:10.1080/00958964.2019.1710444.

Lang, Daniel J., Arnim Wiek, Matthias Bergmann, Michael Stauffacher, Pim Martens, Peter Moll, Mark Swilling, and Christopher J. Thomas. 2012. Transdisciplinary research in sustainability science: Practice, principles, and challenges. *Sustainability Science* 7(S1): 25–43. doi:10.1007/s11625-011-0149-x.

McCormick, John. 1999. The role of environmental NGOs in international regimes. In Regina Axelrod and Stacy Vandeveer (eds.), *The Global Environment: Institutions, Law and Policy*, 20. CQ Press, Washington, DC.
McPhail, Ken. 2004. An emotional response to the state of accounting education: Developing accounting students' emotional intelligence. *Critical Perspectives on Accounting* 15(4–5): 629–648. doi:10.1016/S1045-2354(03)00050-9.
MHRD (Go I). 2020. *National Education Policy 2020*. MHRD, New Delhi, India. https://www.education.gov.in/sites/upload_files/mhrd/files/NEP_Final_English_0.pdf.
Munsif, Rabia, Muhammad Zubair, Ayesha Aziz, and Muhammad Nadeem Zafar. 2021. Industrial air emission pollution: Potential sources and sustainable mitigation. In Richard Viskup (eds.), *Environmental Emissions*. IntechOpen, New York. doi:10.5772/intechopen.93104.
Nassani, Abdelmohsen A., Zahid Yousaf, Magdalena Radulescu, and Mohamed Haffar. 2022. Environmental performance through environmental resources conservation efforts: Does corporate social responsibility authenticity act as mediator? *Sustainability* 14(4): 2330. doi:10.3390/su14042330.
OECD, World Bank , UNEP. 2018. *Financing Climate Futures*. OECD, Paris, France. doi:10.1787/9789264308114-en.
Ogonsola, Sijuwade Adedayo. 2023. *Modelling Values, Entrepreneurial Competencies and Sustainable Development: A Case of Small, Medium and Micro Enterprises in EThekwini Municipal Area*. Durban University of Technology, Durban, South Africa. doi:10.51415/10321/5092.
Ojala, Maria, Ashlee Cunsolo, Charles A. Ogunbode, and Jacqueline Middleton. 2021. Anxiety, worry, and grief in a time of environmental and climate crisis: A narrative review. *Annual Review of Environment and Resources* 46(1): 35–58. doi:10.1146/annurev-environ-012220-022716.
Patil, Shubhangi. 2024. *A New Service Model for Identifying and Improving the Quality of Emergency Department Operations in Tertiary Settings*. Te Herenga Waka-Victoria University of Wellington, Wellington. doi:10.26686/wgtn.26802976.
Pezzey, John C. V., and Michael A. Toman. 2017. *The Economics of Sustainability*. Routledge, London. doi:10.4324/9781315240084.
Ramesh Kumar, K. C., Manual Selvaraj Bexci, Amiya Bhaumik, and Sateesh Kumar Ojha. 2024. Educating and promoting climate sustainability. In Amiya Bhaumik, Sandeep Poddar, Manish Dadhich, Kamal Kant Hiran, Ruchi Doshi (eds.), *Community Resilience and Climate Change Challenges: Pursuit of Sustainable Development Goals (SDGs)*, 183–202. IGI Global, London, UK. doi:10.4018/979-8-3693-6522-9.ch010.
Rickinson, Mark. 2001. Learners and learning in environmental education: A critical review of the evidence. *Environmental Education Research* 7(3): 207–320. doi:10.1080/13504620120065230.
Rustamova, Rano Parpievna. 2023. Important features of the formation of ecological culture in students. *Journal of Economics and Business Management* 6(4): 122. https://sirpublishers.org/index.php/joeabm/article/view/122.
Saatsaz, Masoud. 2020. A historical investigation on water resources management in Iran. *Environment, Development and Sustainability* 22(3): 1749–1785. doi:10.1007/s10668-018-00307-y.
Sanabria, Jorge, Marco Cruz-Sandoval, Athziri Moreno-Romo, Sofía Bosch-Gómez, and María Soledad Ramírez-Montoya. 2024. Research foresight in bridging open science and open innovation: Overview based on the complex thinking paradigm. *International Journal of Innovation Studies* 8(1): 59–75. doi:10.1016/j.ijis.2023.08.002.
Schoene, Berthold. 2024. *Max Porter's Ruderalism, or What Nature Is Now. Textual Practice*. Taylor & Francis Group, New York. doi:10.1080/0950236X.2024.2380254.

Singh, Surya Pratap, Pooja Yadav, Chirag Kumar, Meena Kumari Sharma, and Rakesh Chandra Gaur. 2021. Performance of an integrated settler based anaerobic biofilm reactor as onsite sanitation system. *IETE Journal of Research* 67(5): 603–610. doi:10.1080/03772063.2021.1874840.

Srisathan, Wutthiya A., and Phaninee Naruetharadhol. 2022. A COVID-19 disruption: The great acceleration of digitally planned and transformed behaviors in Thailand. *Technology in Society* 68: 101912. doi:10.1016/j.techsoc.2022.101912.

Sutherland, William J., Peter N. M. Brotherton, Zoe G. Davies, Nancy Ockendon, Nathalie Pettorelli, and Juliet A. Vickery, (eds.) 2020. Making a difference in conservation: Linking science and policy. In *Conservation Research, Policy and Practice*, pp. 3–8. Cambridge University Press, Cambridge. doi:10.1017/9781108638210.

Tõnurist, Piret ; Hanson, Angela. 2020. *Anticipatory Innovation Governance*. Vol. 44. OECD Working Papers on Public Governance, Paris, France. doi:10.1787/cce14d80-en.

UNESCO. 2024. *What You Need to Know about Education for Sustainable Development*. https://www.unesco.org/en/sustainable-development/education/need-know.

van de Sande, Luc. 2024. *Emergence of Organizing Inter-Organizational Collaboration in Addressing Societal Challenges: A Multilayer Paradox Perspective*. Radboud University Press, Nijmegen. doi:10.54195/9789493296787.

Waite, Sue. 2019. Teaching and learning outside the classroom: Personal values, alternative pedagogies and standards- outdoor learning research. *Education* 39(1): 65–82.

Yadav, Shailesh Kumar, Arnab Banerjee, Manoj Kumar Jhariya, Ram Swaroop Meena, Abhishek Raj, Nahid Khan, Sandeep Kumar, and Seema Sheoran. 2022. Environmental education for sustainable development. In Manoj Kumar Jhariya, Ram Swaroop Meena, Arnab Banerjee, Surya Nandan Meena (eds.), *Natural Resources Conservation and Advances for Sustainability*, pp. 415–431. Elsevier, Amsterdam, Netherlands. doi:10.1016/B978-0-12-822976-7.00010-7.

Zhao, Yanzhi, Mingsi Zhao, and Fengyu Shi. 2023. Integrating moral education and educational information technology: A strategic approach to enhance rural teacher training in universities. *Journal of the Knowledge Economy* 15(3): 15053–15093. doi:10.1007/s13132-023-01693-z.

Index

A

B

C

D

E

F

G

H

I

L

M

N

O

P

R

S

T

U

W